VOLUME SIX HUNDRED AND THREE

METHODS IN ENZYMOLOGY

Chemical and Biochemical Approaches for the Study of Anesthetic Function Part B

METHODS IN ENZYMOLOGY

VOLUME SIX HUNDRED AND THREE

METHODS IN ENZYMOLOGY

Chemical and Biochemical Approaches for the Study of Anesthetic Function Part B

Edited by

RODERIC G. ECKENHOFF
University of Pennsylvania Perelman School of Medicine, Philadelphia, PA, United States

IVAN J. DMOCHOWSKI
University of Pennsylvania, Philadelphia, PA, United States

ELSEVIER

ACADEMIC PRESS
An imprint of Elsevier

Academic Press is an imprint of Elsevier
50 Hampshire Street, 5th Floor, Cambridge, MA 02139, United States
525 B Street, Suite 1800, San Diego, CA 92101–4495, United States
The Boulevard, Langford Lane, Kidlington, Oxford OX5 1GB, United Kingdom
125 London Wall, London, EC2Y 5AS, United Kingdom

First edition 2018

ISBN: 978-0-12-814574-6
ISSN: 0076-6879

Publisher: Zoe Kruze
Acquisition Editor: Zoe Kruze
Editorial Project Manager: Alina Cleju
Production Project Manager: Divya Krishna Kumar
Cover Designer: Miles Hitchen

Typeset by SPi Global, India

CONTENTS

Section II
Evolving Biophysical Technologies

CONTRIBUTORS

Mitchell S. Albert
Lakehead University; Thunder Bay Regional Health Research Institute; Northern Ontario School of Medicine, Thunder Bay, ON, Canada

Emilee C. Alkire
The Center for the Neurobiology of Learning and Memory, University of California, Irvine, Irvine, CA, United States

Michael T. Alkire
The Center for the Neurobiology of Learning and Memory, University of California, Irvine, Irvine; VA Long Beach Healthcare System, Long Beach, CA, United States

Hasan Babazada
Perelman School of Medicine at the University of Pennsylvania, Philadelphia, PA, United States

Helen A. Baghdoyan
University of Tennessee, Knoxville; Oak Ridge National Laboratory, Oak Ridge, TN, United States

J. Kent Blasie
University of Pennsylvania, Philadelphia, PA; University of California Irvine, Irvine, CA, United States

Vasyl Bondarenko
University of Pittsburgh School of Medicine, Pittsburgh, PA, United States

Simrun Chahal
Lakehead University, Thunder Bay, ON, Canada

Qiang Chen
University of Pittsburgh School of Medicine, Pittsburgh, PA, United States

Rohan Desai
University of Michigan, Ann Arbor, MI, United States

Ivan J. Dmochowski
University of Pennsylvania, Philadelphia, PA, United States

Daniel J. Emerson
University of Pennsylvania, Philadelphia, PA, United States

Zoe Gerstle
University of Michigan, Ann Arbor, MI, United States

Francis T. Hane
Lakehead University; Thunder Bay Regional Health Research Institute, Thunder Bay, ON, Canada

Andrew E. Hudson
UCLA, Los Angeles, CA, United States

Oskari Kantonen
Turku University Hospital, Turku, Finland; The Center for the Neurobiology of Learning and Memory, University of California, Irvine, Irvine, CA, United States

Max B. Kelz
University of Pennsylvania, Philadelphia, PA, United States

Jaakko Långsjö
Turku PET Centre, University of Turku and the Hospital District of Southwest Finland, Turku; Tampere University Hospital, Tampere, Finland

Renyu Liu
Perelman School of Medicine at the University of Pennsylvania, Philadelphia, PA, United States

Patrick J. Loll
Drexel University College of Medicine, Philadelphia, PA, United States

Ralph Lydic
University of Tennessee, Knoxville; Oak Ridge National Laboratory, Oak Ridge, TN, United States

Amanda L. May
University of Tennessee, Knoxville, TN, United States

Megan McGrath
Massachusetts General Hospital, Boston, MA, United States

Andrew R. McKinstry-Wu
University of Pennsylvania, Philadelphia, PA, United States

Jonathan Z. Pan
University of California at San Francisco, San Francisco, CA, United States

Braedan R.J. Prete
Lakehead University, Thunder Bay, ON, Canada

Alex Proekt
Perelman School of Medicine, Department of Anesthesiology and Critical Care, University of Pennsylvania, Philadelphia, PA, United States

Douglas E. Raines
Massachusetts General Hospital, Boston, MA, United States

Annalotta Scheinin
Turku PET Centre, University of Turku and the Hospital District of Southwest Finland; Turku University Hospital, Turku, Finland

Harry Scheinin
Turku PET Centre, University of Turku and the Hospital District of Southwest Finland; Turku University Hospital; Integrative Physiology and Pharmacology, Institute of Biomedicine, University of Turku, Turku, Finland

Efthimios M.C. Skoulakis
Biomedical Science Research Center "Alexander Fleming", Vari, Greece

Ken Solt
Massachusetts General Hospital; Harvard Medical School, Boston; Massachusetts Institute of Technology, Cambridge, MA, United States

Pei Tang
University of Pittsburgh School of Medicine, Pittsburgh, PA, United States

Luca Turin
Biomedical Science Research Center "Alexander Fleming", Vari, Greece

Christa J. Van Dort
Massachusetts General Hospital; Harvard Medical School, Boston; Massachusetts Institute of Technology, Cambridge, MA, United States

Sarah L. Veatch
University of Michigan, Ann Arbor, MI, United States

Ksenia Vlasov
Massachusetts General Hospital, Boston; Massachusetts Institute of Technology, Cambridge, MA, United States

Alanna Wade
Lakehead University, Thunder Bay, ON, Canada

Marta Wells
University of Pittsburgh School of Medicine, Pittsburgh, PA, United States

Yan Xu
University of Pittsburgh School of Medicine, Pittsburgh, PA, United States

PREFACE

"Anesthesia" is a reversible drug-induced state that most importantly (for the surgeon, at least!), causes an inability to generate a motor response to what would normally be painful stimuli. The general anesthetic state includes unconsciousness and amnesia. These features are considered to be produced in the central nervous system, and in fact, when absent (e.g., during normal "awake" function), they collectively define consciousness. Thus, the study of general anesthesia overlaps with the study of consciousness—probably the most enigmatic state that exists! It should not be surprising then, that the range of experimental and theoretical approaches to understanding the anesthetic state is very large. We have attempted to capture the scope, complexity, and significance of the many approaches to understanding the state of anesthesia in this two-volume set of *Methods in Enzymology*. This is, perhaps, an unusual collection for MIE, in that many of the approaches do not fit the term "enzymology," or for that matter, even chemistry or biochemistry. This second volume: *Chemical and Biochemical Approaches for the Study of Anesthetic Function Part B* details structural approaches (X-ray crystallography, NMR, X-ray interferometry) focused on both soluble and membrane-bound proteins, many incorporating anesthetics. Also included are biophysical approaches for probing anesthetic binding affinity and localization, which include fluorescence, isothermal titration calorimetry, and electron paramagnetic resonance (EPR) techniques, as well as methods for generating and manipulating model membrane systems. We conclude this volume with a description of several cutting-edge in vivo approaches for studying anesthetic function. Novel drug discovery efforts are increasingly guided by light-triggered optoanesthetic and optogenetic approaches. Methods for profiling neuronal activity in space and time monitor changes in calcium secondary messenger, genome, and proteome. And, finally, novel applications of electroencephalography, PET, and MRI are helping to establish the relevant in vivo sites of anesthetic action. We have sought out the recognized experts in each field, to provide a succinct account of what these approaches can contribute and how to employ them most effectively. With such a diverse and intense campaign, one might have thought the mechanism of anesthesia would have succumbed by now! But this could not be further from the truth; despite finding ever-increasing use in modern medicine, anesthesia remains almost as enigmatic as when it was discovered over

170 years ago. Through further deployment and refinement of these approaches, and many more to be developed, we may finally reveal the mechanisms underlying anesthesia, and eventually improve it.

Ivan J. Dmochowski and Roderic G. Eckenhoff

SECTION I

Structural Approaches

CHAPTER ONE

Structural Analysis of Anesthetics in Complex with Soluble Proteins

Patrick J. Loll[1]
Drexel University College of Medicine, Philadelphia, PA, United States
[1]Corresponding author: e-mail address: pjl28@drexel.edu

Contents

Abstract

Anesthetics interact with a broad range of different targets, including both soluble and membrane-bound proteins. Understanding these interactions at the molecular level requires detailed structural knowledge of anesthetic–protein complexes, and one of the most productive routes to such knowledge is X-ray crystallography. In this chapter we discuss the application of this technique to the analysis of complexes of anesthetics with soluble proteins. The model protein apoferritin is highlighted, and protocols are presented for obtaining diffraction-quality crystals of this protein in complex with different general anesthetics.

Methods in Enzymology, Volume 603
ISSN 0076-6879
https://doi.org/10.1016/bs.mie.2018.01.016

1. INTRODUCTION

Many methods can provide structural information that is useful in analyzing protein–ligand interactions. However, only diffraction methods allow for the direct visualization of molecular complexes. For many decades, the premier diffraction technique for analyzing molecular structures has been X-ray crystallography, which provides unbiased images at or near atomic resolution. Recently, electron microscopy has advanced to the stage where it, too, can yield images that approach atomic resolution; however, crystallography continues to offer many advantages and will provide detailed and useful insights for a long time to come.

In particular, crystallography has been a fruitful approach to analyzing how anesthetics interact with protein targets, including both soluble and membrane-bound proteins. The latter category includes many of the clinically relevant targets for general anesthetics, such as ion channels and receptors; such molecules are challenging subjects for crystallographic analysis, owing to the difficulties associated with producing, purifying, and crystallizing membrane proteins (Bill et al., 2011; Loll, 2014). However, decades of effort have overcome many of the hurdles associated with membrane protein structural biology, and structures of crucial biological molecules such as GPCRs and ion channels are now appearing regularly in the literature. This specific topic is discussed in more detail in another chapter within this volume.

There are also good reasons to study the interaction of anesthetics with soluble proteins. It is clear that anesthetics are promiscuous and interact with many different proteins, including soluble ones (Eckenhoff, Chan, & Eckenhoff, 2002; Weiser et al., 2015). Not all of these interactions contribute to general anesthesia, but even those that do not may still contribute to side effects. Also, despite the advances mentioned above, membrane protein structural biology remains challenging, and when experimenters require more tractable model systems, soluble proteins can be accommodating surrogates for their more recalcitrant membrane-bound cousins. For example, well-behaved workhorse proteins such as apoferritin, serum albumin, and luciferase have yielded numerous insights into anesthetic–protein interactions that would have been very difficult to obtain with membrane-bound targets (Bhattacharya, Curry, & Franks, 2000; Franks, Jenkins, Conti, Lieb, & Brick, 1998; Liu, Loll, & Eckenhoff, 2005; Sawas, Pentyala, & Rebecchi, 2004). In this chapter, we discuss the use of X-ray crystallography to analyze interactions of anesthetics with soluble proteins.

2. CRYSTALLOGRAPHIC ISSUES RELEVANT TO ANALYSIS OF ANESTHETIC BINDING

We will not attempt to explain the basics of crystallography here and refer interested students to the wide variety of available texts, which range from basic (Loll & Lattman, 2008; Rhodes, 2006) to more advanced (Drenth, 1994; Rupp, 2010; Wlodawer, Dauter, & Jaskolski, 2017). Rather, we will concentrate on those practical aspects of crystallography that are likely to be of interest to nonspecialists who wish to probe how anesthetics and related ligands interact with macromolecules.

2.1 Protein Crystallization

The first issue to be tackled in any crystallographic experiment is the generation of appropriate crystals; this is frequently also the most difficult issue to resolve. Crystals are required because X-rays interact only weakly with matter, and therefore, the diffraction signal produced by an individual molecule is far too feeble to measure. Crystals allow us to overcome this problem, because even a protein crystal of modest size contains a huge number of molecules, e.g., a protein crystal 50 μm on a side might easily contain 10^{12} molecules, depending on the molecular weight of the protein. Importantly, every molecule in the crystal—or at least every molecule in the irradiated volume—contributes to the final image, which therefore represents an average view of all molecules. In addition to cramming large numbers of scatterers into a confined space, crystals also ensure that the proteins are well aligned with each other. This means that different molecules scatter additively, giving rise to a measurable diffraction signal from the crystal as a whole. Therefore, well-ordered crystals are a sine qua non for any crystallography experiment. In practice, the bulk of the effort in a crystallographic project is frequently devoted to generating such crystals.

Proteins are flexible and irregularly shaped, which are features that tend to work against the formation of well-ordered crystals. Thus, crystallization of even soluble proteins can prove vexing. Many variables potentially affect crystal growth, and so experiments are designed to screen large numbers of different combinations of variables, in the hope of hitting upon one or more conditions favorable for crystal growth. Several hundred to a thousand different solutions are typically screened; these solutions are chosen to sample the available parameter space as efficiently as possible (McPherson & Gavira, 2014). If all goes well, initial crystals are obtained from such screens. If this happens, the key variables are manipulated so to improve crystal quality, using well-described approaches (McPherson & Cudney, 2014). Sometimes, the optimization process involves seeding, a process in which very small crystals or fragments of crystals are placed into fresh protein solutions in the hope of nucleating new and larger crystals (Bergfors, 2003; Stewart, Kolek, Briggs, Chayen, & Baldock, 2011).

However, certain proteins may defy efforts at optimization or may fail to yield any crystals at all, necessitating other approaches. One such approach is to introduce small changes into the protein's surface features. Such changes can have tremendous impact on crystallization behavior, because the protein–protein interactions that form crystal contacts are exquisitely sensitive to the details of a protein's surface. A time-honored (but still useful) way to accomplish this is to use orthologous proteins from different species. The expectation is that orthologs will adopt similar structures, but will have slightly different surface features, and hence will pack more or less readily into a crystal lattice (Kendrew, Parrish, Marrack, & Orlans, 1954). A more targeted approach is to identify and alter surface amino acids likely to prove problematic for crystallization. For example, highly flexible residues such as lysine tend to resist participating in crystal contacts, since this would require locking the side chain into a single conformation (Derewenda, 2011; Derewenda & Vekilov, 2006). Mutation of such residues to amino acids with smaller, more rigid (i.e., lower entropy) side chains can vastly improve crystallization behavior. Online servers are available that will analyze the protein's primary sequence, seeking putative surface regions that would be amenable to such treatment (Goldschmidt, Eisenberg, & Derewenda, 2014).

Small changes in surface residues can significantly improve crystallization outcomes for compact protein domains, but are unlikely to be sufficient if the protein to be crystallized contains entire regions that are highly flexible. This is a particularly significant issue for eukaryotic proteins, as they are very likely to contain either multiple domains connected by flexible linkers or intrinsically disordered regions (Pancsa & Tompa, 2012). Therefore, when considering a hitherto uncrystallized eukaryotic protein, the first step should be a careful analysis of the primary sequence, looking for regions of predicted disorder. Once such regions are identified, protein engineering approaches are applied to remove the offending sections of the sequence, thereby generating constructs that are likely to be more crystallization-friendly (Ruggiero, Smaldone, Squeglia, & Berisio, 2012).

2.2 Crystallizing the Protein–Ligand Complex

Growing crystals of the protein of interest is not enough, of course; one must obtain crystals of the desired complex, namely, protein bound to the anesthetic of interest. Two different approaches are commonly taken to achieve this, namely, cocrystallization and soaking. In the former approach, the ligand of interest is included in the solution during the crystallization experiment, in the expectation that it will bind to the protein and the species actually crystallizing is the protein–ligand complex. In the latter approach, preformed crystals are placed in a solution containing the ligand; with luck, the ligand will diffuse through solvent channels within the crystal and thus find and occupy its binding site. Both approaches have advantages and disadvantages. If the protein of interest can be crystallized readily, then the soaking experiment is easy to try and requires relatively little material. Typically, the crystal is transferred to a so-called artificial mother liquor (this is a solution that matches as closely as possible the solution in which the crystal originally grew; sometimes, the artificial mother liquor contains slightly higher precipitant concentrations than the original experiment, to maintain the stability of the crystal). This mother liquor also contains the anesthetic. Soak times vary widely, ranging from minutes to weeks. The kinetics of binding depend critically on the accessibility of the binding site(s), which could potentially be occluded by neighboring molecules in the crystal lattice. Success also depends upon whether binding is accompanied by any conformational change in the protein, which could disrupt the integrity of the lattice and cause the crystal to crack or dissolve.

Cocrystallization experiments require more material and more time than soaking experiments; however, they should, in principle, allow one to avoid problems related to conformational changes and binding site accessibility, as the protein is already bound to the anesthetic when it is incorporated into the lattice. In such experiments, knowledge of the approximate binding affinity of the ligand is invaluable, as it allows the experimenter to calculate whether or not the concentrations of ligand used are adequate to saturate the binding site. It should be noted that addition of a ligand frequently alters a protein's crystallization behavior, either by engendering conformational changes (even subtle changes can disfavor packing in a specific lattice) or by altering the physicochemical properties of the crystallization solution. Therefore, even if one knows how to crystallize the protein alone, the protein–anesthetic complex may not crystallize under exactly the same conditions; hence it is prudent to explore a range of values for key parameters such as pH and precipitant concentration, bracketing the values identified for the protein alone.

Commonly used crystallization solutions can deviate substantially from physiological conditions, and may include high ionic strength, extremes of pH, and high concentrations of polymers such as polyethylene glycol (PEG). Therefore, it is possible that the affinity of the anesthetic for the protein will change under the conditions used to crystallize the protein. One should not simply assume that, because the anesthetic binds the protein at neutral pH and moderate salt concentrations, it will also bind under the conditions used to produce crystals. Once crystallization conditions have been identified, it is possible to measure the affinity under those exact conditions; however, in actuality it may prove easier to determine the structure in order to establish whether or not the anesthetic is bound.

The issues described above are generally applicable when crystallizing any protein–ligand complex. However, anesthetic ligands bring special challenges. To begin, most anesthetics are hydrophobic, and simply achieving high enough concentrations to significantly populate their binding sites can be difficult. Use of organic cosolvents can help in this regard, but will typically have profound effects on crystallization behavior. Interestingly, hydrophobic solutes are frequently much more soluble in PEG solutions than they are in water. Since PEGs are some of the most frequently used precipitants in protein crystallization experiments, this can provide a fortuitous boost in solubility.

Another issue associated with anesthetics is volatility. This makes it difficult to maintain appropriate concentrations of the anesthetic throughout the crystallization experiment, which can easily span days or weeks. The most volatile anesthetics can easily pass through plastic crystallization plates, and so evaporation can quickly deplete the anesthetic in a protein solution. To reduce this problem, glass crystallization vessels can be used, rather than plastic; glass is not a panacea, however, because even glass experimental setups have seams and joints, and anesthetics will still escape from these points at a nonzero rate (see Fig. 1). Also, the overwhelming majority of commercially available crystallization plates are plastic, so if one chooses to use glass vessels, one should expect some do-it-yourself work. A different approach is to use commercially available plastic crystallization plates, but to enclose these plates in larger containment vessels. The larger container might be glass (e.g., a large TLC chamber laid on its side), or it might simply be a plastic food storage container. A small vial of neat

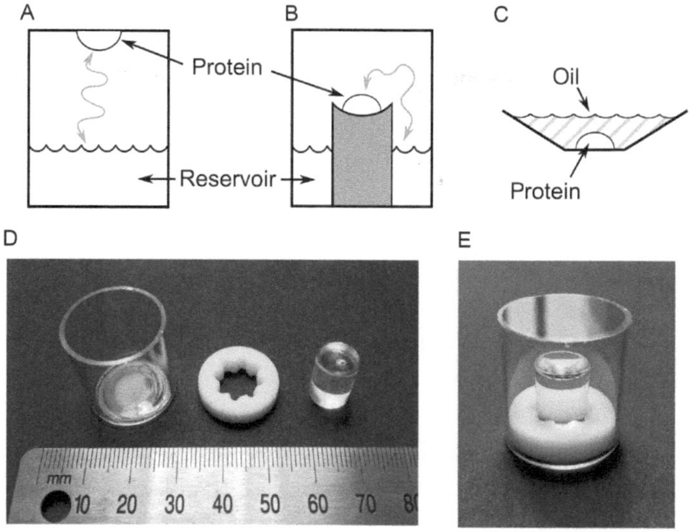

Fig. 1 Typical approaches to crystallizing soluble proteins. (A) The hanging drop geometry for vapor diffusion experiments. Protein solution is mixed with the precipitant (found in the reservoir) and pipetted onto a cover slip, which is then inverted over a reservoir containing precipitant. Diffusion of water vapor (and other volatile components) then allows for the drop to slowly equilibrate with the reservoir so that the precipitant concentration in the drop ultimately approaches that in the reservoir. (B) The sitting drop geometry for vapor diffusion experiments. (C) The microbatch-under-oil experiment. Protein solution is mixed with the precipitant, pipetted into a well, and then covered with oil. Depending on the nature of the oil used, either no water evaporates from the drop (making this a batch experiment—i.e., one where the solution conditions do not change over time), or else water can escape slowly, leading the gradual concentration of all components. (D) and (E) A homemade container for sitting drop vapor diffusion experiments, suitable for use with anesthetics. Panel (D) shows the components: From *left* to *right*, a 2-cm diameter glass vial that was cut down to reduce its height; a Teflon washer that was machined to hold the central rod; and a glass sitting drop rod (Hampton Research), which is a glass cylinder with a polished indentation in one end. The rod is siliconized before use. Panel (E) shows the assembled apparatus. After the reservoir and protein solutions are added, the chamber is capped with a cover slip and sealed with vacuum grease. Because the optical path is entirely glass, progress of the crystallization experiment can be monitored without opening the vessel.

anesthetic is included in the secondary container; this vial has a small hole in its lid, to maintain saturating levels of anesthetic vapor in the air around the crystallization plates and thereby slow the loss of anesthetic from the crystallization experiments. Depending on how permeable the secondary container is, this vial may need to be replenished every few days. (*Note*: Do not allow neat anesthetic to come into direct contact with plasticware, as many anesthetics will attack or dissolve plastics.) If vapor diffusion experiments are being used to grow crystals (Fig. 1), then ideally anesthetic should be included in both the protein drop and the reservoir; if microbatch-under-oil experiments are being used (Chayen, Stewart, & Blow, 1992), the oil used to cover the crystallization drops should be doped with anesthetic (the anesthetic might actually be significantly more soluble in the oil than in the aqueous protein solution; sufficient anesthetic should be included in the oil to ensure that the oil serves as a source, and not a sink, for the anesthetic in the crystallization drop).

Volatility poses special problems during crystal mounting. During mounting, a crystallization container is opened to expose the crystal; the experimenter then extracts it by capturing it in a small plastic or fiber loop, and next flash-cools it in either a stream of cold gas or a cryogen such as liquid nitrogen. Once the container is opened to allow access to the crystal, the mounting procedure can take anywhere from a few seconds to a few minutes. The time required depends on the skill of the experimenter, but is also affected by factors outside of the experimenter's control, such as whether a skin has formed on the drop, the size of the crystals, and the viscosity and surface tension of the specific crystallization solution being used. Even when working as quickly as possible, the crystal will be exposed to the atmosphere long enough that the loss of volatile anesthetic from the crystal is potentially concerning. One possibility is to enclose the microscope used to harvest the crystals in a glove bag filled with anesthetic vapor, but this is both cumbersome and carries the risk of exposing the experimenter to undesirable levels of anesthetic.

A different approach is to forego flash-cooling the crystal and instead collect diffraction data at room temperature. While this will hasten the rate at which radiation damage accumulates, it allows for the possibility of enclosing the crystal in a glass or quartz capillary, which can be sealed in order to maintain appropriate concentrations of anesthetic around the crystal. The practice of mounting protein crystals in sealed capillaries was the norm before ~1990, although it has fallen out of favor since (consult a senior colleague for assistance).

For the specific case of noble gases, devices have been designed to allow crystals to be exposed to high pressures of the gas, and then flash-cooled without ever removing the crystal from the gas environment (note that this represents a special case of the crystal "soaking" experiment described above) (Djinovic-Carugo, Everitt, & Tucker, 1998; Sauer, Schmidt, & Kratky, 1997). Commercial versions of such apparatus have appeared in the past, but to the author's knowledge, none are being marketed at the time of this writing.

Having listed all these caveats, it is necessary to stress that, in fact, it IS possible to observe volatile anesthetics bound specifically to proteins in crystal structures, even in cases where no extraordinary precautions were taken to minimize loss of ligand during crystal mounting (Soltis, Stowell, Wiener, Phillips, & Rees, 1997). Hence, if you are contemplating such an experiment, the best advice is to simply try it, and to use whatever common-sense measures you can easily implement to preserve ligand binding.

3. CRYSTALLIZING APOFERRITIN WITH ISOFLURANE AND PROPOFOL

The soluble iron-sequestering protein ferritin forms spherical 24-mer assemblies. Within these spherical shells, cavities can be found at the interface between two protomers. These cavities are buried and largely hydrophobic in character, and bind a variety of general anesthetic molecules (Vedula et al., 2009). Apoferritin (the iron-free form) is stable and easily obtained, making it an excellent model system for studying protein–anesthetic interactions. The affinities with which apoferritin recognizes a wide range of anesthetics correlate well with their anesthetic potencies (Butts et al., 2009), suggesting that insights about protein–anesthetic recognition gleaned from this system are likely to prove applicable to other, more clinically relevant (and less tractable) targets.

3.1 Equipment

1. FPLC or low-pressure chromatography system with UV detection
2. HiPrep Sephacryl S-300 HR size-exclusion column, 26 × 600 mm (GE Healthcare Life Sciences)
3. Centrifugal filtration units (30,000 Da nominal MW cutoff)
4. UV/visible spectrophotometer
5. Analytical balance
6. Bath sonicator
7. VDX™ crystallization plate with sealant (Hampton Research)
8. Siliconized 12 × 0.22- or 12 × 0.96-mm circular cover slides
9. 0.5- or 1-μL Hamilton syringe

3.2 Chemicals

1. Apoferritin from equine spleen (0.2 μm filtered; Sigma-Aldrich)
2. Sodium acetate
3. Acetic acid
4. Cadmium sulfate 8/3 hydrate (Sigma-Aldrich)

5. Ammonium sulfate
6. HEPES
7. Isoflurane (Sigma-Aldrich)
8. Propofol (2,6-diisopropylphenol; Sigma-Aldrich)
9. Methanol

4. PROTOCOL

4.1 Purification of Commercial Apoferritin via Size-Exclusion Chromatography

1.1 Working at 4°C, equilibrate the S-300 column with at least three column volumes (≥1 L) of 0.2 *M* sodium acetate, pH 5.0.

1.2 Load 0.5 mL of apoferritin on the S-300 column and elute with 0.2 *M* sodium acetate buffer, pH 5.0, at 1 mL/min. Collect 4-mL fractions.

1.3 As illustrated in Fig. 2, the chromatogram should reveal a large monomer peak, along with smaller, earlier eluting peaks that correspond to dimers and trimers (Thomas, Carter, & Rosenberger, 1998). Pool the fractions from the center of the monomer peak.

1.4 Measure the absorbance of the pooled fractions at 280 nm and estimate the concentration of the protein sample using an extinction coefficient $E_{0.1\%} = 0.723$.

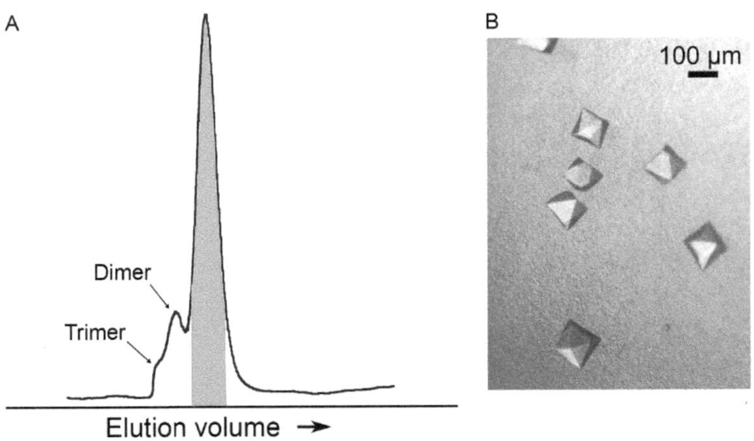

Fig. 2 (A) Purification of apoferritin by size-exclusion chromatography. The elution profile (OD at 280 nm) shows evidence for dimer and trimer species. The *shaded gray area* corresponds to the fractions chosen for crystallization. (B) Apoferritin crystals.

1.5 Adjust the concentration of the protein sample (by ultrafiltration or dilution with 0.2 *M* sodium acetate, pH 5.0) to 12 mg/mL. Store on ice until used.

Tip: Apoferritin from Sigma-Aldrich is supplied in saline solution at a concentration greater than 50 mg/mL. If starting with a more dilute protein solution, use a centrifugal concentrator to achieve the desired concentration.

Tip: Protein concentration is a critical parameter in crystallization. To ensure reproducibility, take great care when using absorbance to estimate concentration. The suggested approach is to dilute a small sample of the protein solution to an OD_{280} value between 0.2 and 1.0. When performing the dilution, determine the degree of dilution gravimetrically, rather than volumetrically (with a good analytical balance, measuring milligram masses is more precise than measuring microliter volumes). Perform dilutions and absorbance measurements in triplicate.

4.2 Cocrystallization with Isoflurane

2.1 Prepare the following stock solutions
250 m*M* $CdSO_4$ (641 mg $CdSO_4 \cdot 8/3H_2O$ in 10 mL)
2.5 *M* sodium acetate, pH 7.5
1 *M* Na·HEPES, pH 7.5

2.2 Calculate the dilutions required to prepare 1 mL of reservoir solution in each well of a 24-well ZDX™ plate. Rows A–D should contain 25, 50, 75, and 100 m*M* $CdSO_4$, respectively, while columns 1–6 should contain 0.2, 0.4, 0.6, 0.8, 0.9, and 1.0 *M* sodium acetate. All wells should contain 0.1 *M* HEPES buffer.

2.3 Place the required volumes of the three stock solutions in glass vials. For one 24-well plate, using 1 mL of reservoir per well, you will require 6 mL of the cadmium sulfate solution, 2.4 mL of the HEPES buffer solution, and 8.4 mL of the sodium acetate solution. Also, prepare a glass vial containing 10 mL of water. To each of the four vials, add 5 μL isoflurane per mL of liquid (this is sufficient to saturate the solution). Cap the vials, vortex to disperse the isoflurane, and then place the vials into a bath sonicator for 30 min to dissolve the isoflurane.

2.4 While the reservoir stock solutions are in the sonicator, add 1 μL isoflurane to 25 μL of apoferritin solution in a 0.5-mL centrifuge tube. Pipet to mix, cap, and incubate on ice for 30 min.

2.5 Working quickly, dispense the required volumes of water and stock solutions to prepare the dilutions calculated above. Dispense and mix the reservoir solutions one row at a time; then, for each well in that row, mix 1 μL of apoferritin solution and 1 μL of the reservoir solution in the center of a siliconized cover slide. Invert the cover slide and place over the well, pressing gently on the cover slide to bed it into the grease and seal the well. Repeat for all four rows.

2.6 Place the finished crystallization plate into a secondary container (e.g., a plastic food storage container with a tight-fitting lid). Place 2 mL of isoflurane into a small glass vial, cover tightly with aluminum foil, and poke several small pinholes into the foil. Use tape to secure the vial to the inside of the secondary container, to prevent it from tipping over. Place the lid on the secondary container and store at cool room temperature (ideally 18–22°C).

2.7 Crystals should appear within 1–2 days and achieve full size within 1 week. The plate may be briefly removed from the secondary container to examine the drops under a microscope. Monitor the isoflurane levels in the glass vial within the secondary container, replenishing as necessary.

2.8 If the crystals are to be mounted for X-ray diffraction analysis, they should be passed quickly through a cryoprotectant solution containing 30% sucrose (w/v) in the corresponding reservoir buffer and then flash-cooled in liquid nitrogen. Before use, the cryoprotectant should be saturated with isoflurane as described above.

4.3 Cocrystallization with Propofol

3.1 Prepare the following stock solutions
500 m*M* $CdSO_4$ (1.28 g $CdSO_4 \cdot 8/3H_2O$ in 10 mL)
2.0 *M* ammonium sulfate

3.2 Calculate the dilutions required to prepare 1 mL of reservoir solution in each well of a 24-well ZDX™ plate. Rows A–D should contain 150, 175, 200, and 225 m*M* $CdSO_4$, respectively, while columns 1–6 should contain 0.6, 0.7, 0.8, 0.9, 1.0, and 1.1 *M* ammonium sulfate. No buffer is used in these reservoir solutions.

3.3 As described in step 2.3, prepare propofol-saturated solutions of water, cadmium sulfate, and ammonium sulfate, using a nominal propofol concentration of at least 2 m*M*.

3.4 Prepare a 200 m*M* stock solution of propofol in methanol. Add 0.18 μL of this solution to 36 μL of apoferritin and mix well.

3.5 As described above (step 2.5), prepare the reservoir solutions using the propofol-saturated stocks. Mix 1.5 μL of the apoferritin solution with 1.5 μL of the reservoir solution to prepare the crystallization drops.

3.6 Incubate the tray at cool room temperature. No secondary container is necessary.

3.7 For diffraction analysis, cryoprotect the crystals before cooling by dipping briefly into a propofol-saturated solution containing 7 volumes of reservoir buffer plus 3 volumes of glycerol.

5. ANALYSIS OF THE STRUCTURE

5.1 Relevance of the Crystalline Environment (and What Does It Mean if I Do Not See the Anesthetic?)

In order to crystallize, proteins must adopt well-populated, stable conformations (Deller, Kong, & Rupp, 2016). After more than half a century of protein crystallography, we can be very confident that the weak forces imposed by crystal packing do not significantly distort these conformations; hence, crystal structures are revealing the true minimum-energy states defined by the protein's amino acid sequence. However, in the case of proteins that have evolved to have multiple minimum-energy states (e.g., a channel that toggles between open and closed forms), then the specific conditions of a crystallization experiment may drive the protein into one particular conformer; or, alternatively, even if several conformers are present, only one might be selected to join the crystal lattice. Obviously, if the anesthetic binds preferentially to one of several possible conformers, then it is necessary to crystallize that conformer to visualize the protein–anesthetic interaction. Thus, while the conformation seen in a crystal structure is almost certainly biologically relevant, it may not be the one you want to see. In practice, crystallizing the "correct" conformer requires some luck, since it is difficult or impossible to predict in advance which conformation will crystallize.

5.2 Dynamics

Proteins are dynamic molecules, a fact that has implications for visualizing proteins via X-ray crystallography. The issue of dynamics is particularly relevant for anesthetics, since these molecules are generally bound only weakly to their target proteins. Dynamic behavior gives rise to disorder, which is a departure from the ideal notion that all copies of a molecule in a crystal are identical to each other. Disorder may be dynamic or static. Dynamic disorder occurs because the crystallographic experiment takes place on a timescale from seconds to hours, which is much slower than the timescale of molecular motions. Thus, individual atoms are constantly moving throughout the experiment—vibrating, undergoing ring flips, etc. (this is analogous to a photographer's subject breathing or blinking during a photo). Diffraction experiments are typically conducted at low temperatures (~100 K), which slows atomic motions and hence reduces dynamic disorder. However, static disorder is also present, and in protein crystals, static disorder is generally more significant than dynamic disorder. Static disorder results when different copies of the molecules that are present in the crystal adopt different orientations or ground-state conformations. Both dynamic and static disorders are present, to varying degrees, in all crystal structures. Because the images obtained from crystallography are averaged both in time (over the course of the diffraction experiment) and in space (over all molecules in the crystal), both types of disorder have similar effects on our ability to visualize molecules via crystallography. In simple terms, more disorder (either dynamic or static) gives rise to more blurring and a weaker image.

It is not straightforward to rigorously account for the effects of disorder. Crystallographers have historically used the atomic displacement parameter, or "*B* value," to model motion in crystal structures. The *B* value derives from a simple disorder model in which an atom vibrates isotropically about its central position. *B* is defined as being proportional to the mean square amplitude of the atom's motion. This model fails to account for anisotropic motions (which abound in proteins) and does not accurately represent the effects of static disorder. However, a better alternative is not readily available, and therefore, the *B* value is still used routinely in protein crystallography. However, it should not be taken as direct measure of motion.

The other parameter relevant to this discussion is occupancy. If we are visualizing the complex of a protein bound to an anesthetic, we can safely assume that the protein is present at an occupancy of 100%–i.e., every position in the crystal lattice is occupied by a protein molecule. However, not every one of these protein molecules necessarily contains an anesthetic molecule in its binding site; if the binding affinity is weak, it is unlikely that all sites are saturated, and, as discussed above, there are ample opportunities for weakly bound anesthetics to escape from the crystal during handling. Therefore, a priori the occupancy of the anesthetic will be unknown and may range anywhere from 0% (no anesthetic present) to 100% (every binding site contains an anesthetic molecule).

Both the occupancy and *B* value are model parameters that are optimized during the refinement process. However, these numbers should be interpreted with caution. Importantly, the two parameters are highly correlated. This may not be intuitive, since from a molecular standpoint, changing an atom's occupancy and changing the amplitude of its motion are two very different things. However, decreasing occupancy and increasing *B* value both have the effect of reducing the atom's electron density, and therefore in practice it is difficult to separate the two effects (Fig. 3). The confounding effects of B values on occupancy make it difficult to quantify ligand binding in crystals. Some investigators have tackled this problem for gaseous ligands, deducing affinities from the concentration dependence of occupancies (Marassio et al., 2011); however, this is not a trivial undertaking, requiring as it does collection of many different data sets at different anesthetic gas pressures, and careful control of crystal-to-crystal variation.

In anesthetic–protein complex structures, anesthetics tend to have higher *B* values than the proteins to which they bind, suggesting that the anesthetics may be moving within their binding cavities. This tends to blur out the

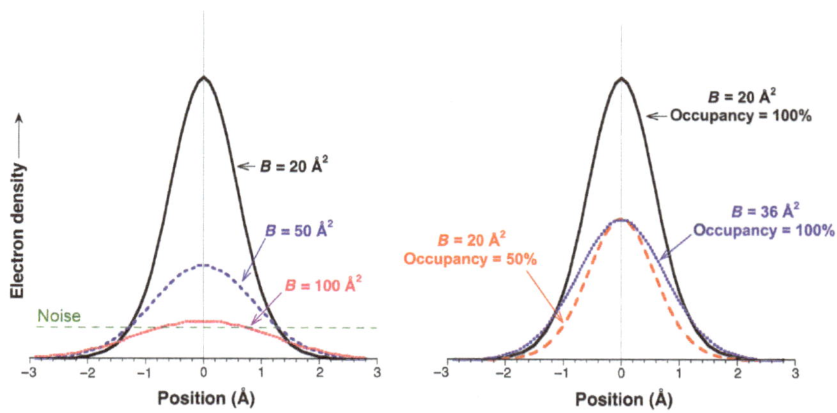

Fig. 3 The effects of *B* value and occupancy on electron density. *Left panel* shows how the theoretical electron density for a single atom changes with the *B* value. The peak height drops and the peak broadens as the *B* value increases; in practice, atoms with too high a *B* value become invisible, as the height of the electron density peak falls below the noise value in the map (hypothetical noise value shown as a *dotted line*). *Right panel* demonstrates the linkage between the *B* value and occupancy. The *upper, solid line* shows the electron density for a single atom with full occupancy and a low *B* value. Either reducing the occupancy (*dashed line*) or increasing the *B* value (*dotted line*) will reduce the peak height. One can see that there are subtle differences in the theoretical peak shapes associated with the two effects, but in actual experimental electron density maps, these differences are too small to be noticed.

electron density, as discussed above, which can make it difficult to unambiguously assign the orientation of the ligand, particularly for structures at modest resolution (lower resolution = less detail). In this case, it can be very useful to have a reporter that highlights a specific atom on the anesthetic molecule. Heavy atoms or anomalous scatterers are ideal for this purpose. Heavy atoms are electron-rich, and so have stronger signals in the electron density map than the C, O, and N atoms found in proteins and most organic molecules; however, most truly heavy atoms, such as mercury and platinum, cannot be included into an anesthetic or anesthetic analog (iodine is a possible exception). Anomalous scatterers exhibit changes in their diffraction profiles that are sharply wavelength-dependent, and thus, when probed with adjustable wavelength X-ray sources, can very precisely report on the identity and location of specific atoms. Since even relatively light elements such as Br exhibit significant anomalous scattering, inclusion of anomalous scatters into an anesthetic-like molecule is considerably more feasible than including a heavy atom. In particular, the anomalous scattering properties of bromine have proven useful in analyzing protein complexes with halothane and bromoform (Liu et al., 2005; Spurny et al., 2013). While bromine is not a common constituent of anesthetics, sulfur and chlorine are already present in some anesthetics and also have weak anomalous scattering signals, which can be exploited to provide reporters in protein–anesthetic complexes (Fourati et al., 2017).

6. SUMMARY

Investigators who seek to use crystallography to analyze how soluble proteins recognize anesthetics may encounter a number of specific challenges. This chapter describes these challenges, attempting to explain the underlying mechanisms and to suggest constructive approaches for overcoming them. While the hydrophobic character and volatility of anesthetic ligands can make them taxing to work with, careful attention to basic biophysical and biochemical principles will minimize problems and render structural analysis feasible, even when working with these formidable molecules.

ACKNOWLEDGMENT

This work was supported by NIH P01 GM055876.

REFERENCES

Bergfors, T. (2003). Seeds to crystals. *Journal of Structural Biology, 142*, 66–76.

Bhattacharya, A. A., Curry, S., & Franks, N. P. (2000). Binding of the general anesthetics propofol and halothane to human serum albumin. *Journal of Biological Chemistry, 275*, 38731–38738.

Bill, R. M., Henderson, P. J., Iwata, S., Kunji, E. R., Michel, H., Neutze, R., et al. (2011). Overcoming barriers to membrane protein structure determination. *Nature Biotechnology, 29*, 335–340.

Butts, C. A., Xi, J., Brannigan, G., Saad, A. A., Venkatachalan, S. P., Pearce, R. A., et al. (2009). Identification of a fluorescent general anesthetic, 1-aminoanthracene. *Proceedings of the National Academy of Sciences of the United States of America, 106*, 6501–6506.

Chayen, N. E., Stewart, P. D. S., & Blow, D. M. (1992). Microbatch crystallization under oil—A new technique allowing many small-volume crystallization trials. *Journal of Crystal Growth, 122*, 176–180.

Deller, M. C., Kong, L., & Rupp, B. (2016). Protein stability: A crystallographer's perspective. *Acta Crystallographica Section F, 72*, 72–95.

Derewenda, Z. S. (2011). It's all in the crystals. *Acta Crystallographica Section D, 67*, 243–248.

Derewenda, Z. S., & Vekilov, P. G. (2006). Entropy and surface engineering in protein crystallization. *Acta Crystallographica Section D, 62*, 116–124.

Djinovic-Carugo, K., Everitt, P., & Tucker, P. A. (1998). A cell for producing xenon-derivative crystals for cryocrystallographic analysis. *Journal of Applied Crystallography, 31*, 812–814.

Drenth, J. (1994). *Principles of protein X-ray crystallography*. New York: Springer-Verlag.

Eckenhoff, M. F., Chan, K., & Eckenhoff, R. G. (2002). Multiple specific binding targets for inhaled anesthetics in the mammalian brain. *The Journal of Pharmacology and Experimental Therapeutics, 300*, 172–179.

Fourati, Z., Ruza, R. R., Laverty, D., Drege, E., Delarue-Cochin, S., Joseph, D., et al. (2017). Barbiturates bind in the GLIC ion channel pore and cause inhibition by stabilizing a closed state. *Journal of Biological Chemistry, 292*, 1550–1558.

Franks, N. P., Jenkins, A., Conti, E., Lieb, W. R., & Brick, P. (1998). Structural basis for the inhibition of firefly luciferase by a general anesthetic. *Biophysical Journal, 75*, 2205–2211.

Goldschmidt, L., Eisenberg, D., & Derewenda, Z. S. (2014). Salvage or recovery of failed targets by mutagenesis to reduce surface entropy. *Methods in Molecular Biology, 1140*, 201–209.

Kendrew, J. C., Parrish, R. G., Marrack, J. R., & Orlans, E. S. (1954). The species specificity of myoglobin. *Nature, 174*, 946–949.

Liu, R., Loll, P. J., & Eckenhoff, R. G. (2005). Structural basis for high-affinity volatile anesthetic binding in a natural 4-helix bundle protein. *FASEB Journal, 19*, 567–576.

Loll, P. J. (2014). Membrane proteins, detergents and crystals: What is the state of the art? *Acta Crystallographica Section F, 70*, 1576–1583.

Loll, P. J., & Lattman, E. E. (2008). *Protein crystallography: A concise guide*. Baltimore: Johns Hopkins University Press.

Marassio, G., Prange, T., David, H. N., Santos, J. S., Gabison, L., Delcroix, N., et al. (2011). Pressure-response analysis of anesthetic gases xenon and nitrous oxide on urate oxidase: A crystallographic study. *FASEB Journal, 25*, 2266–2275.

McPherson, A., & Cudney, B. (2014). Optimization of crystallization conditions for biological macromolecules. *Acta Crystallographica Section F, 70*, 1445–1467.

McPherson, A., & Gavira, J. A. (2014). Introduction to protein crystallization. *Acta Crystallographica Section F, 70*, 2–20.

Pancsa, R., & Tompa, P. (2012). Structural disorder in eukaryotes. *PLoS One*, 7. e34687.

Rhodes, G. (2006). *Crystallography made crystal clear* (3rd ed.). New York: Academic Press.

Ruggiero, A., Smaldone, G., Squeglia, F., & Berisio, R. (2012). Enhanced crystallizability by protein engineering approaches: A general overview. *Protein & Peptide Letters*, *19*, 732–742.

Rupp, B. (2010). *Biomolecular crystallography: Principles, practice, and application to structural biology*. New York: Garland Science.

Sauer, O., Schmidt, A., & Kratky, C. (1997). Freeze-trapping isomorphous xenon derivatives of protein crystals. *Journal of Applied Crystallography*, *30*, 476–486.

Sawas, A. H., Pentyala, S. N., & Rebecchi, M. J. (2004). Binding of volatile anesthetics to serum albumin: Measurements of enthalpy and solvent contributions. *Biochemistry*, *43*, 12675–12685.

Soltis, S. M., Stowell, M. H. B., Wiener, M. C., Phillips, G. N., & Rees, D. C. (1997). Successful flash-cooling of xenon-derivatized myoglobin crystals. *Journal of Applied Crystallography*, *30*, 190–194.

Spurny, R., Billen, B., Howard, R. J., Brams, M., Debaveye, S., Price, K. L., et al. (2013). Multisite binding of a general anesthetic to the prokaryotic pentameric Erwinia chrysanthemi ligand-gated ion channel (ELIC). *Journal of Biological Chemistry*, *288*, 8355–8364.

Stewart, P. D. S., Kolek, S. A., Briggs, R. A., Chayen, N. E., & Baldock, P. F. M. (2011). Random microseeding: A theoretical and practical exploration of seed stability and seeding techniques for successful protein crystallization. *Crystal Growth and Design*, *11*, 3432–3441.

Thomas, B. R., Carter, D., & Rosenberger, F. (1998). Effect of microheterogeneity on horse spleen apoferritin crystallization. *Journal of Crystal Growth*, *187*, 499–510.

Vedula, L. S., Brannigan, G., Economou, N. J., Xi, J., Hall, M. A., Liu, R., et al. (2009). A unitary anesthetic binding site at high resolution. *Journal of Biological Chemistry*, *284*, 24176–24184.

Weiser, B. P., Hall, M. A., Weinbren, N. L., Woll, K. A., Dailey, W. P., Eckenhoff, M. F., et al. (2015). Macroscopic and macromolecular specificity of alkylphenol anesthetics for neuronal substrates. *Scientific Reports*, *5*, 9695.

Wlodawer, A., Dauter, Z., & Jaskolski, M. (Eds.), (2017). *Protein crystallography: Methods and protocols*. New York: Humana Press.

CHAPTER TWO

X-Ray Crystallographic Studies for Revealing Binding Sites of General Anesthetics in Pentameric Ligand-Gated Ion Channels

Qiang Chen, Yan Xu, Pei Tang[1]
University of Pittsburgh School of Medicine, Pittsburgh, PA, United States
[1]Corresponding author: e-mail address: tangp@anes.upmc.edu

Contents

Abstract

X-ray crystallography is a powerful tool in structural biology and can offer insight into structured-based understanding of general anesthetic action on various relevant molecular targets, including pentameric ligand-gated ion channels (pLGICs). In this chapter, we outline the procedures for expression and purification of pLGICs. Optimization of crystallization conditions, especially to achieve high-resolution structures of pLGICs bound with general anesthetics, is also presented. Case studies of pLGICs bound with the volatile general anesthetic isoflurane, 2-bromoethanol, and the intravenous general anesthetic ketamine are revisited.

Methods in Enzymology, Volume 603
ISSN 0076-6879
https://doi.org/10.1016/bs.mie.2018.01.017

1. INTRODUCTION

Pentameric ligand-gated ion channels (pLGICs) are targets of general anesthetics (Campagna, Miller, & Forman, 2003; Franks & Lieb, 1994; Hemmings et al., 2005). However, there is no consensus on the underlying molecular mechanisms of how general anesthetics modulate pLGICs. A number of experimental methods and structural tools have been developed and utilized to identify the binding sites of general anesthetics and understand their modulation on pLGICs, including site-directed mutagenesis in combination with electrophysiology measurements (Mihic et al., 1997), photoaffinity labeling (Chen et al., 2012; Chiara, Dangott, Eckenhoff, & Cohen, 2003; Chiara et al., 2012, 2014; Eckenhoff & Shuman, 1993; Eckenhoff et al., 2010; Husain et al., 2010; Jayakar, Dailey, Eckenhoff, & Cohen, 2013; Jayakar et al., 2014; Woll, Dailey, Brannigan, & Eckenhoff, 2016; Yip et al., 2013), nuclear magnetic resonance (Bondarenko, Mowrey, Liu, Xu, & Tang, 2013; Bondarenko et al., 2014; Bondarenko, Yushmanov, Xu, & Tang, 2008; Kinde et al., 2016), and X-ray crystallography (Chen et al., 2015, 2017; Nury et al., 2011; Pan, Chen, Willenbring, Mowrey, et al., 2012; Sauguet, Howard, et al., 2013; Spurny et al., 2013). Among these methods, X-ray crystallography is particularly well suited to determine atomic-resolution structures of large channel proteins in complex with anesthetics.

In the last decade, remarkable milestones have been reached in solving crystal structures of pLGICs, beginning with the prokaryotic channels ELIC (Hilf & Dutzler, 2008) and GLIC (Bocquet et al., 2009; Hilf & Dutzler, 2009) and more recently with the eukaryotic channels GluCl (Hibbs & Gouaux, 2011), $5HT_{3A}R$ (Hassaine et al., 2014), $\beta3GABA_AR$ (Miller & Aricescu, 2014), α3GlyR (Huang, Chen, Michelsen, Schneider, & Shaffer, 2015), and α4β2nAChR (Morales-Perez, Noviello, & Hibbs, 2016). The crystallographic approach has also successfully led to the discovery of multiple binding sites of general anesthetics in pLGICs (Chen et al., 2015, 2017; Nury et al., 2011; Pan, Chen, Willenbring, Mowrey, et al., 2012; Spurny et al., 2013). These studies provide not only structural insights into the molecular recognition of general anesthetics and the mechanisms underlying anesthetic modulation of pLGICs but also a structural framework for the discovery of novel general anesthetics.

In spite of several recent successes (Chen et al., 2015, 2017; Nury et al., 2011; Pan, Chen, Willenbring, Mowrey, et al., 2012; Spurny et al., 2013), the application of X-ray crystallography to study the interactions between general anesthetics and pLGICs is by no means straightforward. pLGICs are ~200 kDa (prokaryotic) or ~300 kDa (eukaryotic) integral membrane proteins composed of five identical or homologous subunits. They assemble into pentamers under optimal purification conditions, but they also have a tendency to aggregate or to dissociate into smaller oligomers. Although most of the extracellular domain (ECD) and the transmembrane helices are well ordered, there are flexible regions that interfere with crystallization, including the flexible N- and C-termini and the highly variable intracellular domain of eukaryotic pLGICs (Hibbs & Gouaux, 2011; Huang et al., 2015; Miller & Aricescu, 2014; Morales-Perez et al., 2016). Posttranslational modifications of proteins expressed in eukaryotic systems may also interfere with monodispersity and the ability to form high-quality crystals (Hassaine et al., 2013). Formation of high-quality cocrystals with anesthetics is further complicated by the nature of these relative small and mostly hydrophobic molecules. They often bind to multiple sites on pLGICs with low affinities (μ*M*–m*M*). All these factors may affect the crystal quality and structural resolution.

In this chapter, we provide protocols for crystallization of pLGICs in the absence and presence of general anesthetics, and for the identification and functional validation of general anesthetic-binding sites in pLGICs. Although these protocols were developed for prokaryotic pLGICs, similar principles should apply to the eukaryotic channels.

2. PRODUCTION OF MONODISPERSE pLGICs WITH HIGH PURITY

X-ray crystallography requires an adequate amount of monodisperse protein with high purity, often on the order of milligrams. Successful protein production normally includes three major steps: making proper constructs, finding a suitable expression system, and optimizing purification protocols. Sometimes, this procedure may need to be iteratively optimized based on the crystal diffraction quality of the proteins. pLGICs for X-ray crystallographic studies have been obtained as recombinant proteins overexpressed in *Escherichia coli* (*E. coli*) cells for prokaryotic pLGICs (Bocquet et al., 2009, 2007; Hilf & Dutzler, 2008, 2009) and from insect (Du, Lu, Wu, Cheng, & Gouaux, 2015; Hibbs & Gouaux, 2011; Huang et al., 2015) or human embryotic kidney (HEK) (Hassaine et al., 2014, 2013; Miller & Aricescu, 2014; Morales-Perez et al., 2016) cells for eukaryotic pLGICs.

The lower cost, shorter timeline, and versatile strains make *E. coli* an attractive candidate for protein expression not only for prokaryotic proteins but also for eukaryotic membrane proteins. Recently, we have successfully produced full-length wild-type human α7nAChR using *E. coli* and demonstrated that the purified α7nAChR retains functionality and pharmacological properties of native α7nAChR (Tillman et al., 2016). The success with α7nAChR paves a new path for the production of eukaryotic pLGICs using *E. coli*. Other examples of eukaryotic membrane protein expression in *E. coli* can be found in a recent review (Snijder & Hakulinen, 2016).

2.1 Expression of pLGICs

All pLGICs crystallized to date have been modified to remove flexible regions or posttranslational modifications that may interfere with crystal formation. All eukaryotic pLGICs structures obtained so far have removed most or all residues in the intracellular domain because it contains flexible disordered regions (Hassaine et al., 2014; Hibbs & Gouaux, 2011; Huang et al., 2015; Miller & Aricescu, 2014; Morales-Perez et al., 2016). Other highly flexible regions, including the N- and C-termini, may also need to be modified for crystallizations (Hassaine et al., 2014; Hibbs & Gouaux, 2011; Huang et al., 2015; Miller & Aricescu, 2014; Morales-Perez et al., 2016). In principle, nonhomogeneous glycosylation of pLGICs expressed from insect or HEK cells can interfere with crystallization and this heterogeneity can be reduced with deglycosylation enzymes or by point mutations of the glycosylation site (Hassaine et al., 2013; Miller & Aricescu, 2014). However, this step can be skipped in some cases with few glycosylation sites (Du et al., 2015; Hibbs & Gouaux, 2011; Huang et al., 2015; Morales-Perez et al., 2016). A cleavable purification tag, such as polyhistidine (Bocquet et al., 2009; Hibbs & Gouaux, 2011; Hilf & Dutzler, 2008, 2009) and strep (Hassaine et al., 2013; Huang et al., 2015; Morales-Perez et al., 2016), or antibody recognition motif (Miller & Aricescu, 2014) can be added to the N- or C-terminus to facilitate purification. For expressions in *E. coli*, fusion with a soluble protein (e.g., maltose-binding protein (MBP)) may aid in the expression of well-folded channel proteins (Bocquet et al., 2007). Table 1 summarizes fusion proteins and expression tags that have been used for overexpression of pLGICs. Fluorescence-detection size-exclusion chromatography (FSEC) (Kawate & Gouaux, 2006) is an efficient tool for construct screening (Bird et al., 2015), which has been used in the design of crystal constructs

Table 1 Host Cells, Tags, and Detergents Used for Expression and Purification of pLGICs

pLGICs	Host Cells	Tags	Detergents	References
ELIC	*E. coli* BL21(DE3)	N-His_{10}-MBP	UDM	Bertozzi, Zimmermann, Engeler, Hilf, and Dutzler (2016), Hilf and Dutzler (2008), Zimmermann and Dutzler (2011), and Zimmermann, Marabelli, Bertozzi, Sivilotti, and Dutzler (2012)
	E. coli BL21-Gold (DE3)			Gonzalez-Gutierrez et al. (2012)
	E. coli C43			Spurny et al. (2013, 2012) and Ulens et al. (2014)
	E. coli Rosetta(DE3) pLysS			Chen et al. (2015, 2017) and Pan, Chen, Willenbring, Yoshida, et al. (2012)
GLIC	*E. coli* BL21(DE3)	N-His_{10}-MBP	DDM	Bertozzi et al. (2016), Hilf et al. (2010), Hilf and Dutzler (2009), Velisetty and Chakrapani (2012), and Velisetty, Chalamalasetti, and Chakrapani (2012)
	E. coli BL21-Gold (DE3)			Gonzalez-Gutierrez, Cuello, Nair, and Grosman (2013)
	E. coli C43			Basak, Schmandt, Gicheru, and Chakrapani (2017), Bocquet et al. (2009), Fourati, Sauguet, and Delarue (2015), Laurent et al. (2016), Menny et al. (2017), Nury et al. (2011), Prevost et al. (2012), Sauguet, Fourati, Prange, Delarue, and Colloc'h (2016), Sauguet, Poitevin, et al. (2013), and Sauguet et al. (2014)
	E. coli Rosetta(DE3) pLysS			Mowrey et al. (2013) and Pan, Chen, Willenbring, Mowrey, et al. (2012)
	Drosophila Schneider (S2)	C-His_{10}		Sauguet et al. (2014)
GLIC-GlyR				Moraga-Cid et al. (2015)
ELIC-GLIC	*E. coli* BL21(DE3)	N-His_{10}-MBP		Schmandt et al. (2015)

Continued

Table 1 Host Cells, Tags, and Detergents Used for Expression and Purification of pLGICs—cont'd

pLGICs	Host Cells	Tags	Detergents	References
α7nAChR	Rosetta2(DE3)pLysS	C-His_8	DPC	Tillman et al. (2016)
GluCl	Sf9		DDM	Althoff, Hibbs, Banerjee, and Gouaux (2014) and Hibbs and Gouaux (2011)
α1GlyR				Du et al. (2015)
α3GlyR		C-Strep		Huang et al. (2015), Huang, Chen, and Shaffer (2017), and Huang, Shaffer, et al. (2017)
α4β2nAChR	GnTI-HEK		DDM +CHS	Morales-Perez et al. (2016)
β3$GABA_A$R	HEK293S-$GnTI^-$	TETSQVAPA	DMNG +CHS	Miller and Aricescu (2014)
$5HT_{3A}$R	T-REx-293	N-Strep	C12E9	Hassaine et al. (2014)

of pLGICs, including GluCl (Hibbs & Gouaux, 2011), β3GABA$_A$R (Miller & Aricescu, 2014), and α4β2nAChR (Morales-Perez et al., 2016).

A brief generalized protocol for overexpression of pLGICs in *E. coli* is given as follows:

1. Transform the expression plasmid into the chosen *E. coli* strain.
2. Pick a few colonies into 10 mL rich media, such as Luria–Bertani (LB) broth supplemented with the selective antibiotics. Grow cells overnight at 37°C.
3. Inoculate (1:100) overnight culture into 1 L LB supplemented with antibiotics. Grow cells at 37°C in an environmental shaker at 250 rpm until optical density (OD, measured at 600 nm) reaches 0.6–0.8.
4. Lower the environmental shaker temperature to 15°C. After 1-h temperature equilibration, induce the culture by adding 0.2 m*M* isopropyl β-D-1-thiogalactopyranoside (IPTG).
5. Harvest cells after expression for ~16 h. Store cell pellets at −80°C.

Note 1: Expression media, OD_{600} at induction, IPTG concentration, temperature, and expression time should be optimized for individual proteins.

Note 2: For the expression of pLGICs in insect or HEK cells, please refer to the similar protein expression protocols referenced in Table 1.

2.2 Purification of pLGICs

The process of protein purification isolates the desired protein from other cell components in the expression system. In general, the purification of pLGICs involves up to four steps: (1) isolating the cell membrane that contains the expressed protein, (2) solubilizing the membrane with a proper detergent, (3) purifying the expressed pLGICs from other membrane fractions, and (4) cleaving the fusion protein or tags whenever necessary. General guides on purifying membrane proteins have been recently reviewed (Roy, 2015). Specific purification protocols for pLGICs used in crystallizations can be found in the references for Table 1.

One of the key factors affecting pLGIC purification is the choice of detergent, which is crucial not only for solubilizing the membrane but also for maintaining the correct folding of pLGICs (Hovius et al., 1998; Miller & Aricescu, 2014). Table 1 lists the detergents used for successful purification of pLGICs. *n*-Dodecyl-β-D-maltopyranoside (DDM) is a mild detergent commonly used for solubilizing and purifying pLGICs. FSEC can be applied to identify suitable detergents for a given pLGIC (Miller & Aricescu, 2014). The purification is normally performed at 4°C to prevent aggregation.

A brief purification protocol for pLGICs expressed in *E. coli* is given as follows:

1. Lyse cells and collect the membrane with ultracentrifugation at least 100,000 × *g* for ∼1 h.
2. Solubilize the collected membrane using ∼2% of the chosen detergent.
3. Purify the pLGIC with immobilized metal ion affinity chromatography or antibody resin based on the designed purification tag. Check protein purity with sodium dodecyl sulfate polyacrylamide gel electrophoresis (SDS-PAGE).
4. Cleave and separate the fusion protein and/or tags from the pLGIC if necessary.
5. Collect the purified pentameric fraction of the protein by size-exclusion chromatography (SEC).
6. Check protein purity with SDS-PAGE.

Note 3: The amount of detergent to solubilize the membrane is about 20% in weight and should be optimized according to the quantity of collected membrane.

2.3 Functional Validation of pLGIC Constructs Used for Crystallizations

As pLGICs are often modified to improve their monodispersity and fit crystallization requirements, it is necessary to check if modifications to the recombinant protein construct have resulted in adverse effects on pharmacological features and channel functions, particularly the characteristic response to general anesthetics. Most studies have performed functional measurements in *Xenopus* oocytes or HEK cells heterologously expressing the pLGIC constructs (Bocquet et al., 2007; Du et al., 2015; Hibbs & Gouaux, 2011; Hilf & Dutzler, 2009; Huang et al., 2015; Miller & Aricescu, 2014; Morales-Perez et al., 2016). The purified pLGIC can also be directly tested for channel function, either directly measuring function of reconstituted pLGICs in the lipid bilayer (Hilf & Dutzler, 2008) or by injecting vesicles containing purified pLGICs into *Xenopus* oocytes for functional measurements (Labriola et al., 2013; Tillman et al., 2016).

3. CRYSTALLIZATION OF pLGICs IN THE ABSENCE AND PRESENCE OF GENERAL ANESTHETICS

Protein crystallization can be achieved through vapor diffusion (Delmar, Bolla, Su, & Yu, 2015), microbatch (Chayen, 1998), or free-interface diffusion (Segelke, 2005). For a new soluble protein, the standard crystallization procedure involves screening the purified protein against sparse-matrix conditions (Luft, Newman, & Snell, 2014; Newman et al., 2005), including pH, temperature, and the choice of salt and type of polyethylene glycol (PEG) at varying concentrations. Crystallization of membrane proteins, such as pLGICs, involves additional variables due to the requirement of using detergents or other membrane mimetics (Bill et al., 2011; Landau & Rosenbusch, 1996). Cocrystallizing pLGICs with anesthetics offers a reliable way to reveal anesthetic-binding sites and pLGIC conformational changes induced by the anesthetic binding. However, the inclusion of general anesthetics may add extra complications to crystallization. The choice of anesthetics and the method to introduce anesthetics to crystallization samples can make a profound difference in outcomes. A comprehensive consideration of multiple factors increases success in obtaining pLGIC crystal structures.

3.1 Crystallization of pLGICs

Crystallization of pLGICs begins with systematic screening of multiple variables, including pH, temperature, the type of buffer solution, and the type and concentration of precipitants and salts. The screening can use commercially available membrane protein crystallization screening kits, such as MemGold, MemGold2 (Molecular Dimensions), and MembFac (Hampton Research). Table 2 summarizes the crystallization conditions for pLGICs structures that have been determined so far, including protein concentration, reservoir solution, additives, crystallization method, and cryoprotectants. The same detergent used in purification (Table 1) is typically also used in crystallization. The concentration of pLGICs used for crystallization spans a wide range, from as low as 1–2 mg/mL for GluCl to as high as 8–12 mg/mL for GLIC. Thus, it requires optimization for individual pLGICs. Additives, such as agonists, antagonist, channel modulators, lipids, and nanobodies, can be used to improve the quality of pLGIC crystals. High-quality diffracting crystals of pLGICs have been obtained using both sitting-drop and hanging-drop methods at temperatures ranging from 4°C to 20°C. The crystals often need one additional week or even longer after the initial formation to reach their peak size (Sauguet et al., 2014). The final crystals are normally collected with cryoprotection using 20%–30% glycerol, ethylene glycol, or lower-molecular-weight PEG, such as PEG350MME or PEG400 (Table 2). The harvested crystals are flash frozen and stored in liquid N_2.

The crystals of pLGICs bound with anesthetics can be obtained in two different ways: grown in the presence of anesthetics (cocrystallization) or soaked with an anesthetic-containing solution after crystals form. It appears

Table 2 Conditions of Crystallization and Cryoprotection for pLGICs

pLGIC	PDB Code	Protein (mg/mL)	Reservoir Solution	Additive	Crystallization Method	Cryoprotectant
ELIC	2VL0	10	200 m*M* $(NH_4)_2SO_4$, 50 m*M* ADA (pH 6.5), 10%–15% PEG 4000	0.5 mg/mL *E. coli* lipids	Sitting drop (4°C)	30% Glycerol or ethylene glycol
	2YKS		200 m*M* $NaLiSO_4$, 50 m*M* ADA (pH 6.5), 10% PEG 4000			30% Glycerol
	2YN6		50 m*M* $BaAc_2$, 50 m*M* ADA (pH 6.5), 10% PEG 4000			30% Ethylene glycol
	2YOE, 4A97, 4A98, 3ZKR, 4TWD, 4TWF, 4TWH, 5LG3, 5LID		200 m*M* $(NH_4)_2SO_4$, 50 m*M* ADA (pH 6.5), 9%–12% PEG 4000			30% Glycerol
	3RQU, 3RQW		200 m*M* $(NH_4)_2SO_4$, 100 m*M* MES (pH 6.1–6.3), 10%–12% PEG 4000			20% Glycerol
	5HEJ, 5HEO, 5HEU, 5HEW		200 m*M* $(NH_4)_2SO_4$, 50 m*M* ADA (pH 6.5), 10%–13% PEG 4000			30% Ethylene glycol
	4Z90, 4Z91, 5SXU, 5SXV	5–6	200 m*M* $(NH_4)_2SO_4$, 100 m*M* MES (pH 6.1), 10%–12% PEG 4000	0.01–0.02 mg/mL *E. coli* lipids		20% Glycerol
	3UQ4, 3UQ5, 3UQ7	9–10	200 m*M* $(NH_4)_2SO_4$, 50 m*M* ADA (pH 6.5–6.9), 10%–12% PEG 4000	0.5 mg/mL *E. coli* lipids	Hanging drop (4°C)	30% Ethylene glycol

GLIC	3EHZ, 3EI0	10	225 m*M* $(NH_4)_2SO_4$, 50 m*M* NaAc (pH 4.0), 9%–12% PEG 4000	0.5 mg/mL *E. coli* lipids	Sitting drop (4°C)	30% Glycerol
	5HEG, 5HEH, 5HEJ, 5HEO, 5HEU, 5HEW					30% Ethylene glycol
	2XQ3, 2XQ4, 2XQ5, 2XQ6, 2XQ7, 2XQ8, 2XQ9, 2XQA		200–500 m*M* $(NH_4)_2SO_4$, 50 m*M* NaAc (pH 4.0), 9%–12% PEG 4000			30% Ethylene glycol
	4F8H		225 m*M* $(NH_4)_2SO_4$, 50 m*M* NaAc (pH 3.9–4.1), 10%–12% PEG 4000			20% Glycerol
	5J0Z	9–10	225 m*M* $(NH_4)_2SO_4$, 50 m*M* NaAc (pH 3.9–4.2), 10%–12% PEG 4000			30% Ethylene glycol
	4IRE	~10	225 m*M* $(NH_4)_2SO_4$, 50 m*M* NaAc (pH 3.9–4.3), 10%–12% PEG 4000		Hanging drop (4°C)	20% Glycerol
	4LMK, 4LMJ, 4LML	9–10	200–250 m*M* $(NH_4)_2SO_4$, 50 m*M* NaAc (pH 3.9–4.3), 10%–12% PEG 4000			30% Ethylene glycol
	3EAM, 3TLS, 3TLT, 3TLU, 3TLV, 3YLW, 3UU3, 3UU4, 3UU5, 3UU6, 3UU8, 3UUB	8–12	400 m*M* NH_4SCN, 100 m*M* NaAc (pH 4.6), 15% PEG 4000	—	Hanging drop (20°C)	20% Glycerol
	3P4W, 3P50, 4HFB, 4HFC, 4HFD, 4HFE, 4HFH		400 m*M* NaSCN, 100 m*M* NaAc (pH 4.0), 12%–16% PEG 4000			
	4HFI, 4IL4, 4IL9, 4ILA, 4ILB, 4ILC	—	400 m*M* NaSCN, 100 m*M* NaAc (pH 4.0), 12%–15% PEG 4000			

Continued

Table 2 Conditions of Crystallization and Cryoprotection for pLGICs—cont'd

pLGIC	PDB Code	Protein (mg/mL)	Reservoir Solution	Additive	Crystallization Method	Cryoprotectant
	5IUX					No further cryoprotection
	4NPP	6–8	50 m*M* $NiCl_2$, 100 m*M* NaAc (pH 4.0), 15%–20% PEG 2000MME		Hanging drop (20°C), microseeding	18% Glycerol
	4NPQ	10	200 m*M* KSCN, 10 m*M* $CaCl_2$, 3% TMAO 100 m*M* NaHEPES (pH 7.5), 13%–15% PEG 4000			
	4QH1, 4QH5		400 m*M* NaSCN, 100 m*M* NaAc or $NaH_2/Na_2H\text{-}PO_4$ (pH 4.0), 2% DMSO, 10% glycerol, 12%–14.5% PEG 4000			No further cryoprotection
	5HCJ, 5HCM	6–8	400 m*M* NaSCN, 100 m*M* NaAc (pH 4.0), 17% glycerol, 12%–16% PEG 4000			
	4ZZB, 4ZZC	—	400 m*M* NaSCN, 100 m*M* NaAc (pH 4.0), 12%–15% PEG 4000		Hanging drop (18°C), microseeding	20% Glycerol
	5L47, 5L4E, 5L4H	—	400 m*M* NaSCN, 100 m*M* NaAc (pH 4.0), 3% DMSO, 16% glycerol, 12%–14% PEG 4000			No further cryoprotection
ELIC-GLIC	4YEU	9–10	200 m*M* $(NH_4)_2SO_4$, 50 m*M* ADA (pH 6.7–7.6), 7.5%–10% PEG 4000	0.5 mg/mL *E. coli* lipids	Sitting drop (4°C)	30% Ethylene glycol
GLIC-GlyR	4X5T	6–8	50 m*M* $NiCl_2$, 0.1 *M* NaAc (pH 3.0), 16%–20% (vol/vol) PEG 2000MME, 4% (vol/vol) DMSO, 11% (vol/vol) ethylene glycol		Hanging drop (20°C), microseeding	

GluCl	3RHW, 3RI5, 3RIA, 3RIF	1–2	70 m*M* NaCl, 50 m*M* sodium citrate (pH 4.5), 21%–23% PEG400	Fab antibody, POPC or DPPC	Hanging drop (4°C)	30% PEG400
	4TNV, 4TNW	1–2	100 m*M* KCl, 50 m*M* sodium citrate (pH 5.5), 35%–36% pentaerythritol propoxylate (5/4 PO/OH)	Fab antibody, POPC	Hanging drop (4°C)	No further cryoprotection
β3GABA$_A$R	4COF	3	100 m*M* NaCl, 100 m*M* Li_2SO_4, 100 m*M* ADA (pH 6.5), 11.5% PEG 4000	2% (w/v) benzamidine	Sitting drop (6.5°C)	20% Glycerol
5HT$_{3A}$R	4PIR	2.7	0.1 *M* Na_2SO_4, 0.1 *M* Tris (pH 8.5), 20%–25% PEG 10000	1:2 VHH5 nanobody, 0.56 m*M* Cymal-6	Hanging drop (4°C)	20% Glycerol or 40% PEG 10000
α3GlyR	5CFB	3	100 m*M* KCl, 200 m*M* $MgCl_2$, 25 m*M* sodium citrate (pH 4.0), 30%–33% PEG400	0.2 m*M* strychnine	Hanging drop (4°C)	No further cryoprotection
	5VDH, 5VDI	~3	200 m*M* $CaCl_2$, 100 m*M* MES (pH 6.5), 22.5%–27.5% PEG350MME	0.5 m*M* glycine		30% PEG350MME
	5TIO, 5TIN	~3	200 m*M* $CaCl_2$, 100 m*M* MES (pH 6.5), 22.5%–27.5% PEG350MME	0.5 m*M* glycine		
α4β2nAChR	5KXI	1.5–2.5	50 m*M* ADA (pH 6.8), 12.5% PEG 1500 and 10% PEG 1000	1 m*M* nicotine	Hanging drop (14°C)	PEG 1000, PEG 1500, and ethylene glycol

that the cocrystallization technique has outperformed the soaking method based on the currently available crystal structures of pLGICs bound with anesthetics (Table 3). Once a favorable crystallization condition is identified for a given pLGIC, it can be used as a starting point to optimize further the pLGIC crystallization in the presence of anesthetics or other ligands. In the presence of anesthetics, the original optimized condition for apo-pLGICs often needs to be adjusted by tweaking pH, salt compositions, and concentrations of precipitants. Additives, such as glycerol and DMSO, have been shown useful in cocrystallizing pLGICs with anesthetics (Fourati et al., 2017). Most general anesthetics have limited solubility in the aqueous phase; the anesthetic concentration chosen for cocrystallization must be balanced, as a low concentration may lead to low incorporation, while a high concentration may degrade crystal diffraction. Likewise, soaking anesthetics into pLGIC crystals also requires a delicate balance between permeation of sufficient anesthetics into the crystals and potential disruption of the crystal lattice. Nontraditional soaking methods have proved successful for incorporating anesthetics into pLGICs, including adding saturated volatile desflurane to the reservoir well after GLIC crystals are observed (Nury et al., 2011) and pressuring xenon into GLIC crystals in a gas chamber (Sauguet et al., 2016).

Cryoprotection conditions for the crystallization of apo-pLGICs normally also work for pLGICs bound with general anesthetics, but may need some modification. For cryoprotection, the cryoprotectant should contain a slightly higher concentration of anesthetics than that used for cocrystallization or soaking.

3.2 Data Collection and Processing

Synchrotron radiation facilities are usually used for collecting high-resolution diffraction data of pLGIC crystals. BioSync (http://biosync.sbkb.org/index.jsp) keeps up-to-date information about synchrotron facilities around the world. Crystals of pLGICs are often small, fragile, and prone to radiation damage, so a microfocus beamline with a fast readout detector is highly preferred. It enables data collection from small crystals or from the selected better-diffracted regions in heterogeneous crystals (Riekel, Burghammer, & Schertler, 2005) and allows for complete data collection before severe radiation damage. Data collection for the anesthetic-containing crystals is similar to that for apo crystals, unless anesthetic molecules contain anomalous diffracting

Table 3 Anesthetics Identified in Crystal Structures of pLGICs

pLGICs	Anesthetics	PDB Code (Resolution), Functional States	Reservoir Solution	Crystallization Conditions
ELIC	Isofluorane	4Z91 (3.39 Å), resting 4Z90 (3.0 Å), desensitized	200 m*M* $(NH_4)_2SO_4$, 100 m*M* MES (pH 6.1), 10%–12% PEG 4000	Cocrystallization[a], sitting drop (4°C)
	2-Bromoethanol	5SXU (3.1 Å), desensitized 5SXV (3.4 Å), resting		
	Bromoform	3ZKR (3.65 Å), resting	200 m*M* $(NH_4)_2SO_4$, 50 m*M* ADA (pH 6.5), 9%–12% PEG 4000	
GLIC	Desflurane	3P4W (3.2 Å), open	400 m*M* NaSCN, 100 m*M* NaAc (pH 4.0), 12%–16% PEG 4000	Saturating amount of desflurane added in the reservoir well after crystals were formed, hanging drop (20°C)
	Propofol	3P50 (3.3 Å), open		Pure propofol added in crystallization mixture, hanging drop (20°C)
	Ketamine	4F8H (2.99 Å), desensitized	225 m*M* $(NH_4)_2SO_4$, 50 m*M* NaAc (pH 3.9–4.1), 10%–12% PEG 4000	Cocrystallization[a], sitting drop (4°C)
	2-Bromoethanol	4HFC (3.05 Å), desensitized F14′A mutant	400 m*M* NaSCN, 100 m*M* NaAc (pH 4.0), 12%–16% PEG 4000	Cocrystallization[a], hanging drop (20°C)
	Ethanol	4HFE (2.8 Å), desensitized F14′A mutant		
	Bromoform	4HFD (3.1 Å), desensitized F14′A mutant 4HFH (2.65 Å), desensitized		
		5HCM (3.15 Å), desensitized locally closed mutant	400 m*M* NaSCN, 100 m*M* NaAc (pH 4.0), 17% glycerol 2% bromoform, 12%–16% PEG 4000	

Continued

Table 3 Anesthetics Identified in Crystal Structures of pLGICs—cont'd

pLGICs	Anesthetics	PDB Code (Resolution), Functional States	Reservoir Solution	Crystallization Conditions
	Xenon	4ZZB (3.4 Å), desensitized locally closed mutant 4ZZC (3.1 Å), open	400 m*M* NaSCN, 100 m*M* NaAc (pH 4.0), 12%–15% PEG 4000	Crystals pressured under xenon, hanging drop (18°C)
	Seleniated barbiturate	5L47 (3.3 Å), desensitized locally closed mutant	400 m*M* NaSCN, 100 m*M* NaAc (pH 4.0), 3% DMSO, 16% glycerol, 12%–14% PEG 4000	Cocrystallization, hanging drop (18°C)
	Thiopental	5L4E (3.5 Å), desensitized locally closed mutant		
	Brominated barbiturate	5L4H (3.3 Å), desensitized locally closed mutant		

[a]Anesthetics were cocrystallized with proteins.

C12E9, α-dodecyl-w-hydroxy-poly(oxy-1,2-ethanediyl); *DDM*: *n*-dodecyl-β-D-maltopyranoside; *DMNG*, decyl maltose neopentyl glycol; *DPC*, *n*-dodecylphosphocholine; *POPC*, 1-palmitoyl-2-oleoyl-*sn*-glycero-3-phosphocholine; *UDM*, *n*-undecyl-β-D-maltopyranoside.

atoms. In such a case, data collection must be performed at or above the X-ray absorption edge so that the anomalous signal is maximized or close to the maximum. The collected diffraction data can be indexed, integrated, and scaled using software such as XDS (Kabsch, 2010a, 2010b), HKL2000 (Otwinowski & Minor, 1997), MOSFLM (Battye, Kontogiannis, Johnson, Powell, & Leslie, 2011), CCP4 (Winn et al., 2011), and d*TREK (Pflugrath, 1999). Initial structures are normally obtained by molecular replacement with the program PHASER (McCoy, 2007; McCoy et al., 2007) using known structures of homologous channels as search models. Structure refinements are performed with PHENIX (Adams et al., 2010) or BUSTER (Blanc et al., 2004). COOT (Emsley & Cowtan, 2004) is used for manual examination and rebuilding of the coordinates based on the electron density maps of 2Fo-Fc and Fo-Fc. Structure analyses are performed using PHENIX (Adams et al., 2010), and stereochemistry is evaluated using MolProbity (Chen et al., 2010). PyMOL (DeLano, 2002) and VMD (Humphrey, Dalke, & Schulten, 1996) programs can be used for structural analysis and visualization.

3.3 Identification of Anesthetics Bound to pLGICs

Identification of anesthetics bound to pLGICs cannot begin until the structural model of a given pLGIC fits well to the X-ray electron density. This often requires an iterative building–refinement process. In addition to the pLGIC electron density, extra electron density will be observed if anesthetic molecules bound to the pLGIC and diffracted well. How well the shape and size of this extra electron density matches with the molecular features of the anesthetic determines the certainty of anesthetic assignment. Comparing the electron densities from apo and anesthetic-containing crystals obtained at otherwise the same conditions can boost the certainty, as demonstrated in the previous studies of ELIC and GLIC bound with ketamine (Pan, Chen, Willenbring, Mowrey, et al., 2012), isoflurane (Chen et al., 2015), thiopental (Fourati et al., 2017), desflurane, or propofol (Nury et al., 2011). Anomalous signals from xenon, bromide, or selenium atoms, if they exist in the chosen anesthetics or anesthetic analogs, will also boost the certainty of the assignment of anesthetics. Xenon (Sauguet et al., 2016), 2-bromoethanol (Chen et al., 2017; Sauguet, Howard, et al., 2013), bromoform (Laurent et al., 2016; Spurny et al., 2013), and brominated or seleniated barbiturates (Fourati et al., 2017) have all been successfully identified based on anomalous signals.

4. FUNCTIONAL VALIDATION OF THE BOUND ANESTHETICS

Due to the low-binding affinity (μ*M*–m*M*) of general anesthetics, multiple binding sites may be observed in the crystal structures of pLGICs. However, the functional impact of anesthetic binding to each site may not be equal. Therefore, it is necessary to evaluate the roles of individual binding sites in triggering functional changes of pLGICs. Site-directed mutagenesis combined with functional measurements, such as two-electrode voltage clamp experiments, can fulfill this purpose (Chen et al., 2015, 2017; Nury et al., 2011; Pan, Chen, Willenbring, Mowrey, et al., 2012; Sauguet, Howard, et al., 2013).

5. CASE STUDIES

5.1 Crystallization of ELIC with the Volatile General Anesthetic Isofluorane or 2-Bromoethanol (Chen et al., 2015, 2017)

ELIC, a cation-conducting pLGIC from *Erwinia chrysanthemi*, is functionally inhibited by the volatile anesthetic isoflurane and ethanol. Elucidating the molecular interactions between ELIC and isoflurane or ethanol aids in our understanding of inhibitory mechanisms of general anesthetics and alcohols on pLGICs.

As a prokaryotic pLGIC, ELIC expresses well in *E. coli*, especially when fused to the C-terminus of MBP. A *Herpes simplex* (HRV) 3C protease site was inserted between MBP and ELIC for their subsequent separation. The pelB signal sequence and His_{10} purification tag were placed at the N-terminus of the MBP–ELIC fusion protein. The *E. coli* strain Rosetta(DE3)pLysS provided the best expression from the T7 promoter in the expression plasmid. The pLysS and expression plasmids were selected by chloramphenicol (35 μg/mL) and kanamycin (50 μg/mL), respectively. The expression began with an overnight preculture, which was diluted 1:100 into LB media containing antibiotics and grown at 37°C. When OD reached 0.9, the culture was harvested via centrifugation at 5000 × *g* for 20 min at 15°C. The cell pellet was suspended in M9 media and incubated at 15°C for 1–2 h before induction with 0.2 m*M* IPTG. After expression for 24 h, the cells were harvested AND suspended in buffer A (150 m*M* NaCl, 50 m*M* sodium phosphate, pH 8) with a protease inhibitor cocktail.

The cells were lysed with a M-110Y microfluidizer, and the membrane fraction was collected at 4°C by ultracentrifugation at 235,000 × *g* for ~1 h. The fusion protein His_{10}-MBP–ELIC was extracted from the membrane with 3.5% (wt/vol) *n*-undecyl-β-D-maltoside (Anatrace) in buffer A, clarified by centrifugation at 5000 × *g* for 20 min, and purified using Ni-NTA chromatography at 4°C. The purified fusion protein was cleaved overnight with HRV3C protease, and ELIC was separated from MBP using Ni-NTA chromatography. The pentameric fraction of ELIC was obtained by SEC using a Superdex 200 10/300GL column. The buffer was exchanged to 0.025% (wt/vol) *n*-dodecyl-β-D-maltoside (Anatrace) in 10 m*M* sodium phosphate (pH 8) and 150 m*M* NaCl at this step. ELIC was concentrated to ~5–6 mg/mL for crystallization.

ELIC crystals were obtained using the vapor-diffusion method in sitting drops at 4°C. *E. coli* polar lipids (0.01–0.02 mg/mL) and 2-bromoethanol (100 m*M*) were mixed and equilibrated with ELIC for at least half an hour before setting up for crystallization trays. The equilibrated protein was mixed in a 1:1 ratio with the reservoir solution and resulted in a 2-bromoethanol concentration of 50 m*M*. The reservoir solution contained 10%–12% PEG 4000, 200 m*M* ammonium sulfate, and 100 m*M* MES buffer at pH 6.1. The crystallization of ELIC with isoflurane followed a similar procedure except that isoflurane was not directly added to the protein and was instead added to the reservoir solution at its saturated concentration (14.4 m*M*) (Xu, Tang, Firestone, & Zhang, 1996). The amount of *E. coli* polar lipids, 2-bromoethanol, and isoflurane were adjusted and optimized for the best diffracting crystals. The crystals typically formed in 1 or 2 weeks and were harvested after briefly soaking in the reservoir solution supplemented with 20% glycerol and 100 m*M* 2-bromoethanol or a saturated isoflurane solution. The harvested crystals were flash-frozen in liquid nitrogen for short storage before shipping to a synchrotron facility for data collection.

X-ray crystal structures of ELIC bound with isoflurane were determined with a resolution up to 3.0 Å (Chen et al., 2015). The high electron density in the Fo-Fc density map revealed double occupancies of isoflurane inside the pore of ELIC, near T237(6′) and A244(13′). A contraction of the pore radius was observed near the extracellular entrance upon isoflurane binding. Electrophysiology measurements of oocytes expressing ELIC with a single-point mutation at either the T237(6′) or A244(13′) position validated that binding to these sites are responsible for the inhibitory action of isoflurane. The study presents compelling evidence for a direct pore-binding mechanism of isoflurane inhibition.

The X-ray crystal structure of ELIC bound with 2-bromoethanol was solved at 3.1 Å (Chen et al., 2017) and showed multiple 2-bromoethanol-binding sites, including inside the pore near T237(6′) in the transmembrane domain (TMD) and at three different locations in the ECD. The strong anomalous signal from the bromide atom of 2-bromoethanol facilitated the determination of these binding sites, which would be difficult, if not impossible, to directly determine for the small, low-affinity ethanol molecule. Mutations to the three binding sites in the ECD did not alter the functional response of ELIC to 2-bromoethanol or ethanol, suggesting a negligible contribution of ethanol binding in the ECD to the functional modulation ofELIC. In contrast, both 2-bromoethanol and ethanol showed reduced potencies on the T237(6′) A mutant, indicating that the binding to the pore predominately contributes to the inhibition ofELIC by ethanol. This conclusion was further substantiated using an ELIC-α1β3GABA$_A$R chimera, which has the ECD of ELIC but the TMD of α1β3GABA$_A$R. While 2-bromoethanol and ethanol inhibit ELIC, both 2-bromoethanol and ethanol potentiate the ELIC-α1β3GABA$_A$R chimera. Replacing the TMD of ELIC reverses the functional effect of 2-bromoethanol and ethanol, supporting a pore-blocking mechanism of short-chain alcohols in ELIC, and likely also in other cation-conducting pLGICs.

5.2 Crystallization of GLIC with the Intravenous General Anesthetic Ketamine (Pan, Chen, Willenbring, Yoshida, et al., 2012)

GLIC, a prokaryotic cation-conducting pLGIC from *Gloeobacter violaceus*, is inhibited by ketamine, an anesthetic commonly known as an antagonist of NMDA receptors. The crystal structure of ketamine bound to GLIC provides novel insight into the molecular recognition and allosteric anesthetic modulation of pLGICs.

The expression and purification of GLIC were conducted in a similar manner as done for ELIC, including the use of the fusion protein MBP and the *E. coli* strain Rosetta(DE3)pLysS. The purified GLIC was concentrated, equilibrated with 0.5 mg/mL *E. coli* polar lipids in the presence of ketamine (1 m*M*) for at least 1 h at 4°C, and then mixed in a 1:1 ratio with the reservoir solution (10%–12% PEG 4000, 225 m*M* ammonium sulfate, and 50 m*M* sodium acetate buffer at pH 3.9–4.1). Crystals were obtained using the sitting-drop method at 4°C and usually appeared within 1 week. The crystals were cryoprotected by soaking in the reservoir solution supplemented with 20% glycerol and 10 m*M* ketamine for 30 min.

The harvested crystals were flash frozen in liquid nitrogen for short storage before data collection at a synchrotron facility.

The X-ray structure of GLIC bound with ketamine was determined at 2.99 Å. Ketamine was identified in a single intersubunit cavity in the ECD in each of five equivalent subunits of GLIC. The binding site partially overlaps with the antagonist-binding site in homologous pLGICs. The functional impact of ketamine binding to the site revealed in the crystal structure was further investigated by additional electrophysiology experiments. A point mutation at N152C inside the ketamine-binding site greatly altered the activation of GLIC by protons and decreased the potency of ketamine inhibition on GLIC. In addition, chemical labeling of N152C with 8-(chloromercuri)-2-dibenzofuransulfonic acid (CBFS) resulted in the inhibition of GLIC, mimicking the effect of ketamine binding at this site. Treatment with 10 m*M* dithiothreitol (DTT) removed the CBFS label at N152C and brought the channel back to a normal function. The success in identifying the ketamine-binding site by crystallography and functional validation of the relevance of this site for ketamine inhibition of GLIC provides a useful template for future studies of anesthetic binding to pLGICs.

ACKNOWLEDGMENTS

The authors would like to thank Dr. Tommy Tillman, Ms. Marta Wells, and Ms. Sandy Hirsch for their valuable comments and suggestions. Use of the Stanford Synchrotron Radiation Lightsource, SLAC National Accelerator Laboratory, is supported by the U.S. Department of Energy, Office of Science, Office of Basic Energy Sciences under Contract No. DE-AC02-76SF00515, the National Institutes of Health, and National Institute of General Medical Sciences (including P41GM103393). The research is supported by NIH (R01GM056257). The authors declare no conflicts of interest with the contents of this chapter.

REFERENCES

Adams, P. D., Afonine, P. V., Bunkoczi, G., Chen, V. B., Davis, I. W., Echols, N., et al. (2010). PHENIX: A comprehensive python-based system for macromolecular structure solution. *Acta Crystallographica. Section D, Biological Crystallography*, *66*(Pt. 2), 213–221. https://doi.org/10.1107/S0907444909052925.

Althoff, T., Hibbs, R. E., Banerjee, S., & Gouaux, E. (2014). X-ray structures of GluCl in apo states reveal a gating mechanism of Cys-loop receptors. *Nature*, *512*(7514), 333–337. https://doi.org/10.1038/nature13669.

Basak, S., Schmandt, N., Gicheru, Y., & Chakrapani, S. (2017). Crystal structure and dynamics of a lipid-induced potential desensitized-state of a pentameric ligand-gated channel. *Elife 6*, e23886. https://doi.org/10.7554/eLife.23886.

Battye, T. G., Kontogiannis, L., Johnson, O., Powell, H. R., & Leslie, A. G. (2011). iMOSFLM: A new graphical interface for diffraction-image processing with MOSFLM. *Acta Crystallographica. Section D, Biological Crystallography*, *67*(Pt. 4), 271–281. https://doi.org/10.1107/S0907444910048675.

Bertozzi, C., Zimmermann, I., Engeler, S., Hilf, R. J., & Dutzler, R. (2016). Signal transduction at the domain interface of prokaryotic pentameric ligand-gated ion channels. *PLoS Biology*, *14*(3), e1002393. https://doi.org/10.1371/journal.pbio.1002393.

Bill, R. M., Henderson, P. J., Iwata, S., Kunji, E. R., Michel, H., Neutze, R., et al. (2011). Overcoming barriers to membrane protein structure determination. *Nature Biotechnology*, *29*(4), 335–340. https://doi.org/10.1038/nbt.1833.

Bird, L. E., Rada, H., Verma, A., Gasper, R., Birch, J., Jennions, M., et al. (2015). Green fluorescent protein-based expression screening of membrane proteins in Escherichia coli. *Journal of Visualized Experiments*, *95*, e52357. https://doi.org/10.3791/52357.

Blanc, E., Roversi, P., Vonrhein, C., Flensburg, C., Lea, S. M., & Bricogne, G. (2004). Refinement of severely incomplete structures with maximum likelihood in BUSTER-TNT. *Acta Crystallographica. Section D, Biological Crystallography*, *60*(Pt. 12 Pt. 1), 2210–2221. https://doi.org/10.1107/S0907444904016427.

Bocquet, N., Nury, H., Baaden, M., Le Poupon, C., Changeux, J. P., Delarue, M., et al. (2009). X-ray structure of a pentameric ligand-gated ion channel in an apparently open conformation. *Nature*, *457*(7225), 111–114. https://doi.org/10.1038/nature07462.

Bocquet, N., Prado de Carvalho, L., Cartaud, J., Neyton, J., Le Poupon, C., Taly, A., et al. (2007). A prokaryotic proton-gated ion channel from the nicotinic acetylcholine receptor family. *Nature*, *445*(7123), 116–119. https://doi.org/10.1038/nature05371.

Bondarenko, V., Mowrey, D., Liu, L. T., Xu, Y., & Tang, P. (2013). NMR resolved multiple anesthetic binding sites in the TM domains of the alpha4beta2 nAChR. *Biochimica et Biophysica Acta*, *1828*(2), 398–404. https://doi.org/10.1016/j.bbamem.2012.09.014.

Bondarenko, V., Mowrey, D. D., Tillman, T. S., Seyoum, E., Xu, Y., & Tang, P. (2014). NMR structures of the human alpha7 nAChR transmembrane domain and associated anesthetic binding sites. *Biochimica et Biophysica Acta*, *1838*(5), 1389–1395. https://doi.org/10.1016/j.bbamem.2013.12.018.

Bondarenko, V., Yushmanov, V. E., Xu, Y., & Tang, P. (2008). NMR study of general anesthetic interaction with nAChR beta2 subunit. *Biophysical Journal*, *94*(5), 1681–1688. https://doi.org/10.1529/biophysj.107.116772.

Campagna, J. A., Miller, K. W., & Forman, S. A. (2003). Mechanisms of actions of inhaled anesthetics. *The New England Journal of Medicine*, *348*(21), 2110–2124. https://doi.org/10.1056/NEJMra021261.

Chayen, N. E. (1998). Comparative studies of protein crystallization by vapour-diffusion and microbatch techniques. *Acta Crystallographica. Section D, Biological Crystallography*, *54*(Pt. 1), 8–15.

Chen, V. B., Arendall, W. B., 3rd, Headd, J. J., Keedy, D. A., Immormino, R. M., Kapral, G. J., et al. (2010). MolProbity: All-atom structure validation for macromolecular crystallography. *Acta Crystallographica. Section D, Biological Crystallography*, *66*(Pt. 1), 12–21. https://doi.org/10.1107/S0907444909042073.

Chen, Q., Kinde, M. N., Arjunan, P., Wells, M. M., Cohen, A. E., Xu, Y., et al. (2015). Direct pore binding as a mechanism for isoflurane inhibition of the pentameric ligand-gated ion channel ELIC. *Scientific Reports*, *5*, 13833. https://doi.org/10.1038/srep13833.

Chen, Z. W., Manion, B., Townsend, R. R., Reichert, D. E., Covey, D. F., Steinbach, J. H., et al. (2012). Neurosteroid analog photolabeling of a site in the third transmembrane domain of the beta3 subunit of the GABA(A) receptor. *Molecular Pharmacology*, *82*(3), 408–419. https://doi.org/10.1124/mol.112.078410.

Chen, Q., Wells, M. M., Tillman, T. S., Kinde, M. N., Cohen, A., Xu, Y., et al. (2017). Structural basis of alcohol inhibition of the pentameric ligand-gated ion channel ELIC. *Structure*, *25*(1), 180–187. https://doi.org/10.1016/j.str.2016.11.007.

Chiara, D. C., Dangott, L. J., Eckenhoff, R. G., & Cohen, J. B. (2003). Identification of nicotinic acetylcholine receptor amino acids photolabeled by the volatile anesthetic halothane. *Biochemistry*, *42*(46), 13457–13467. https://doi.org/10.1021/bi0351561.

Chiara, D. C., Dostalova, Z., Jayakar, S. S., Zhou, X., Miller, K. W., & Cohen, J. B. (2012). Mapping general anesthetic binding site(s) in human alpha1beta3 gamma-aminobutyric acid type A receptors with [(3)H]TDBzl-etomidate, a photoreactive etomidate analogue. *Biochemistry*, *51*(4), 836–847. https://doi.org/10.1021/bi201772m.

Chiara, D. C., Gill, J. F., Chen, Q., Tillman, T., Dailey, W. P., Eckenhoff, R. G., et al. (2014). Photoaffinity labeling the propofol binding site in GLIC. *Biochemistry*, *53*(1), 135–142. https://doi.org/10.1021/bi401492k.

DeLano, W. L. (2002). *The PyMOL molecular graphics system, Schrödinger, LLC*. Palo Alto, CA: Delano Scientific LLC.

Delmar, J. A., Bolla, J. R., Su, C. C., & Yu, E. W. (2015). Crystallization of membrane proteins by vapor diffusion. *Methods in Enzymology*, *557*, 363–392. https://doi.org/10.1016/bs.mie.2014.12.018.

Du, J., Lu, W., Wu, S., Cheng, Y., & Gouaux, E. (2015). Glycine receptor mechanism elucidated by electron cryo-microscopy. *Nature*, *526*(7572), 224–229. https://doi.org/10.1038/nature14853.

Eckenhoff, R. G., & Shuman, H. (1993). Halothane binding to soluble proteins determined by photoaffinity labeling. *Anesthesiology*, *79*(1), 96–106.

Eckenhoff, R. G., Xi, J., Shimaoka, M., Bhattacharji, A., Covarrubias, M., & Dailey, W. P. (2010). Azi-isoflurane, a photolabel analog of the commonly used inhaled general anesthetic isoflurane. *ACS Chemical Neuroscience*, *1*(2), 139–145. https://doi.org/10.1021/cn900014m.

Emsley, P., & Cowtan, K. (2004). Coot: Model-building tools for molecular graphics. *Acta Crystallographica. Section D, Biological Crystallography*, *60*(Pt. 12 Pt. 1), 2126–2132. https://doi.org/10.1107/S0907444904019158.

Fourati, Z., Ruza, R. R., Laverty, D., Drege, E., Delarue-Cochin, S., Joseph, D., et al. (2017). Barbiturates bind in the GLIC ion channel pore and cause inhibition by stabilizing a closed state. *The Journal of Biological Chemistry*, *292*(5), 1550–1558. https://doi.org/10.1074/jbc.M116.766964.

Fourati, Z., Sauguet, L., & Delarue, M. (2015). Genuine open form of the pentameric ligand-gated ion channel GLIC. *Acta Crystallographica. Section D, Biological Crystallography*, *71*(Pt. 3), 454–460. https://doi.org/10.1107/S1399004714026698.

Franks, N. P., & Lieb, W. R. (1994). Molecular and cellular mechanisms of general anaesthesia. *Nature*, *367*(6464), 607–614. https://doi.org/10.1038/367607a0.

Gonzalez-Gutierrez, G., Cuello, L. G., Nair, S. K., & Grosman, C. (2013). Gating of the proton-gated ion channel from Gloeobacter violaceus at pH 4 as revealed by X-ray crystallography. *Proceedings of the National Academy of Sciences of the United States of America*, *110*(46), 18716–18721. https://doi.org/10.1073/pnas.1313156110.

Gonzalez-Gutierrez, G., Lukk, T., Agarwal, V., Papke, D., Nair, S. K., & Grosman, C. (2012). Mutations that stabilize the open state of the Erwinia chrysanthemi ligand-gated ion channel fail to change the conformation of the pore domain in crystals. *Proceedings of the National Academy of Sciences of the United States of America*, *109*(16), 6331–6336. https://doi.org/10.1073/pnas.1119268109.

Hassaine, G., Deluz, C., Grasso, L., Wyss, R., Tol, M. B., Hovius, R., et al. (2014). X-ray structure of the mouse serotonin 5-HT3 receptor. *Nature*, *512*(7514), 276–281. https://doi.org/10.1038/nature13552.

Hassaine, G., Deluz, C., Tol, M. B., Li, X. D., Graff, A., Vogel, H., et al. (2013). Large scale expression and purification of the mouse 5-HT3 receptor. *Biochimica et Biophysica Acta*, *1828*(11), 2544–2552. https://doi.org/10.1016/j.bbamem.2013.05.028.

Hemmings, H. C., Jr., Akabas, M. H., Goldstein, P. A., Trudell, J. R., Orser, B. A., & Harrison, N. L. (2005). Emerging molecular mechanisms of general anesthetic action. *Trends in Pharmacological Sciences*, *26*(10), 503–510. https://doi.org/10.1016/j.tips.2005.08.006.

Hibbs, R. E., & Gouaux, E. (2011). Principles of activation and permeation in an anion-selective cys-loop receptor. *Nature*, *474*(7349), 54–60. https://doi.org/10.1038/nature10139.

Hilf, R. J., Bertozzi, C., Zimmermann, I., Reiter, A., Trauner, D., & Dutzler, R. (2010). Structural basis of open channel block in a prokaryotic pentameric ligand-gated ion channel. *Nature Structural & Molecular Biology*, *17*(11), 1330–1336. https://doi.org/10.1038/nsmb.1933.

Hilf, R. J., & Dutzler, R. (2008). X-ray structure of a prokaryotic pentameric ligand-gated ion channel. *Nature*, *452*(7185), 375–379. https://doi.org/10.1038/nature06717.

Hilf, R. J., & Dutzler, R. (2009). Structure of a potentially open state of a proton-activated pentameric ligand-gated ion channel. *Nature*, *457*(7225), 115–118. https://doi.org/10.1038/nature07461.

Hovius, R., Tairi, A. P., Blasey, H., Bernard, A., Lundstrom, K., & Vogel, H. (1998). Characterization of a mouse serotonin 5-HT3 receptor purified from mammalian cells. *Journal of Neurochemistry*, *70*(2), 824–834.

Huang, X., Chen, H., Michelsen, K., Schneider, S., & Shaffer, P. L. (2015). Crystal structure of human glycine receptor-alpha3 bound to antagonist strychnine. *Nature*, *526*(7572), 277–280. https://doi.org/10.1038/nature14972.

Huang, X., Chen, H., & Shaffer, P. L. (2017). Crystal structures of human GlyRalpha3 bound to Ivermectin. *Structure*, *25*(6), 945–950, e942. https://doi.org/10.1016/j.str.2017.04.007.

Huang, X., Shaffer, P. L., Ayube, S., Bregman, H., Chen, H., Lehto, S. G., et al. (2017). Crystal structures of human glycine receptor alpha3 bound to a novel class of analgesic potentiators. *Nature Structural & Molecular Biology*, *24*(2), 108–113. https://doi.org/10.1038/nsmb.3329.

Humphrey, W., Dalke, A., & Schulten, K. (1996). VMD: Visual molecular dynamics. *Journal of Molecular Graphics*, *14*(1), 33–38, (27–38).

Husain, S. S., Stewart, D., Desai, R., Hamouda, A. K., Li, S. G., Kelly, E., et al. (2010). p-Trifluoromethyldiazirinyl-etomidate: A potent photoreactive general anesthetic derivative of etomidate that is selective for ligand-gated cationic ion channels. *Journal of Medicinal Chemistry*, *53*(17), 6432–6444. https://doi.org/10.1021/jm100498u.

Jayakar, S. S., Dailey, W. P., Eckenhoff, R. G., & Cohen, J. B. (2013). Identification of propofol binding sites in a nicotinic acetylcholine receptor with a photoreactive propofol analog. *The Journal of Biological Chemistry*, *288*(9), 6178–6189. https://doi.org/10.1074/jbc.M112.435909.

Jayakar, S. S., Zhou, X., Chiara, D. C., Dostalova, Z., Savechenkov, P. Y., Bruzik, K. S., et al. (2014). Multiple propofol-binding sites in a gamma-aminobutyric acid type A receptor (GABAAR) identified using a photoreactive propofol analog. *The Journal of Biological Chemistry*, *289*(40), 27456–27468. https://doi.org/10.1074/jbc.M114.581728.

Kabsch, W. (2010a). Integration, scaling, space-group assignment and post-refinement. *Acta Crystallographica. Section D, Biological Crystallography*, *66*(Pt. 2), 133–144. https://doi.org/10.1107/S0907444909047374.

Kabsch, W. (2010b). Xds. *Acta Crystallographica. Section D, Biological Crystallography*, *66*(Pt. 2), 125–132. https://doi.org/10.1107/S0907444909047337.

Kawate, T., & Gouaux, E. (2006). Fluorescence-detection size-exclusion chromatography for precrystallization screening of integral membrane proteins. *Structure*, *14*(4), 673–681. https://doi.org/10.1016/j.str.2006.01.013.

Kinde, M. N., Bondarenko, V., Granata, D., Bu, W., Grasty, K. C., Loll, P. J., et al. (2016). Fluorine-19 NMR and computational quantification of isoflurane binding to the voltage-gated sodium channel NaChBac. *Proceedings of the National Academy of Sciences of the United States of America*, *113*(48), 13762–13767. https://doi.org/10.1073/pnas.1609939113.

Labriola, J. M., Pandhare, A., Jansen, M., Blanton, M. P., Corringer, P. J., & Baenziger, J. E. (2013). Structural sensitivity of a prokaryotic pentameric ligand-gated ion channel to its membrane environment. *The Journal of Biological Chemistry*, *288*(16), 11294–11303. https://doi.org/10.1074/jbc.M113.458133.

Landau, E. M., & Rosenbusch, J. P. (1996). Lipidic cubic phases: A novel concept for the crystallization of membrane proteins. *Proceedings of the National Academy of Sciences of the United States of America*, *93*(25), 14532–14535.

Laurent, B., Murail, S., Shahsavar, A., Sauguet, L., Delarue, M., & Baaden, M. (2016). Sites of anesthetic inhibitory action on a cationic ligand-gated ion channel. *Structure*, *24*(4), 595–605. https://doi.org/10.1016/j.str.2016.02.014.

Luft, J. R., Newman, J., & Snell, E. H. (2014). Crystallization screening: The influence of history on current practice. *Acta Crystallographica. Section F, Structural Biology Communications*, *70*(Pt. 7), 835–853. https://doi.org/10.1107/S2053230X1401262X.

McCoy, A. J. (2007). Solving structures of protein complexes by molecular replacement with phaser. *Acta Crystallographica. Section D, Biological Crystallography*, *63*(Pt. 1), 32–41. https://doi.org/10.1107/S0907444906045975.

McCoy, A. J., Grosse-Kunstleve, R. W., Adams, P. D., Winn, M. D., Storoni, L. C., & Read, R. J. (2007). Phaser crystallographic software. *Journal of Applied Crystallography*, *40*, 658–674. https://doi.org/10.1107/S0021889807021206.

Menny, A., Lefebvre, S. N., Schmidpeter, P. A., Drege, E., Fourati, Z., Delarue, M., et al. (2017). Identification of a pre-active conformation of a pentameric channel receptor. *eLife 6*, e23955. https://doi.org/10.7554/eLife.23955.

Mihic, S. J., Ye, Q., Wick, M. J., Koltchine, V. V., Krasowski, M. D., Finn, S. E., et al. (1997). Sites of alcohol and volatile anaesthetic action on GABA(A) and glycine receptors. *Nature*, *389*(6649), 385–389. https://doi.org/10.1038/38738.

Miller, P. S., & Aricescu, A. R. (2014). Crystal structure of a human GABAA receptor. *Nature*, *512*(7514), 270–275. https://doi.org/10.1038/nature13293.

Moraga-Cid, G., Sauguet, L., Huon, C., Malherbe, L., Girard-Blanc, C., Petres, S., et al. (2015). Allosteric and hyperekplexic mutant phenotypes investigated on an alpha1 glycine receptor transmembrane structure. *Proceedings of the National Academy of Sciences of the United States of America*, *112*(9), 2865–2870. https://doi.org/10.1073/pnas.1417864112.

Morales-Perez, C. L., Noviello, C. M., & Hibbs, R. E. (2016). X-ray structure of the human alpha4beta2 nicotinic receptor. *Nature*, *538*(7625), 411–415. https://doi.org/10.1038/nature19785.

Mowrey, D., Chen, Q., Liang, Y., Liang, J., Xu, Y., & Tang, P. (2013). Signal transduction pathways in the pentameric ligand-gated ion channels. *PLoS One*, *8*(5), e64326. https://doi.org/10.1371/journal.pone.0064326.

Newman, J., Egan, D., Walter, T. S., Meged, R., Berry, I., Ben Jelloul, M., et al. (2005). Towards rationalization of crystallization screening for small- to medium-sized academic laboratories: The PACT/JCSG+ strategy. *Acta Crystallographica. Section D, Biological Crystallography*, *61*(Pt. 10), 1426–1431. https://doi.org/10.1107/S0907444905024984.

Nury, H., Van Renterghem, C., Weng, Y., Tran, A., Baaden, M., Dufresne, V., et al. (2011). X-ray structures of general anaesthetics bound to a pentameric ligand-gated ion channel. *Nature*, *469*(7330), 428–431. https://doi.org/10.1038/nature09647.

Otwinowski, Z., & Minor, W. (1997). Processing of X-ray diffraction data collected in oscillation mode. *Methods in Enzymology*, *276*, 307–326.

Pan, J., Chen, Q., Willenbring, D., Mowrey, D., Kong, X. P., Cohen, A., et al. (2012). Structure of the pentameric ligand-gated ion channel GLIC bound with anesthetic ketamine. *Structure*, *20*(9), 1463–1469. https://doi.org/10.1016/j.str.2012.08.009.

Pan, J., Chen, Q., Willenbring, D., Yoshida, K., Tillman, T., Kashlan, O. B., et al. (2012). Structure of the pentameric ligand-gated ion channel ELIC cocrystallized with its competitive antagonist acetylcholine. *Nature Communications*, *3*, 714. https://doi.org/10.1038/ncomms1703.

Pflugrath, J. W. (1999). The finer things in X-ray diffraction data collection. *Acta Crystallographica. Section D, Biological Crystallography*, *55*(Pt. 10), 1718–1725.

Prevost, M. S., Sauguet, L., Nury, H., Van Renterghem, C., Huon, C., Poitevin, F., et al. (2012). A locally closed conformation of a bacterial pentameric proton-gated ion channel. *Nature Structural & Molecular Biology*, *19*(6), 642–649. https://doi.org/10.1038/nsmb.2307.

Riekel, C., Burghammer, M., & Schertler, G. (2005). Protein crystallography microdiffraction. *Current Opinion in Structural Biology*, *15*(5), 556–562. https://doi.org/10.1016/j.sbi.2005.08.013.

Roy, A. (2015). Membrane preparation and solubilization. *Methods in Enzymology*, *557*, 45–56. https://doi.org/10.1016/bs.mie.2014.11.044.

Sauguet, L., Fourati, Z., Prange, T., Delarue, M., & Colloc'h, N. (2016). Structural basis for xenon inhibition in a cationic pentameric ligand-gated ion channel. *PLoS One*, *11*(2), e0149795. https://doi.org/10.1371/journal.pone.0149795.

Sauguet, L., Howard, R. J., Malherbe, L., Lee, U. S., Corringer, P. J., Harris, R. A., et al. (2013). Structural basis for potentiation by alcohols and anaesthetics in a ligand-gated ion channel. *Nature Communications*, *4*, 1697. https://doi.org/10.1038/ncomms2682.

Sauguet, L., Poitevin, F., Murail, S., Van Renterghem, C., Moraga-Cid, G., Malherbe, L., et al. (2013). Structural basis for ion permeation mechanism in pentameric ligand-gated ion channels. *The EMBO Journal*, *32*(5), 728–741. https://doi.org/10.1038/emboj.2013.17.

Sauguet, L., Shahsavar, A., Poitevin, F., Huon, C., Menny, A., Nemecz, A., et al. (2014). Crystal structures of a pentameric ligand-gated ion channel provide a mechanism for activation. *Proceedings of the National Academy of Sciences of the United States of America*, *111*(3), 966–971. https://doi.org/10.1073/pnas.1314997111.

Schmandt, N., Velisetty, P., Chalamalasetti, S. V., Stein, R. A., Bonner, R., Talley, L., et al. (2015). A chimeric prokaryotic pentameric ligand-gated channel reveals distinct pathways of activation. *The Journal of General Physiology*, *146*(4), 323–340. https://doi.org/10.1085/jgp.201511478.

Segelke, B. (2005). Macromolecular crystallization with microfluidic free-interface diffusion. *Expert Review of Proteomics*, *2*(2), 165–172. https://doi.org/10.1586/14789450.2.2.165.

Snijder, H. J., & Hakulinen, J. (2016). Membrane protein production in E. coli for applications in drug discovery. *Advances in Experimental Medicine and Biology*, *896*, 59–77. https://doi.org/10.1007/978-3-319-27216-0_5.

Spurny, R., Billen, B., Howard, R. J., Brams, M., Debaveye, S., Price, K. L., et al. (2013). Multisite binding of a general anesthetic to the prokaryotic pentameric Erwinia chrysanthemi ligand-gated ion channel (ELIC). *The Journal of Biological Chemistry*, *288*(12), 8355–8364. https://doi.org/10.1074/jbc.M112.424507.

Spurny, R., Ramerstorfer, J., Price, K., Brams, M., Ernst, M., Nury, H., et al. (2012). Pentameric ligand-gated ion channel ELIC is activated by GABA and modulated by benzodiazepines. *Proceedings of the National Academy of Sciences of the United States of America*, *109*(44), E3028–3034. https://doi.org/10.1073/pnas.1208208109.

Tillman, T. S., Alvarez, F. J., Reinert, N. J., Liu, C., Wang, D., Xu, Y., et al. (2016). Functional human alpha7 nicotinic acetylcholine receptor (nAChR) generated from Escherichia coli. *The Journal of Biological Chemistry*, *291*(35), 18276–18282. https://doi.org/10.1074/jbc.M116.729970.

Ulens, C., Spurny, R., Thompson, A. J., Alqazzaz, M., Debaveye, S., Han, L., et al. (2014). The prokaryote ligand-gated ion channel ELIC captured in a pore blocker-bound conformation by the Alzheimer's disease drug memantine. *Structure*, *22*(10), 1399–1407. https://doi.org/10.1016/j.str.2014.07.013.

Velisetty, P., & Chakrapani, S. (2012). Desensitization mechanism in prokaryotic ligand-gated ion channel. *The Journal of Biological Chemistry*, *287*(22), 18467–18477. https://doi.org/10.1074/jbc.M112.348045.

Velisetty, P., Chalamalasetti, S. V., & Chakrapani, S. (2012). Conformational transitions underlying pore opening and desensitization in membrane-embedded Gloeobacter violaceus ligand-gated ion channel (GLIC). *The Journal of Biological Chemistry*, *287*(44), 36864–36872. https://doi.org/10.1074/jbc.M112.401067.

Winn, M. D., Ballard, C. C., Cowtan, K. D., Dodson, E. J., Emsley, P., Evans, P. R., et al. (2011). Overview of the CCP4 suite and current developments. *Acta Crystallographica. Section D, Biological Crystallography*, *67*(Pt. 4), 235–242. https://doi.org/10.1107/S0907444910045749.

Woll, K. A., Dailey, W. P., Brannigan, G., & Eckenhoff, R. G. (2016). Shedding light on anesthetic mechanisms: Application of photoaffinity ligands. *Anesthesia and Analgesia*, *123*(5), 1253–1262. https://doi.org/10.1213/ANE.0000000000001365.

Xu, Y., Tang, P., Firestone, L., & Zhang, T. T. (1996). 19F nuclear magnetic resonance investigation of stereoselective binding of isoflurane to bovine serum albumin. *Biophysical Journal*, *70*(1), 532–538. https://doi.org/10.1016/S0006-3495(96)79599-1.

Yip, G. M., Chen, Z. W., Edge, C. J., Smith, E. H., Dickinson, R., Hohenester, E., et al. (2013). A propofol binding site on mammalian GABAA receptors identified by photolabeling. *Nature Chemical Biology*, *9*(11), 715–720. https://doi.org/10.1038/nchembio.1340.

Zimmermann, I., & Dutzler, R. (2011). Ligand activation of the prokaryotic pentameric ligand-gated ion channel ELIC. *PLoS Biology*, *9*(6), e1001101. https://doi.org/10.1371/journal.pbio.1001101.

Zimmermann, I., Marabelli, A., Bertozzi, C., Sivilotti, L. G., & Dutzler, R. (2012). Inhibition of the prokaryotic pentameric ligand-gated ion channel ELIC by divalent cations. *PLoS Biology*, *10*(11), e1001429. https://doi.org/10.1371/journal.pbio.1001429.

CHAPTER THREE

Solution NMR Studies of Anesthetic Interactions with Ion Channels

Vasyl Bondarenko, Marta Wells, Yan Xu, Pei Tang[1]
University of Pittsburgh School of Medicine, Pittsburgh, PA, United States
[1]Corresponding author: e-mail address: tangp@anes.upmc.edu

Contents

Abstract

NMR spectroscopy is one of the major tools to provide atomic resolution protein structural information. It has been used to elucidate the molecular details of interactions between anesthetics and ion channels, to identify anesthetic binding sites, and to characterize channel dynamics and changes introduced by anesthetics. In this chapter, we present solution NMR methods essential for investigating interactions between ion channels and general anesthetics, including both volatile and intravenous anesthetics. Case studies are provided with a focus on pentameric ligand-gated ion channels and the voltage-gated sodium channel NaChBac.

Methods in Enzymology, Volume 603
ISSN 0076-6879
https://doi.org/10.1016/bs.mie.2018.01.018

ABBREVIATIONS

5-HT$_{3A}$R	5-hydroxytryptamine, or serotonin receptor
DDM	*n*-dodecyl-β-D-maltopyranoside
DSS	2,2-dimethyl-2-silapentane-5-sulfonic acid
ECD	extracellular domain
GABA$_A$R	type-A gamma-aminobutyric acid receptor
GluCl	glutamate-gated chloride channel
GlyR	glycine receptor
ICD	intracellular domain
LDAO	lauryldimethylamine-*N*-oxide
NaChBac	bacterial voltage-gated sodium channel
nAChR	nicotinic acetylcholine receptors
NMR	nuclear magnetic resonance
pLGICs	pentameric ligand-gated ion channels
STD NMR	saturation transfer difference NMR
TET	2,2,2-trifluoroethanethiol
TMD	transmembrane domain
UDM	*n*-undecyl-β-D-maltopyranoside

1. INTRODUCTION

Nuclear magnetic resonance (NMR) spectroscopy has a wide range of applications, including the determination of protein structures, characterization of protein–ligand interactions, and detection of changes in protein conformations and dynamics upon ligand binding (Kay, 2016; Kim, Howell, Van Horn, Jeon, & Sanders, 2009; Kitevski-LeBlanc & Prosser, 2012; Liang & Tamm, 2016; Liu, Horst, Katritch, Stevens, & Wuthrich, 2012; Oxenoid & Chou, 2013, 2016; Rosenzweig & Kay, 2016; Ye, Van Eps, Zimmer, Ernst, & Prosser, 2016; Zhuang et al., 2013). For the specific purpose of understanding anesthetic interactions with proteins, NMR has been used to determine the structures of anesthetic targets (Bondarenko et al., 2012, 2014; Bondarenko, Tillman, Xu, & Tang, 2010; Cui et al., 2012; Ma, Brandon, et al., 2008; Ma, Liu, Li, Tang, & Xu, 2005; Ma, Tillman, et al., 2008; Mowrey, Cui, et al., 2013; Mowrey, Kinde, Xu, & Tang, 2015; Tang, Mandal, & Xu, 2002), identify anesthetic binding sites in proteins (Bondarenko, Mowrey, Liu, Xu, & Tang, 2013; Bondarenko et al., 2014; Bondarenko, Yushmanov, Xu, & Tang, 2008; Kinde, Bondarenko, et al., 2016; Kinde, Bu et al., 2016; Mowrey, Liu, et al., 2013; Tang, Eckenhoff, & Xu, 2000), characterize the direct interactions between proteins and anesthetics over a broad range of binding affinities (Bondarenko et al., 2013, 2008; Canlas, Cui, Li, Xu, & Tang, 2008; Cui et al., 2008; Ma, Brandon, et al., 2008; Tang et al., 2000; Tang, Hu, Liachenko, & Xu, 1999; Xu, Seto, Tang, & Firestone, 2000; Xu, Tang, Firestone, & Zhang, 1996), and determine how anesthetic binding affects protein structures and dynamics (Canlas et al., 2008; Cui et al., 2008; Cui, Canlas, Xu, & Tang, 2010; Mowrey, Liu, et al., 2013).

Pentameric ligand-gated ion channels (pLGICs) are important targets of general anesthetics (Campagna, Miller, & Forman, 2003; Eckenhoff, 1996; Franks & Lieb, 1994; Hemmings et al., 2005). Each pLGIC subunit consists of an extracellular domain (ECD) and a transmembrane domain (TMD) composed of four helices (TM1–TM4) with TM2 forming the pore. Eukaryotic pLGICs also contain an intracellular domain (ICD) composed of a large loop between TM3 and TM4. General anesthetics also interact with tetrameric voltage-gated sodium channels by stabilizing the inactivated state or altering the inactivation kinetics (Barber, Carnevale, Klein, Eckenhoff, & Covarrubias, 2014; Covarrubias, Barber, Carnevale, Treptow, & Eckenhoff, 2015; Herold & Hemmings, 2012; Ouyang, Jih, Zhang, Correa, & Hemmings, 2007; Purtelll, Gingrich, Ouyang, Herold, & Hemmings, 2015). Although solution NMR has been used to obtain high-resolution TMD structures of the human pLGICs α4β2nAChR, α7nAChR, and α1GlyR (Bondarenko et al., 2012, 2014, 2010; Mowrey, Cui, et al., 2013), its real strength lies in characterizing the interactions between anesthetics and ion channels, revealing anesthetic binding sites with atomic resolution, and measuring the influence of anesthetics on the dynamics of specific regions in a channel. X-ray crystallography has been successfully employed to identify anesthetic binding sites in pLGICs (Chen et al., 2015, 2017; Nury et al., 2011; Pan et al., 2012; Sauguet et al., 2013; Spurny et al., 2013); however, solution NMR provides more flexibility in sample preparation and measurements at more physiologically relevant conditions. Overall, solution NMR is complementary to other tools in providing structural insight into anesthetic interactions with ion channels.

This chapter outlines solution NMR methods used to investigate the interactions of anesthetics with ion channels and the effects of anesthetic binding on the dynamics of these channels. Expression and purification protocols for isotope-labeled ion channels and sample preparation for solution NMR are also presented.

2. PREPARATION OF ION CHANNELS FOR SOLUTION NMR

2.1 Design Proper Constructs

Recent advances in solution NMR spectroscopy have made it possible to determine structures of various ion channels with molecular weights less than 80 kDa (Mowrey et al., 2015; OuYang et al., 2013; Oxenoid & Chou, 2016; Schnell & Chou, 2008). With an increased molecular mass, both the sensitivity and resolution of NMR spectra worsen dramatically; NMR signals broaden significantly due to slower rotational tumbling of the larger protein. The use of transverse relaxation-optimized spectroscopy (TROSY) has considerably improved the limit of protein mass in solution NMR (Pervushin, Riek, Wider, & Wuthrich, 1997). However, a typical full-length eukaryotic pLGIC has a molecular weight of ~280 kDa and also carries extra weight from membrane mimetics used to stabilize the channel. Similarly, voltage-gated sodium channels have a molecular weight of ~125 kDa. Thus, the full-length ion channel challenges the limit of conventional solution NMR for structure determination. One convenient approach to solving the protein-size problem for ion channels is to truncate the large full-length protein into smaller individual domains, which can retain functionality in isolation. For example, the TMDs of nAChRs and GlyRs have been shown to conduct current in the absence of an ECD or ICD (Bondarenko et al., 2012, 2014; Mowrey, Cui, et al., 2013). Conversely, determination of anesthetic binding sites can be conducted using full-length channels, for which isotopic labeling for NMR observation is performed only on selected residues rather than uniformly labeling the entire protein. Thus, the protein-size issue becomes less problematic, as shown in studies of anesthetic binding to full-length ELIC and NaChBac (Kinde, Bondarenko, et al., 2016; Kinde, Bu, et al., 2016).

2.2 Choose Membrane Mimetics

The choice of membrane mimetic is an important factor in the stability of ion channel samples for solution NMR. Although a native membrane environment is the most ideal choice, the large size of lipid vesicles is problematic for solution NMR. Alternative choices include detergent micelles, phospholipid bicelles, and nanodiscs (Etzkorn et al., 2013; Hagn, Etzkorn, Raschle, & Wagner, 2013; Kucharska, Edrington, Liang, & Tamm, 2015; Laguerre et al., 2016; Liang & Tamm, 2016; Piai, Fu, Dev, & Chou, 2017; Sanders & Mittendorf, 2011; Susac, Horst, & Wuthrich, 2014). Detergent micelles are often used for solution NMR due to their small size. In most cases, detergent micelles provide better quality of NMR spectra and a greater number of resolved residues due to the faster tumbling compared to bicelles and nanodiscs. However, bicelles and nanodiscs more closely resemble the native membrane environment. The NMR spectra for samples in bicelles and nanodiscs, even with fewer resolved residues, can be used at least as a reference to judge whether the protein in micelles is folded properly.

2.3 Prepare Uniformly Isotope-Labeled Ion Channels

Ion channel samples for high-resolution solution NMR require a protein concentration of ~200–300 μM to achieve a good signal to noise ratio of spectra within a reasonable period of time. Ion channel samples must be stable during the period of data acquisition, typically lasting hours to days, depending on the designed experiments. In addition to the selection of membrane mimetic, choices of buffer type, salt concentration, and pH are extremely important for sample stability. Most NMR experiments are required to be performed at pH $\leq$ 7 to prevent exchange-induced signal weakening. Thus, the protein samples must be stable at this pH condition. The ion channel construct for NMR experiments should be designed to fulfill this requirement. For example, in order to stabilize the TMDs of α4, β2, and α7nAChRs at pH $\leq$ 7, a TEV protease recognition site and a polyhistidine tag were placed at the N-terminus of the TMD. The ICD was replaced with a short glycine linker. More importantly, insertion of 2–3 glutamate residues at the N-terminus and mutation of two hydrophobic residues within the TM2–TM3 linker to serine greatly prevented protein aggregation and extended the lifetime of the samples at low pH (Bondarenko et al., 2012, 2014).

Expression and purification of ion channels in *E. coli* follow the same generalized protocols as described in the chapter "X-ray Crystallographic Studies for Revealing Binding Sites of General Anesthetics in Pentameric Ligand-Gated Ion Channels" by Chen, Xu, and Tang in this volume. For ^{15}N and ^{13}C NMR, isotope labeling is achieved by replacing LB broth with M9 media containing ^{15}N-$(NH_4)_2SO_4$ and/or ^{13}C-glucose during expression. After purification, the sample is prepared to fit the requirements of NMR, including pH below 7. To make pH adjustment easier, the buffer concentration should be reduced to <5 mM using a diafiltration or desalting column. The protein should be concentrated to ~200–300 μM, and the detergent concentration should be minimized.

2.4 Prepare Selectively ^{19}F-labeled Ion Channels

^{19}F NMR is a popular tool to study protein conformations, dynamics, and interactions with ligands. ^{19}F has the advantage of high natural abundance, large chemical shift dispersion, and the absence of native ^{19}F signals in proteins, offering a clean background for NMR experiments. ^{19}F NMR chemical shifts are sensitive to electrostatic and van der Waals interactions, and the linewidth of a ^{19}F resonance signal can be affected predominately by protein dynamics (Kinde et al., 2015; Kitevski-LeBlanc & Prosser, 2012; Liu et al., 2012). The high sensitivity to the chemical and dynamic environment around the ^{19}F probe makes ^{19}F NMR particularly suitable for detecting channel conformational changes and identifying anesthetic binding sites. In principle, a ^{19}F probe can be placed in a designed residue position through either biosynthetic incorporation (Sharaf & Gronenborn, 2015) or posttranslationally introduced by chemical labeling small ^{19}F-containing molecules to purified proteins with reactive –SH or –NH moieties. However, in practice with ion channels, chemical labeling is preferred to provide a sufficient amount of ^{19}F-labeled channels for NMR. 2,2,2-Trifluoroethanethiol (TET) and 3-bromo-1,1,1-trifluoroacetone (BTFA) have been used as ^{19}F-labeling reagents for ion channels (Kinde, Bondarenko, et al., 2016; Kinde, Bu, et al., 2016; Kinde et al., 2015). 4-(Perfluoro-*tert*-butyl)-phenyliodoacetamide (PFP) and *S*-ethyl-trifluorothioacetate (SETFA) have also been used for ^{19}F-labeling proteins (Kalbitzer et al., 1992; Mehta, Kulkarni, Mason, Constantinescu, & Antich, 1994). In order to have site-specific labeling, it is ideal to use an ion channel construct that contains a single cysteine residue. Thus, mutations are often required if there are multiple cysteines in the native channel. ^{19}F chemical labeling of ion channels typically begins with a brief treatment of a reducing agent, followed by the addition of the selective labeling reagent. A generalized labeling protocol is provided below:

1. Add a 30-fold molar excess of the reducing agent tris(2-carboxyethyl) phosphine (TCEP) to the purified protein and incubate at 4°C for 1 h.
2. Remove TCEP using a desalting column.
3. Label protein with a ~70-fold molar excess of ^{19}F-labeling reagents (TET or BTFA) at room temperature for 3 h, and then 4°C overnight.
4. Remove free ^{19}F-labeling reagent using dialysis (3 changes of 100 × volume, 1 h each), followed by size exclusion chromatography (Superdex 200 10/300GL column (GE Healthcare)).

Note: A few m*M* of lipids (asolectin or DMPC) can be added to the final protein sample to improve stability.

3. NMR METHODS TO IDENTIFY ANESTHETICS BINDING SITES AND DETECT ANESTHETIC-INDUCED DYNAMICS CHANGES

3.1 Chemical Shift Perturbations

The chemical shift is the resonant frequency of a nucleus relative to a standard in NMR spectroscopy. If anesthetics bind to an ion channel, they will alter the chemical environment of contacting residues, causing changes in chemical shifts. Thus, perturbations to ^{1}H, ^{15}N, ^{13}C, or ^{19}F chemical shifts are an initial indication of anesthetic binding. Anesthetic-induced chemical shift perturbation can be measured in 1D, 2D, or 3D NMR spectra by titrating different concentrations of anesthetics into a sample. In general, only ^{1}H and ^{19}F are detected in 1D spectra because of their high sensitivity. 1D ^{1}H spectra of the protein region are usually crowded and are poorly resolved. In contrast, 1D ^{19}F spectra of selectively labeled samples provide distinct signals for characterizing anesthetic binding. 3D ^{15}N- or ^{13}C-filtered NMR spectra provide much better resolution, but data acquisition requires long times. 2D ^{1}H–^{15}N or ^{1}H–^{13}C NMR spectra, such as HSQC, provide adequate resolution within a reasonable time for data collection. Therefore, they have been commonly used to identify drug-binding sites, including for anesthetics (Bondarenko et al., 2013, 2014, 2008; Mowrey, Liu, et al., 2013). In addition to chemical shift changes, drug binding may also introduce changes in the protein dynamics that can be reflected in the 2D NMR spectra. To prevent artificial chemical shift perturbations, one should use the same buffer for both protein and anesthetics during titration experiments. If any additional chemicals have to be introduced to the sample, a control spectrum must be acquired. The concentration of volatile anesthetics titrated to the protein sample may not be stable due to evaporation. Therefore, the volatile anesthetic concentration after each titration step needs to be calibrated using a reference signal, such as a known concentration of DSS in the 1D ^{1}H spectrum or trifluoroacetate (TFA) in the ^{19}F spectrum (Xu et al., 2000, 1996).

Chemical shift perturbations can be used to estimate the dissociation constant (K_D) for anesthetics from fitting the data as a function of anesthetic concentration in Eq. (1) (Fielding, 2003; Morton et al., 1996) below:

$$\Delta\delta_{obs} = \frac{\Delta\delta_{max}}{2[P]_0}\left\{ \left(K_D + [A]_0 + [P]_0\right) - \sqrt{\left(K_D + [A]_0 + [P]_0\right)^2 - 4[A]_0[P]_0} \right\} \tag{1}$$

$\Delta\delta_{obs}$ is the observed chemical shift change in the presence and absence of anesthetics at a given anesthetic concentration, $\Delta\delta_{max}$ is the maximum chemical shift change in the presence and absence of anesthetics, and $[P]_0$ and $[A]_0$ are the protein and anesthetic concentration, respectively.

Although residues showing large chemical shift changes in the presence of anesthetics are likely directly involved in binding, indirect allosteric effects by anesthetics can also produce changes in the spectra. Additional NMR experiments (see Section 3.2) are required to distinguish between the direct and indirect anesthetic effects.

3.2 Saturation Transfer Difference Spectroscopy

Saturation transfer difference (STD) NMR has been used effectively to identify direct binding of small molecules to proteins with a K_D between 10^{-8} and $10^{-3}M$ (Klein, Meinecke, Mayer, & Meyer, 1999; Mayer & Meyer, 1999). It is particularly suitable for investigations of anesthetic–protein interactions due to the low-affinity nature of the anesthetic binding (Bondarenko et al., 2013, 2014; Cui et al., 2008; Mowrey, Liu, et al., 2013). Recent improvements in the STD NMR pulse sequences have allowed for better information on the most significant molecular contacts and interactions that relate to ligand binding modes (Megy et al., 2005; Sorge, Wagstaff, Rowe, Williamson, & Howard, 2015; Vogtherr & Peters, 2000; Wagstaff, Vallath, Marshall, Williamson, & Howard, 2010).

STD experiments collect a pair of NMR spectra in an interleaved fashion, with on- and off-resonance frequencies to saturate selected ^1H or ^{19}F of ion channels. Effective spin diffusion saturation can quickly propagate across the protein as well as bound anesthetics. Thus, the signal from bound anesthetics will be attenuated in the on-resonance spectrum. Subtracting the off- and on-resonance saturation spectra provides the STD spectrum, where only signals from the bound anesthetics are observed. Fitting the STD data collected at different saturation times to Eq. (2) (Mayer & James, 2004) below offers an estimate of several key parameters, including the asymptotic maximum of the STD buildup curve STD_{max}, the observed saturation rate constant k_{sat}, and the crossrelaxation rate constant between the anesthetic and ion channel spins σ ($\sigma = STD_{max}k_{sat}$).

$$STD = STD_{max}\left[1 - e^{(-k_{sat}t)}\right] \qquad (2)$$

where $STD = 100\% \frac{I_{off} - I_{on}}{I_{off}}$, I_{off} and I_{on} are the anesthetic peak intensity from the off- and on-resonance STD spectra, respectively, and t is the saturation time.

If the aforementioned STD NMR experiments are performed over several concentrations of a given anesthetic, the anesthetic binding affinity, K_D, can measured based on Eqs. (3) and (4) (Mayer & Meyer, 2001; Meyer & Peters, 2003),

$$STD_{Ampl.} = \frac{(I_{off} - I_{on})}{I_{off}} * \text{ligand excess} \tag{3}$$

$$STD_{Ampl.} = \frac{STD_{max}[L]}{K_D + [L]} \tag{4}$$

where K_D is calculated by fitting the $STD_{Ampl.}$ data as a function of the anesthetic concentration $[L]$.

3.3 Relaxation Measurements for Anesthetic Modulation of Protein Dynamics

Protein dynamics modulated by anesthetics can be measured both qualitatively and quantitatively. The linewidth and intensity of NMR peaks are related to protein motion. Within the same protein, residues in more flexible regions often experience faster motion and have higher intensities and narrower linewidths in NMR spectra than rigid residues. Anesthetics can cause line broadening and reduced intensity of residues in one region of a protein, but result in narrowed linewidths and higher intensities for residues in another region, indicative of changes in dynamics (Bondarenko et al., 2013, 2014; Canlas et al., 2008; Cui et al., 2008; Mowrey, Liu, et al., 2013). It is not unusual to observe multiple conformations of a given protein in the NMR spectra; the addition of anesthetics can shift the populations of conformations. To quantify changes in protein dynamics introduced by anesthetics, one can perform systematic measurements of spin–lattice (R_1) and spin–spin (R_2) ^{15}N relaxation rate constants and ^{1}H–^{15}N heteronuclear *NOE*. In addition, the ^{15}N R_2 relaxation dispersion data are particularly well suited to determine conformational exchange due to anesthetic binding (Canlas et al., 2008; Cui et al., 2008). In regions where conformational exchange is slow (micro- to millisecond time scale), more than one peak per residue is observed. Slow conformational exchange is especially important because of the relevance to protein function, such as channel gating. Anesthetic binding can not only change the populations of different conformations but also shift the exchange rate between conformations, from the fast to slow regime or vice versa.

4. CASE STUDIES

Two representative studies are presented here to show how solution NMR can be used to identify anesthetic binding sites and characterize the effects of anesthetic binding on the dynamics of ion channels.

4.1 Anesthetic Interactions with nAChR Transmembrane Domains

α4β2nAChR is inhibited by the volatile anesthetic halothane (Mori et al., 2001), and both α4β2 and α7nAChRs are inhibited by the intravenous anesthetic ketamine (Coates & Flood, 2001; Flood & Krasowski, 2000; Yamakura, Chavez-Noriega, & Harris, 2000). To identify specific binding sites for these anesthetics in the TMDs of α4β2 and α7nAChRs, we performed ^{1}H–^{15}N TROSY–HSQC (chemical shift perturbation) and 2D STD NMR experiments (Bondarenko et al., 2013, 2014). nAChR TMD samples for NMR experiments were expressed using plasmids containing N-terminal His-tagged constructs from a T7 promoter, transformed into the *E. coli* strain Rosetta2(DE3)pLysS (Novagen) and plated on an LB-agar plate with ampicillin and chloramphenicol selection. Selected colonies were grown overnight at 37°C in an environmental shaker at 250 rpm. 5 mL of the overnight culture was added to each liter of LB media with antibiotics and grown at 37°C to an optical density of 0.7 at 600 nm. Cells were harvested by centrifugation at 5000 × *g* for 20 min. The cell pellet from 2 L LB culture was resuspended in 0.5 L of M9 media containing ^{15}N-$(NH_4)_2SO_4$ and/or ^{13}C-glucose and grown at 15°C for 1.5 h. Expression was induced by adding 0.2 m*M* IPTG, and cells were grown at 15°C for 3 days. The harvested cell pellet was resuspended in 30 mL lysis buffer (50 m*M* Tris, pH 8.0, 150 m*M* NaCl), and cells were digested at 4°C with lysozyme and benzonase for 1 h. Cells were lysed using a microfluidizer, and the membranes were collected by centrifugation at 100,000 × *g* at 4°C for 1 h. The membranes were homogenized in 10 mL lysis buffer. 2.5% LDAO detergent was subsequently added, and the sample was incubated at 4°C for 1 h with inversion. The solubilized protein was clarified by centrifugation at 13,000 × *g* at 4°C for 10 min.

To purify nAChR TMDs, the solubilized protein was loaded at 0.5 mL/min into a 1-mL Ni-NTA column that was preequilibrated with 20 m*M* imidazole in 96% Buffer A (50 m*M* Tris, pH 8.0, 150 m*M* NaCl, and 0.5% LDAO) and 4% Buffer B (500 m*M* imidazole, pH 8.0, 500 m*M* NaCl, and 0.5% LDAO). The Ni-NTA column was washed with 20 m*M* imidazole until a flat baseline was observed, and then further washed with 10 mL of 50 m*M* imidazole. The protein was eluted with 250 m*M* imidazole. The purity of the protein samples was assessed to determine if additional purification was required.

To prepare the nAChR TMD samples for NMR, the concentration of Tris was reduced to <5 m*M* using diafiltration or desalting columns. The pH was reduced from 8.0 to 4.7 using 1 *M* acetic acid. The NaCl concentration was reduced from 100 to 10 m*M*. Finally, the protein samples were concentrated to 250–300 μ*M*. The concentration of LDAO in the samples was ~1% as measured by 1D ^{1}H NMR. Size exclusion chromatography-multiangle light scattering analysis (SEC-MALS) indicated that the nAChR TMDs spontaneously assembled into pentamers in micelles (Bondarenko et al., 2012, 2014). Furthermore, the purified α7nAChR TMDs formed ion-conducting channels when injected as reconstituted asolectin vesicles into *Xenopus* oocytes. The TMD does not possess the orthosteric agonist-binding site of full-length α7nAChR, but channel current can be elicited by ivermectin (Bondarenko et al., 2014), a known positive allosteric modulator that acts through the TMD (Collins & Millar, 2010; Krause et al., 1998). Moreover, ketamine inhibited ivermectin-induced current in the TMD, consistent with the effect of ketamine on native human α7nAChR (Bondarenko et al., 2014). No ivermectin-elicited current was observed in control oocytes injected with empty asolectin vesicles. These data validate the relevance of using truncated channels for NMR studies.

NMR spectra were acquired on Bruker Avance 600 and 800 MHz spectrometers using triple-resonance inverse-detection cryoprobes (Bruker Instruments, Billerica, MA). For ^{1}H, ^{15}N, and ^{13}C chemical shift assignment and the protein structure determination, a suite of NMR experiments was performed, including HNCA, HN(CO)CA, HNCO, ^{15}N-edited NOESY, ^{1}H–^{13}C HSQC, and ^{1}H–^{15}N TROSY–HSQC. HSQC spectra were collected at 35°C, 40°C, and 45°C to optimize the temperature condition (Bondarenko et al., 2013, 2012, 2014).

In the α4β2nAChR TMD, both halothane and ketamine were found in an intrasubunit cavity near the extracellular end of the β2 transmembrane helices, homologous to a common anesthetic binding site observed in

X-ray structures of GLIC (Nury et al., 2011) and photolabeling of ELIC (Kinde, Bu, et al., 2016). Halothane, but not ketamine, was also found in intersubunit cavities at the interface of α4 and β2. In addition, both anesthetics bound to intrasubunit cavities at the intracellular end of α4 and β2 TMDs. In α7nAChR, ketamine bound to the intracellular end of the TMD and perturbed the channel-gate residue L9′.

Halothane and ketamine binding profoundly altered conformational exchanges of α4β2nAChR TMDs. A number of residues, close to and remote from the binding sites, showed resonance signal splitting from single to double peaks, indicating a decreased conformation exchange rates in the presence of anesthetics. The populations of two conformations were also shifted by anesthetics. Ketamine also introduced notable changes in peak intensity to residues in the α7nAChR TMD, particularly in TM4. Perturbations to TM4 are expected to have functional consequences (Lasalde et al., 1996; Lee et al., 1994; Li, Lee, Pappone, Palma, & McNamee, 1992), and the role of TM4 in pLGIC–lipid interactions as well as in nAChR function has been established (daCosta & Baenziger, 2009; Lasalde et al., 1996; Lee et al., 1994; Li et al., 1992). These studies provided compelling NMR evidence of anesthetic-induced changes in protein dynamics.

4.2 Isoflurane Interaction with NaChBac

The volatile anesthetic isoflurane inhibits voltage-gated sodium channels, including the bacterial channel NaChBac (Ouyang et al., 2007). We used ^{19}F NMR to identify isoflurane binding sites in NaChBac (Kinde, Bondarenko, et al., 2016). NaChBac was expressed in *E. coli* C41 cells (Lucigen) using the in-house expression vector pETHT at 18°C for 24 h. A six-histidine tag and a TEV protease recognition site were placed in the N-terminus of NaChBac. *n*-dodecyl-β-D-maltoside (DDM, Anatrace) detergent was used for membrane solubilization and purification. Protein expression and purification were otherwise similar to the protocols described earlier for nAChR TMDs. Single cysteine mutations were performed on selected residues based on docking predictions (Raju, Barber, LeBard, Klein, & Carnevale, 2013). Multiple binding sites were predicted, including sites in the S4–S5 linker, the base of the S5 helix, the extracellular interface near the hinge of the P1 helix, near the central pore at the bottom of the selectivity filter, and at the N-terminal hinge of the S6 helix. ^{19}F-labeling with TET to cysteine residues was conducted using the protocol described in Section 2.4.

^{19}F NMR STD spectra with on- and off-resonance saturation were acquired at 10°C using five different saturation times (0.1, 0.4, 1, 2, and 4 s) and a 5-s recycle time (Kinde, Bondarenko, et al., 2016). To quantify the isoflurane STD signal in each NaChBac mutant, the $-CF_2$ and $-CF_3$ peaks were integrated in each spectrum and plotted as a function of the saturation time. The data were fit by Eq. (2) (Mayer & James, 2004) in Section 3.2.

The ^{19}F NMR STD spectra showed strong direct interactions between isoflurane and NaChBac at residue positions S129, T189, and S208. Using selective saturation of the $-CF_3$ and $-CF_2$ moieties in isoflurane, we observed an orientation preference for isoflurane bound to T189 and S208, but not to S129. The data conclude that isoflurane inhibits NaChBac by two distinct mechanisms: (i) as a channel blocker at the base of the selectivity filter, and (ii) as a modulator to restrict the pivot motion at the S4–S5 linker and at a critical hinge that controls the gating and inactivation motion of S6 (Kinde, Bondarenko, et al., 2016).

ACKNOWLEDGMENTS

The authors would like to thank Dr. Tommy Tillman, Dr. Qiang Chen, and Ms. Sandy Hirsch for their valuable comments and suggestions. The research is supported by NIH (R01GM056257, R01GM49202). The authors declare no conflicts of interest with the contents of this chapter.

REFERENCES

Barber, A. F., Carnevale, V., Klein, M. L., Eckenhoff, R. G., & Covarrubias, M. (2014). Modulation of a voltage-gated Na+ channel by sevoflurane involves multiple sites and distinct mechanisms. *Proceedings of the National Academy of Sciences of the United States of America*, *111*(18), 6726–6731. https://doi.org/10.1073/pnas.1405768111.

Bondarenko, V., Mowrey, D., Liu, L. T., Xu, Y., & Tang, P. (2013). NMR resolved multiple anesthetic binding sites in the TM domains of the alpha4beta2 nAChR. *Biochimica et Biophysica Acta*, *1828*(2), 398–404. https://doi.org/10.1016/j.bbamem.2012.09.014.

Bondarenko, V., Mowrey, D., Tillman, T., Cui, T., Liu, L. T., Xu, Y., et al. (2012). NMR structures of the transmembrane domains of the alpha4beta2 nAChR. *Biochimica et Biophysica Acta*, *1818*(5), 1261–1268. https://doi.org/10.1016/j.bbamem.2012.02.008.

Bondarenko, V., Mowrey, D. D., Tillman, T. S., Seyoum, E., Xu, Y., & Tang, P. (2014). NMR structures of the human alpha7 nAChR transmembrane domain and associated anesthetic binding sites. *Biochimica et Biophysica Acta*, *1838*(5), 1389–1395. https://doi.org/10.1016/j.bbamem.2013.12.018.

Bondarenko, V., Tillman, T., Xu, Y., & Tang, P. (2010). NMR structure of the transmembrane domain of the n-acetylcholine receptor beta2 subunit. *Biochimica et Biophysica Acta*, *1798*(8), 1608–1614. https://doi.org/10.1016/j.bbamem.2010.04.014.

Bondarenko, V., Yushmanov, V. E., Xu, Y., & Tang, P. (2008). NMR study of general anesthetic interaction with nAChR beta2 subunit. *Biophysical Journal*, *94*(5), 1681–1688. https://doi.org/10.1529/biophysj.107.116772.

Campagna, J. A., Miller, K. W., & Forman, S. A. (2003). Mechanisms of actions of inhaled anesthetics. *The New England Journal of Medicine*, *348*(21), 2110–2124. https://doi.org/10.1056/NEJMra021261.

Canlas, C. G., Cui, T., Li, L., Xu, Y., & Tang, P. (2008). Anesthetic modulation of protein dynamics: Insight from an NMR study. *The Journal of Physical Chemistry B*, *112*(45), 14312–14318. https://doi.org/10.1021/jp805952w.

Chen, Q., Kinde, M. N., Arjunan, P., Wells, M. M., Cohen, A. E., Xu, Y., et al. (2015). Direct pore binding as a mechanism for isoflurane inhibition of the pentameric ligand-gated ion channel ELIC. *Scientific Reports*, *5*, 13833. https://doi.org/10.1038/srep13833.

Chen, Q., Wells, M. M., Tillman, T. S., Kinde, M. N., Cohen, A., Xu, Y., et al. (2017). Structural basis of alcohol inhibition of the pentameric ligand-gated ion channel ELIC. *Structure*, *25*(1), 180–187. https://doi.org/10.1016/j.str.2016.11.007.

Coates, K. M., & Flood, P. (2001). Ketamine and its preservative, benzethonium chloride, both inhibit human recombinant alpha7 and alpha4beta2 neuronal nicotinic acetylcholine receptors in Xenopus oocytes. *British Journal of Pharmacology*, *134*(4), 871–879. https://doi.org/10.1038/sj.bjp.0704315.

Collins, T., & Millar, N. S. (2010). Nicotinic acetylcholine receptor transmembrane mutations convert ivermectin from a positive to a negative allosteric modulator. *Molecular Pharmacology*, *78*(2), 198–204. https://doi.org/10.1124/mol.110.064295.

Covarrubias, M., Barber, A. F., Carnevale, V., Treptow, W., & Eckenhoff, R. G. (2015). Mechanistic insights into the modulation of voltage-gated ion channels by inhalational anesthetics. *Biophysical Journal*, *109*(10), 2003–2011. https://doi.org/10.1016/j.bpj.2015.09.032.

Cui, T., Bondarenko, V., Ma, D., Canlas, C., Brandon, N. R., Johansson, J. S., et al. (2008). Four-alpha-helix bundle with designed anesthetic binding pockets. Part II: Halothane effects on structure and dynamics. *Biophysical Journal*, *94*(11), 4464–4472. https://doi.org/10.1529/biophysj.107.117853.

Cui, T., Canlas, C. G., Xu, Y., & Tang, P. (2010). Anesthetic effects on the structure and dynamics of the second transmembrane domains of nAChR alpha4beta2. *Biochimica et Biophysica Acta*, *1798*(2), 161–166. https://doi.org/10.1016/j.bbamem.2009.08.009.

Cui, T., Mowrey, D., Bondarenko, V., Tillman, T., Ma, D., Landrum, E., et al. (2012). NMR structure and dynamics of a designed water-soluble transmembrane domain of nicotinic acetylcholine receptor. *Biochimica et Biophysica Acta*, *1818*(3), 617–626. https://doi.org/10.1016/j.bbamem.2011.11.021.

daCosta, C. J., & Baenziger, J. E. (2009). A lipid-dependent uncoupled conformation of the acetylcholine receptor. *The Journal of Biological Chemistry*, *284*(26), 17819–17825. https://doi.org/10.1074/jbc.M900030200.

Eckenhoff, R. G. (1996). An inhalational anesthetic binding domain in the nicotinic acetylcholine receptor. *Proceedings of the National Academy of Sciences of the United States of America*, *93*(7), 2807–2810.

Etzkorn, M., Raschle, T., Hagn, F., Gelev, V., Rice, A. J., Walz, T., et al. (2013). Cell-free expressed bacteriorhodopsin in different soluble membrane mimetics: Biophysical properties and NMR accessibility. *Structure*, *21*(3), 394–401. https://doi.org/10.1016/j.str.2013.01.005.

Fielding, L. (2003). NMR methods for the determination of protein-ligand dissociation constants. *Current Topics in Medicinal Chemistry*, *3*(1), 39–53. https://doi.org/10.2174/1568026033392705.

Flood, P., & Krasowski, M. D. (2000). Intravenous anesthetics differentially modulate ligand-gated ion channels. *Anesthesiology*, *92*(5), 1418–1425.

Franks, N. P., & Lieb, W. R. (1994). Molecular and cellular mechanisms of general-anesthesia. *Nature*, *367*(6464), 607–614. https://doi.org/10.1038/367607a0.

Hagn, F., Etzkorn, M., Raschle, T., & Wagner, G. (2013). Optimized phospholipid bilayer nanodiscs facilitate high-resolution structure determination of membrane proteins. *Journal of the American Chemical Society*, *135*(5), 1919–1925. https://doi.org/10.1021/ja310901f.

Hemmings, H. C., Jr., Akabas, M. H., Goldstein, P. A., Trudell, J. R., Orser, B. A., & Harrison, N. L. (2005). Emerging molecular mechanisms of general anesthetic action. *Trends in Pharmacological Sciences*, *26*(10), 503–510. https://doi.org/10.1016/j.tips.2005.08.006.

Herold, K. F., & Hemmings, H. C., Jr. (2012). Sodium channels as targets for volatile anesthetics. *Frontiers in Pharmacology*, *3*, 50. https://doi.org/10.3389/fphar.2012.00050.

Kalbitzer, H. R., Rohr, G., Nowak, E., Goody, R. S., Kuhn, W., & Zimmermann, H. (1992). A new high sensitivity ^{19}F probe for labeling cysteine groups of proteins. *NMR in Biomedicine*, *5*(6), 347–350.

Kay, L. E. (2016). New views of functionally dynamic proteins by solution NMR spectroscopy. *Journal of Molecular Biology*, *428*(2 Pt. A), 323–331. https://doi.org/10.1016/j.jmb.2015.11.028.

Kim, H. J., Howell, S. C., Van Horn, W. D., Jeon, Y. H., & Sanders, C. R. (2009). Recent advances in the application of solution NMR spectroscopy to multi-span integral membrane proteins. *Progress in Nuclear Magnetic Resonance Spectroscopy*, *55*(4), 335–360. https://doi.org/10.1016/j.pnmrs.2009.07.002.

Kinde, M. N., Bondarenko, V., Granata, D., Bu, W., Grasty, K. C., Loll, P. J., et al. (2016). Fluorine-19 NMR and computational quantification of isoflurane binding to the voltage-gated sodium channel NaChBac. *Proceedings of the National Academy of Sciences of the United States of America*, *113*(48), 13762–13767. https://doi.org/10.1073/pnas.1609939113.

Kinde, M. N., Bu, W., Chen, Q., Xu, Y., Eckenhoff, R. G., & Tang, P. (2016). Common anesthetic-binding site for inhibition of pentameric ligand-gated ion channels. *Anesthesiology*, *124*(3), 664–673. https://doi.org/10.1097/ALN.0000000000001005.

Kinde, M. N., Chen, Q., Lawless, M. J., Mowrey, D. D., Xu, J., Saxena, S., et al. (2015). Conformational changes underlying desensitization of the pentameric ligand-gated ion channel ELIC. *Structure*, *23*(6), 995–1004. https://doi.org/10.1016/j.str.2015.03.017.

Kitevski-LeBlanc, J. L., & Prosser, R. S. (2012). Current applications of ^{19}F NMR to studies of protein structure and dynamics. *Progress in Nuclear Magnetic Resonance Spectroscopy*, *62*, 1–33. https://doi.org/10.1016/j.pnmrs.2011.06.003.

Klein, J., Meinecke, R., Mayer, M., & Meyer, B. (1999). Detecting binding affinity to immobilized receptor proteins in compound libraries by HR-MAS STD NMR. *Journal of the American Chemical Society*, *121*(22), 5336–5337. https://doi.org/10.1021/Ja990706x.

Krause, R. M., Buisson, B., Bertrand, S., Corringer, P. J., Galzi, J. L., Changeux, J. P., et al. (1998). Ivermectin: A positive allosteric effector of the alpha7 neuronal nicotinic acetylcholine receptor. *Molecular Pharmacology*, *53*(2), 283–294.

Kucharska, I., Edrington, T. C., Liang, B. Y., & Tamm, L. K. (2015). Optimizing nanodiscs and bicelles for solution NMR studies of two beta-barrel membrane proteins. *Journal of Biomolecular NMR*, *61*(3–4), 261–274. https://doi.org/10.1007/s10858-015-9905-z.

Laguerre, A., Lohr, F., Henrich, E., Hoffmann, B., Abdul-Manan, N., Connolly, P. J., et al. (2016). From nanodiscs to isotropic bicelles: A procedure for solution nuclear magnetic resonance studies of detergent-sensitive integral membrane proteins. *Structure*, *24*(10), 1830–1841. https://doi.org/10.1016/j.str.2016.07.017.

Lasalde, J. A., Tamamizu, S., Butler, D. H., Vibat, C. R., Hung, B., & McNamee, M. G. (1996). Tryptophan substitutions at the lipid-exposed transmembrane segment M4 of Torpedo californica acetylcholine receptor govern channel gating. *Biochemistry*, *35*(45), 14139–14148. https://doi.org/10.1021/bi961583l.

Lee, Y. H., Li, L., Lasalde, J., Rojas, L., McNamee, M., Ortiz-Miranda, S. I., et al. (1994). Mutations in the M4 domain of Torpedo californica acetylcholine receptor dramatically alter ion channel function. *Biophysical Journal, 66*(3 Pt. 1), 646–653.

Li, L., Lee, Y. H., Pappone, P., Palma, A., & McNamee, M. G. (1992). Site-specific mutations of nicotinic acetylcholine receptor at the lipid-protein interface dramatically alter ion channel gating. *Biophysical Journal, 62*(1), 61–63. https://doi.org/10.1016/S0006-3495(92)81779-4.

Liang, B., & Tamm, L. K. (2016). NMR as a tool to investigate the structure, dynamics and function of membrane proteins. *Nature Structural & Molecular Biology, 23*(6), 468–474. https://doi.org/10.1038/nsmb.3226.

Liu, J. J., Horst, R., Katritch, V., Stevens, R. C., & Wuthrich, K. (2012). Biased signaling pathways in beta2-adrenergic receptor characterized by 19F-NMR. *Science, 335*(6072), 1106–1110. https://doi.org/10.1126/science.1215802.

Ma, D., Brandon, N. R., Cui, T., Bondarenko, V., Canlas, C., Johansson, J. S., et al. (2008). Four-alpha-helix bundle with designed anesthetic binding pockets. Part I: Structural and dynamical analyses. *Biophysical Journal, 94*(11), 4454–4463. https://doi.org/10.1529/biophysj.107.117838.

Ma, D., Liu, Z., Li, L., Tang, P., & Xu, Y. (2005). Structure and dynamics of the second and third transmembrane domains of human glycine receptor. *Biochemistry, 44*(24), 8790–8800. https://doi.org/10.1021/bi050256n.

Ma, D., Tillman, T. S., Tang, P., Meirovitch, E., Eckenhoff, R., Carnini, A., et al. (2008). NMR studies of a channel protein without membranes: Structure and dynamics of water-solubilized KcsA. *Proceedings of the National Academy of Sciences of the United States of America, 105*(43), 16537–16542. https://doi.org/10.1073/pnas.0805501105.

Mayer, M., & James, T. L. (2004). NMR-based characterization of phenothiazines as a RNA binding scaffold. *Journal of the American Chemical Society, 126*(13), 4453–4460. https://doi.org/10.1021/ja0398870.

Mayer, M., & Meyer, B. (1999). Characterization of ligand binding by saturation transfer difference NMR spectroscopy. *Angewandte Chemie International Edition, 38*(12), 1784–1788.

Mayer, M., & Meyer, B. (2001). Group epitope mapping by saturation transfer difference NMR to identify segments of a ligand in direct contact with a protein receptor. *Journal of the American Chemical Society, 123*(25), 6108–6117.

Megy, S., Bertho, G., Gharbi-Benarous, J., Evrard-Todeschi, N., Coadou, G., Segeral, E., et al. (2005). STD and TRNOESY NMR studies on the conformation of the oncogenic protein beta-catenin containing the phosphorylated motif DpSGXXpS bound to the beta-TrCP protein. *The Journal of Biological Chemistry, 280*(32), 29107–29116. https://doi.org/10.1074/jbc.M501628200.

Mehta, V. D., Kulkarni, P. V., Mason, R. P., Constantinescu, A., & Antich, P. P. (1994). Fluorinated proteins as potential 19F magnetic resonance imaging and spectroscopy agents. *Bioconjugate Chemistry, 5*(3), 257–261.

Meyer, B., & Peters, T. (2003). NMR spectroscopy techniques for screening and identifying ligand binding to protein receptors. *Angewandte Chemie (International Ed. in English), 42*(8), 864–890. https://doi.org/10.1002/anie.200390233.

Mori, T., Zhao, X., Zuo, Y., Aistrup, G. L., Nishikawa, K., Marszalec, W., et al. (2001). Modulation of neuronal nicotinic acetylcholine receptors by halothane in rat cortical neurons. *Molecular Pharmacology, 59*(4), 732–743.

Morton, C. J., Pugh, D. J., Brown, E. L., Kahmann, J. D., Renzoni, D. A., & Campbell, I. D. (1996). Solution structure and peptide binding of the SH3 domain from human Fyn. *Structure, 4*(6), 705–714.

Mowrey, D. D., Cui, T., Jia, Y., Ma, D., Makhov, A. M., Zhang, P., et al. (2013). Open-channel structures of the human glycine receptor alpha1 full-length transmembrane domain. *Structure, 21*(10), 1897–1904. https://doi.org/10.1016/j.str.2013.07.014.

Mowrey, D. D., Kinde, M. N., Xu, Y., & Tang, P. (2015). Atomistic insights into human Cys-loop receptors by solution NMR. *Biochimica et Biophysica Acta-Biomembranes, 1848*(1), 307–314. https://doi.org/10.1016/j.bbamem.2014.03.014.

Mowrey, D. D., Liu, Q., Bondarenko, V., Chen, Q., Seyoum, E., Xu, Y., et al. (2013). Insights into distinct modulation of alpha7 and alpha7beta2 nicotinic acetylcholine receptors by the volatile anesthetic isoflurane. *The Journal of Biological Chemistry, 288*(50), 35793–35800. https://doi.org/10.1074/jbc.M113.508333.

Nury, H., Van Renterghem, C., Weng, Y., Tran, A., Baaden, M., Dufresne, V., et al. (2011). X-ray structures of general anaesthetics bound to a pentameric ligand-gated ion channel. *Nature, 469*(7330), 428–431. https://doi.org/10.1038/nature09647.

Ouyang, W., Jih, T. Y., Zhang, T. T., Correa, A. M., & Hemmings, H. C., Jr. (2007). Isoflurane inhibits NaChBac, a prokaryotic voltage-gated sodium channel. *The Journal of Pharmacology and Experimental Therapeutics, 322*(3), 1076–1083. https://doi.org/10.1124/jpet.107.122929.

OuYang, B., Xie, S., Berardi, M. J., Zhao, X., Dev, J., Yu, W., et al. (2013). Unusual architecture of the p7 channel from hepatitis C virus. *Nature, 498*(7455), 521–525. https://doi.org/10.1038/nature12283.

Oxenoid, K., & Chou, J. J. (2013). The present and future of solution NMR in investigating the structure and dynamics of channels and transporters. *Current Opinion in Structural Biology, 23*(4), 547–554. https://doi.org/10.1016/j.sbi.2013.03.010.

Oxenoid, K., & Chou, J. J. (2016). A functional NMR for membrane proteins: Dynamics, ligand binding, and allosteric modulation. *Protein Science, 25*(5), 959–973. https://doi.org/10.1002/pro.2910.

Pan, J., Chen, Q., Willenbring, D., Mowrey, D., Kong, X.-P., Cohen, A., et al. (2012). Structure of the pentameric ligand-gated ion channel GLIC bound with anesthetic ketamine. *Structure, 20*(9), 1463–1469. https://doi.org/10.1016/j.str.2012.08.009.

Pervushin, K., Riek, R., Wider, G., & Wuthrich, K. (1997). Attenuated T2 relaxation by mutual cancellation of dipole-dipole coupling and chemical shift anisotropy indicates an avenue to NMR structures of very large biological macromolecules in solution. *Proceedings of the National Academy of Sciences of the United States of America, 94*(23), 12366–12371.

Piai, A., Fu, Q., Dev, J., & Chou, J. J. (2017). Optimal bicelle size q for solution NMR studies of the protein transmembrane partition. *Chemistry, 23*(6), 1361–1367. https://doi.org/10.1002/chem.201604206.

Purtelll, K., Gingrich, K. J., Ouyang, W., Herold, K. F., & Hemmings, H. C. (2015). Activity-dependent depression of neuronal sodium channels by the general anaesthetic isoflurane. *British Journal of Anaesthesia, 115*(1), 112–121. https://doi.org/10.1093/bja/aev203.

Raju, S. G., Barber, A. F., LeBard, D. N., Klein, M. L., & Carnevale, V. (2013). Exploring volatile general anesthetic binding to a closed membrane-bound bacterial voltage-gated sodium channel via computation. *PLoS Computational Biology, 9*(6), e1003090. https://doi.org/10.1371/journal.pcbi.1003090.

Rosenzweig, R., & Kay, L. E. (2016). Solution NMR spectroscopy provides an avenue for the study of functionally dynamic molecular machines: The example of protein disaggregation. *Journal of the American Chemical Society, 138*(5), 1466–1477. https://doi.org/10.1021/jacs.5b11346.

Sanders, C. R., & Mittendorf, K. F. (2011). Tolerance to changes in membrane lipid composition as a selected trait of membrane proteins. *Biochemistry, 50*(37), 7858–7867. https://doi.org/10.1021/bi2011527.

Sauguet, L., Poitevin, F., Murail, S., Van Renterghem, C., Moraga-Cid, G., Malherbe, L., et al. (2013). Structural basis for ion permeation mechanism in pentameric ligand-gated ion channels. *The EMBO Journal, 32*(5), 728–741. https://doi.org/10.1038/emboj.2013.17.

Schnell, J. R., & Chou, J. J. (2008). Structure and mechanism of the M2 proton channel of influenza A virus. *Nature*, *451*(7178), 591–595. https://doi.org/10.1038/nature06531.

Sharaf, N. G., & Gronenborn, A. M. (2015). (19)F-modified proteins and (19)F-containing ligands as tools in solution NMR studies of protein interactions. *Methods in Enzymology*, *565*, 67–95. https://doi.org/10.1016/bs.mie.2015.05.014.

Sorge, J. L., Wagstaff, J. L., Rowe, M. L., Williamson, R. A., & Howard, M. J. (2015). Q2DSTD NMR deciphers epitope-mapping variability for peptide recognition of integrin alphavbeta6. *Organic & Biomolecular Chemistry*, *13*(29), 8001–8007. https://doi.org/10.1039/c5ob01237f.

Spurny, R., Billen, B., Howard, R. J., Brams, M., Debaveye, S., Price, K. L., et al. (2013). Multisite binding of a general anesthetic to the prokaryotic pentameric Erwinia chrysanthemi ligand-gated ion channel (ELIC). *The Journal of Biological Chemistry*, *288*(12), 8355–8364. https://doi.org/10.1074/jbc.M112.424507.

Susac, L., Horst, R., & Wuthrich, K. (2014). Solution-NMR characterization of outer-membrane protein A from E. coli in lipid bilayer nanodiscs and detergent micelles. *Chembiochem*, *15*(7), 995–1000. https://doi.org/10.1002/cbic.201300729.

Tang, P., Eckenhoff, R. G., & Xu, Y. (2000). General anesthetic binding to gramicidin A: The structural requirements. *Biophysical Journal*, *78*(4), 1804–1809. https://doi.org/10.1016/S0006-3495(00)76730-0.

Tang, P., Hu, J., Liachenko, S., & Xu, Y. (1999). Distinctly different interactions of anesthetic and nonimmobilizer with transmembrane channel peptides. *Biophysical Journal*, 77(2), 739–746. https://doi.org/10.1016/S0006-3495(99)76928-6.

Tang, P., Mandal, P. K., & Xu, Y. (2002). NMR structures of the second transmembrane domain of the human glycine receptor alpha(1) subunit: Model of pore architecture and channel gating. *Biophysical Journal*, *83*(1), 252–262. https://doi.org/10.1016/S0006-3495(02)75166-7.

Vogtherr, M., & Peters, T. (2000). Application of NMR based binding assays to identify key hydroxy groups for intermolecular recognition. *Journal of the American Chemical Society*, *122*(25), 6093–6099. https://doi.org/10.1021/ja0001916.

Wagstaff, J. L., Vallath, S., Marshall, J. F., Williamson, R. A., & Howard, M. J. (2010). Two-dimensional heteronuclear saturation transfer difference NMR reveals detailed integrin alphavbeta6 protein-peptide interactions. *Chemical Communications*, *46*(40), 7533–7535. https://doi.org/10.1039/c0cc01846e.

Xu, Y., Seto, T., Tang, P., & Firestone, L. (2000). NMR study of volatile anesthetic binding to nicotinic acetylcholine receptors. *Biophysical Journal*, *78*(2), 746–751. https://doi.org/10.1016/S0006-3495(00)76632-X.

Xu, Y., Tang, P., Firestone, L., & Zhang, T. T. (1996). 19F nuclear magnetic resonance investigation of stereoselective binding of isoflurane to bovine serum albumin. *Biophysical Journal*, *70*(1), 532–538. https://doi.org/10.1016/S0006-3495(96)79599-1.

Yamakura, T., Chavez-Noriega, L. E., & Harris, R. A. (2000). Subunit-dependent inhibition of human neuronal nicotinic acetylcholine receptors and other ligand-gated ion channels by dissociative anesthetics ketamine and dizocilpine. *Anesthesiology*, *92*(4), 1144–1153.

Ye, L., Van Eps, N., Zimmer, M., Ernst, O. P., & Prosser, R. S. (2016). Activation of the A2A adenosine G-protein-coupled receptor by conformational selection. *Nature*, *533*(7602), 265–268. https://doi.org/10.1038/nature17668.

Zhuang, T., Chen, Q., Cho, M. K., Vishnivetskiy, S. A., Iverson, T. M., Gurevich, V. V., et al. (2013). Involvement of distinct arrestin-1 elements in binding to different functional forms of rhodopsin. *Proceedings of the National Academy of Sciences of the United States of America*, *110*(3), 942–947. https://doi.org/10.1073/pnas.1215176110.

CHAPTER FOUR

Time-Resolved Neutron Interferometry and the Mechanism of Electromechanical Coupling in Voltage-Gated Ion Channels

J. Kent Blasie[1]
University of Pennsylvania, Philadelphia, PA, United States
University of California Irvine, Irvine, CA, United States
[1]Corresponding author: e-mail address: jkblasie@sas.upenn.edu

Contents

Abstract

The mechanism of electromechanical coupling for voltage-gated ion channels (VGICs) involved in neurological signal transmission, primarily Nav- and Kv-channels, remains unresolved. Anesthetics have been shown to directly impact this mechanism, at least for Kv-channels. Molecular dynamics computer simulations can now predict the structures of VGICs embedded within a hydrated phospholipid bilayer membrane as a function of the applied transmembrane voltage, but significant assumptions are still

Methods in Enzymology, Volume 603
ISSN 0076-6879
https://doi.org/10.1016/bs.mie.2018.01.019

necessary. Nevertheless, these simulations are providing new insights into the mechanism of electromechanical coupling at the atomic level in 3-D. We show that time-resolved neutron interferometry can be used to investigate directly the profile structure of a VGIC, vectorially oriented within a single hydrated phospholipid bilayer membrane at the solid–liquid interface, as a function of the applied transmembrane voltage in the absence of any assumptions or potentially perturbing modifications of the VGIC protein and/or the host membrane. The profile structure is a projection of the membrane's 3-D structure onto the membrane normal and, in the absence of site-directed deuterium labeling, is provided at substantially lower spatial resolution than the atomic level. Nevertheless, this novel approach can be used to directly test the validity of the predictions from molecular dynamics simulations. We describe the key elements of our novel experimental approach, including why each is necessary and important to providing the essential information required for this critical comparison of "simulation" vs "experiment." In principle, the approach could be extended to higher spatial resolution and to include the effects of anesthetics on the electromechanical coupling mechanism in VGICs.

1. INTRODUCTION

Nav- and Kv-channels are voltage-gated ion channels (VGICs) primarily responsible for the generation and propagation of action potentials involved in neurological signal transmission (Hodgkin, 1951). More recently, it has become increasingly evident that VGICs are potential targets for anesthetic action (Hemmings, 2009; Liang et al., 2015). The mechanism of electromechanical coupling in VGICs, namely, the coupling of conformational changes within the four voltage sensor domains (VSDs) in response to a change in the transmembrane voltage, either depolarizing or polarizing with respect to the resting voltage, to opening or closing the pore domain, respectively, remains unknown (Grizel, Glukov, & Sokolova, 2014; Vargas et al., 2012). Importantly, it has been shown that at least for Kv-channels, anesthetics directly impact the electromechanical coupling mechanism (Liang et al., 2015).

VGICs exist primarily at the resting transmembrane voltage of negative 100–120 mV (electrophysiological convention) and experience the absence of a transmembrane voltage, i.e., 0 mV, only transiently on a few microseconds timescale during the course of an action potential of a few milliseconds duration (Hodgkin & Huxley, 1952). Currently, high-resolution 3-D structures for these Nav- and Kv-channels can only be provided by conventional techniques (e.g., X-ray crystallography or more recently cryo-electron microscopy) necessarily in the absence of a finite (nonzero) transmembrane voltage (Long, Tao, Campbell, & MacKinnon, 2007; Payandeh, Scheurer, Zheng, & Catterall, 2011). This seriously complicates any investigation into the mechanism of electromechanical coupling. For example, in the case of Kv-channels for which the first high-resolution structure was obtained, the pore domain was found to be open and the voltage sensor conformations were assumed to be in an activated conformation since 0 mV would be sufficiently depolarizing with respect to the resting transmembrane voltage (Long, Campbell, & MacKinnon, 2005). Given this structure and the presumed activated, open state of the channel, numerous models were subsequently developed for the deactivated, closed state. These models employed various computational approaches, each necessarily coupled with specific assumptions (as reviewed in Vargas et al., 2012). Some incorporated restraints based on experimental results provided by indirect techniques utilizing, for example, site-directed labeling of the protein with tethered biotin of differing lengths and measurements of its accessibility to avidin from either side of the membrane (Ruta, Chen, & MacKinnon, 2005; Schow, Freites, Nizkorodov, White, & Tobias, 2012) or pairs of site-directed cysteine mutations and subsequent metal crosslinking (Henrion et al., 2012), each measured as a function of the transmembrane voltage. Such studies have collectively led to an overall "consensus structure" for the deactivated, closed state of Kv-channels (Vargas et al., 2012). However, when the first high-resolution 3-D structure of a Nav-channel was determined (Payandeh et al., 2011), the voltage sensors were found to be in an activated conformation, by comparison with Kv-channels in the absence of a transmembrane voltage, but the pore domain was closed contrary to that for the Kv-channels. As a result, the authors had to assume that the state found was either trapped in an activated, preopen state (Payandeh et al., 2011, Supplementary Information) or a stable inactivated state (Payandeh, El-Din, Scheurer, Zheng, & Catterall, 2012).

It has become possible for all-atom molecular dynamics simulations to address the conformational states responsible for voltage gating in a VGIC directly in response to the applied transmembrane voltage. The approach was first applied to the isolated VSD of the prokaryotic KvAP channel embedded within a fully hydrated phospholipid bilayer (Freites, Schow, White, & Tobias, 2012; Schow, Freites, Gogna, White, & Tobias, 2010). More recently, the approach has been applied to a complete, homotetrameric Kv-channel embedded within a hydrated phospholipid bilayer (Jensen et al., 2012). In such simulations, numerous critical assumptions are required. For example, in the case of the complete Kv-channel, despite the very long trajectories of a few hundred microseconds duration, larger than physiological transmembrane voltages together with a reduction in the strength of salt bridges between oppositely charged residues on the interior of the voltage sensors, thought to be essential for stabilizing their various voltage-dependent conformational states, were required to close and open the channel. Herein we show how time-resolved neutron interferometry can be utilized to provide direct experimental evidence, in the absence of any assumptions and/or potentially perturbing site-directed labeling, to test whether the voltage-dependent conformational states of the Kv-channel predicted by the simulation are correct. In principle, it should be straightforward to extend this "simulation" vs "experiment" approach to include investigations into the effects of anesthetics on the mechanism of electromechanical coupling for both Kv- and Nav-channels.

We developed our time-resolved neutron interferometry approach using a pulsed spallation source in a classic "pump–probe" fashion to first investigate the so-called profile structure of the isolated VSD of KvAP vectorially oriented within a phospholipid bilayer at the solid–liquid interface, utilizing a cyclic sequence of polarizing, nonpolarizing, and depolarizing voltage pulses (Tronin et al., 2014). Only recently have we been successful in applying this approach to investigate the profile structure of the complete, homotetrameric KvAP channel, vectorially oriented within a hydrated phospholipid bilayer membrane at a solid–liquid interface, utilizing a cyclic sequence of hyperpolarizing and depolarizing voltage pulses designed to circumvent inactivation of the channel (Schmidt, Cross, & MacKinnon, 2009). This most recent work will be fully reported elsewhere (Tronin et al., 2018). The "profile structure" is the projection of the 3-D structure of the membrane parallel to the membrane plane onto the membrane normal. Averaging the data for each of the voltage pulses collected over many cycles provides upon analysis the dependence of the nSLD (neutron scattering length density) profile structure of the membrane on the applied transmembrane voltage. The difference nSLD profiles for the membrane (polarizing *minus* nonpolarizing and depolarizing *minus* nonpolarizing for the isolated VSD of KvAP, and hyperpolarizing *minus* depolarizing for KvAP itself) were found to be in good agreement with those calculated from the simulations. Despite the relatively low spatial resolution of the experimental nSLD profiles, these experimental results nevertheless provide direct evidence in support of the conformational changes associated with the activation/deactivation of the VSDs, coupled with the opening/closing of the pore domain responsible for voltage gating in the case of the complete KvAP channel, as identified in these particular simulations implementing an applied transmembrane voltage. This is because the predicted conformational changes extended over a sufficiently large region of the membrane profile structure to be detected experimentally at this spatial resolution.

Our goal here is to describe the key elements of our novel experimental approach, including why each is necessary and important to providing the essential information required for this critical comparison of "simulation" vs "experiment."

2. METHODS

2.1 Voltage-Dependent Membrane Structure Probed by X-Ray/Neutron Reflectivity

In order to probe the profile structure of a membrane as a function of the applied transmembrane voltage by either X-ray or neutron reflectivity, we developed two methods for fabricating a lipid bilayer membrane containing a voltage-sensitive integral membrane protein at the solid–liquid interface with a unique vectorial orientation of the protein with respect to the membrane normal (Gupta, Liu, Strzalka, & Blasie, 2011), designated as "Self-Assembly" and "Directed Assembly." Both were shown to be capable of producing a sufficiently high in-plane density of the protein to effectively dominate the membrane's profile structure, with the protein comprising as much as 50% of the membrane area and phospholipid the remainder. Importantly, fabrication of the membrane at the solid–liquid interface facilitates its incorporation into an electrochemical cell designed specifically for either X-ray (Gupta et al., 2011) or neutron reflectivity (Gupta, Dura, Freites, Tobias, & Blasie, 2012) experiments. The simplest of these two methods was termed "Self-Assembly" and was first applied to the isolated VSD of the prokaryotic Kv-channel KvAP, as described schematically in Fig. 1. Since the "profile structure" is the projection of the 3-D structure of the membrane parallel to the membrane plane onto the membrane normal, as described for a small 3×3 ensemble of the VSD within a phospholipid bilayer in Fig. 2, achieving a *unidirectional* vectorial orientation of the voltage-sensitive integral membrane protein within the membrane at high in-plane density was essential to making the identification of the conformational changes in the protein in response to the applied transmembrane voltage unique. For example, more typical membrane reconstitution approaches usually result in a bidirectional vectorial

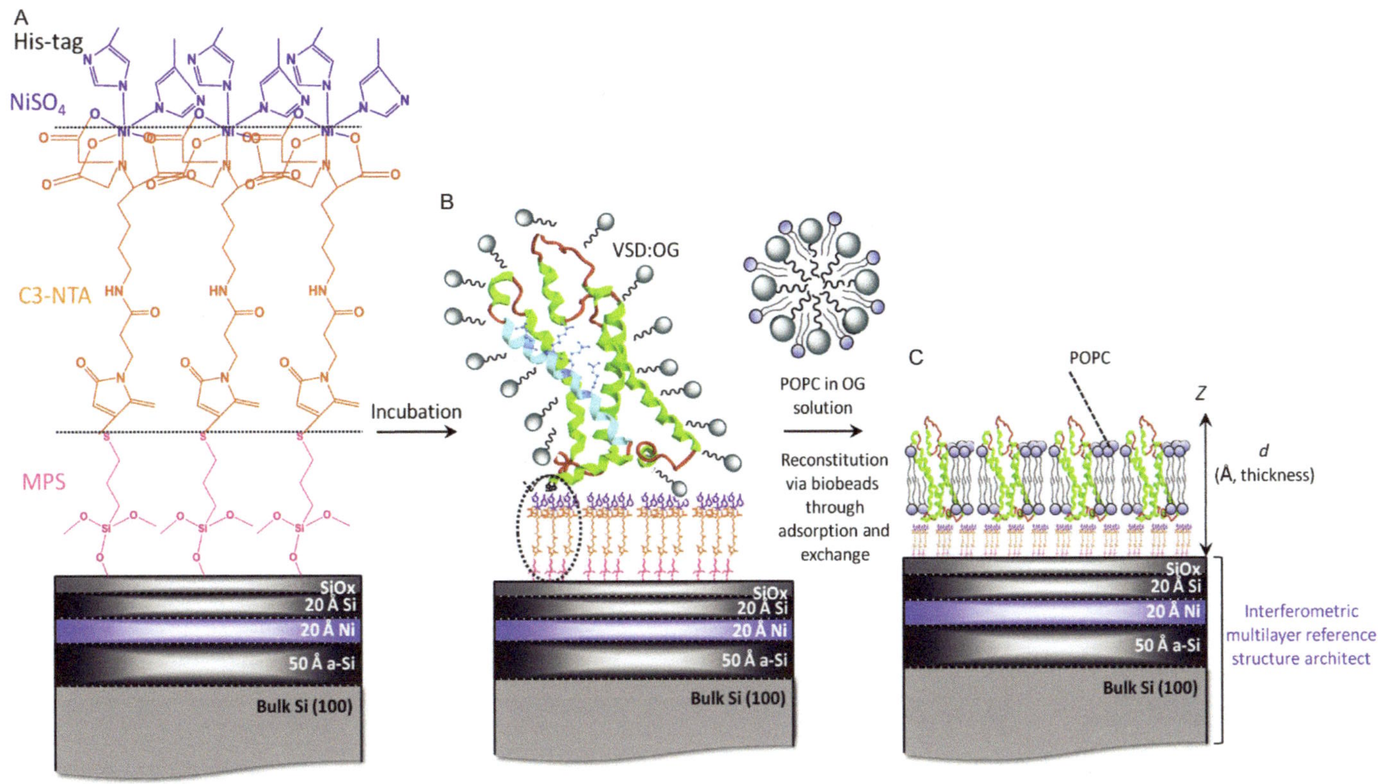

Fig. 1 See legend on opposite page.

orientation in which half of the proteins are oriented in one direction with respect to the membrane normal and the other half in the opposite orientation (Krepkiy et al., 2009). Application of a transmembrane voltage gradient along the membrane normal would then result in one half of the proteins experiencing the voltage gradient in one direction and the other half in the opposite direction. The profile structure of the membrane determined in the reflectivity experiment would then contain a superposition of the two vectorial orientations thereby making it impossible to ascribe any particular voltage-dependent conformation to one particular vectorial orientation of the protein.

2.2 The Choice of Neutrons Over X-Rays

Neutron reflectivity, also enhanced via interferometry as described in Section 2.3, has several particular advantages for the investigation of the profile structure of a membrane at the solid–liquid interface. First, neutrons unlike X-rays readily penetrate most materials thereby greatly simplifying the design of the electrochemical cell as required for the X-ray/neutron reflectivity experiment. Second, as especially important for soft matter, the interaction with neutron radiation is not directly ionizing, unlike the case for X-rays or electrically charged elementary particles, resulting in the absence of radiation damage to the membrane specimen. This aspect is of particular importance for the time-resolved "pump–probe" neutron interferometry approach described herein as the data collection times are very long due to the very low incident neutron fluxes available even with the world's most powerful neutron sources. The third advantage arises from the fact that while X-rays are scattered by atomic electrons thereby varying

Fig. 1 Schematic outlining the Self-Assembly approach for the isolated VSD of KvAP. (A) A Si–Ni–Si multilayer on silicon reference structure, alkylated with mercaptopropylsilane (MPS) to functionalize its surface with thiol groups, is subsequently reacted with the linker maleimido-C_3-NTA to provide nitrilotriacetate endgroups for ligation with Ni^{+2} ions, along with (B) histidines from the protein's C-terminal His_6-tag following incubation with detergent-solubilized VSD. (C) Reconstitution of vectorially oriented VSD in a single phospholipid bilayer environment via exchange against POPC/OG in the presence of biobeads. For aqueous environments, water is expected on both the proximal and distal (with respect to the substrate) sides of the bilayer. The isolated VSD of KvAP is shown as a *ribbon* representation of its X-ray crystal structure (Jiang et al., 2003). *Adapted with permission from Gupta, S., Liu, J., Strzalka, J., & Blasie, J. K. (2011). Profile structures of the VSD and KvAP channel vectorially-oriented in single membranes at solid-vapor or solid-liquid interfaces via X-ray reflectivity.* Physical Review E, *84(3), 031911-1-15; Copyrighted by the American Physical Society.*

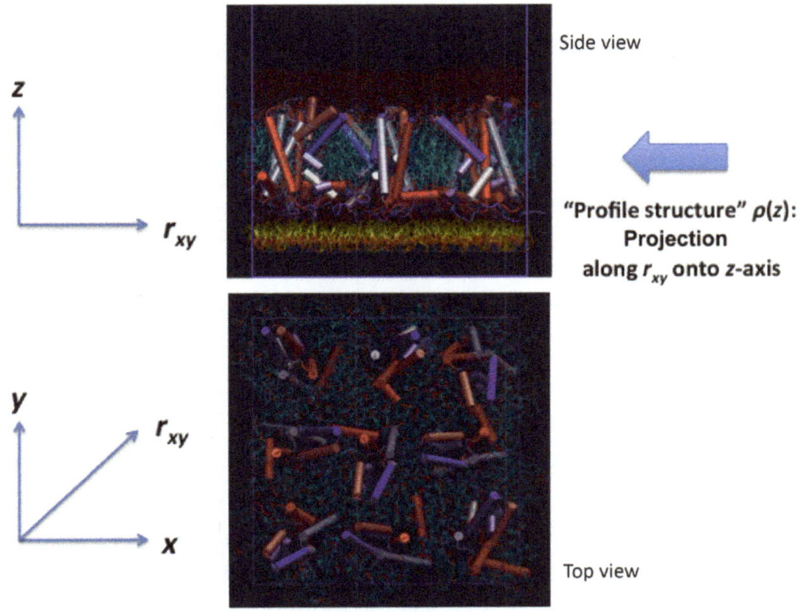

Fig. 2 Molecular graphics representation of a small 3 × 3 ensemble of the vectorially oriented VSD molecules (α-helices represented as *solid cylinders*) solvated by a POPC bilayer (*turquoise*) tethered to the surface of a solid inorganic multilayer substrate (not shown) below the monolayer of tethering chain molecules (*yellow*) in the side view. The "profile structure" arises from the projection of the 3-D structure of the membrane along r_{xy} vectors parallel to the plane onto the *z*-axis normal to the plane of the membrane. The VSD molecules have been rotated randomly with respect to one another about the *z*-axis in the top view. However, unlike in the top view shown, experimentally, the VSDs exhibit only glass-like order in the *x–y* plane, even at this low average area/VSD of ~1400 Å^2 approximating that of the experimental situation. *Adapted with permission from Tronin, A., Strzalka, J. W., Kuzmenko, I., Worcester, D., Lauter, V., Freites, J. A., et al. (2014). Direct evidence of conformational changes associated with voltage-gating in a voltage sensor protein by time-resolved X-ray/neutron interferometry.* Langmuir, 30*(16), 4784–4796; Copyrighted by the American Chemical Society.*

with atomic number, neutrons are scattered by atomic nuclei with most exhibiting so-called potential scattering of comparable amplitude. However, a few isotopes of a few elements exhibit resonance scattering, which changes both the amplitude and phase of the scattered neutrons. Importantly for soft matter, deuterium (^{2}H or D) exhibits potential scattering like most other atomic nuclei of comparable positive scattering length ($b_D = +6.67 \times 10^{-5}$ Å), while hydrogen (^{1}H) exhibits resonance scattering of negative scattering length ($b_H = -3.74 \times 10^{-5}$ Å), by convention the change of sign arising from a change in the relative phase of the scattered neutron for potential vs resonance scattering (Bacon, 1975). Likewise, D_2O- and H_2O-based electrolytes have scattering length densities (SLDs) that are positive ($+6.36 \times 10^{-6}$ Å^{-2}) and negative (-0.56×10^{-6} Å^{-2}), respectively. Importantly, typical membrane molecular components, i.e., proteins and lipids, have average SLDs that lie between these extremes. As a result, the neutron scattering contrast between a structure of interest, a membrane, and its solvation environment, an aqueous medium, can be manipulated by varying either the deuterium/hydrogen ratio of the membrane components or that of the aqueous medium (via the D_2O/H_2O ratio), the latter referred to as "solvent contrast variation." When these two average contrasts are matched, neutron scattering from the structure of interest is either dramatically reduced or vanishes completely. Coupling these approaches allows the investigator to suppress the contrast for a selected membrane component while enhancing that of another, e.g., for the lipid vs the protein or for selected submolecular portions of either. Such isotopic substitution is generally nearly isomorphous for biological structures, thereby allowing the visualization of selected membrane components, or submolecular portions thereof, by comparison of the selectively deuterated structure with that fully hydrogenated via a *difference* structure or profile structure. The same approach is more problematic for the case of nonresonance X-ray scattering because the only way to change the scattering contrast at the atomic level is to change the atom (i.e., its atomic number), which need not be isomorphous, depending on its role in stabilizing the structure of interest, thereby rendering the analogous *difference* structure irrelevant for this purpose. Lastly, the *difference* in the membrane profile structure obtained with neutrons for a solvent contrast at a higher D_2O:H_2O ratio and that obtained at a lower D_2O:H_2O ratio reveals the distribution of both water and water-accessible, exchangeable hydrogen atoms, the latter primarily within the membrane protein.

2.3 X-Ray/Neutron Reflectivity vs Interferometry

An elegant illustration of the substantial advantage provided by X-ray or neutron interferometry over the conventional X-ray or neutron reflectivity experiment was provided by a comparison of the X-ray reflectivity from a Langmuir monolayer of an amphiphilic four-helix bundle peptide compressed at the liquid–gas interface in the absence and presence of the interferometric effect (Krishnan et al., 2010). A schematic representation of the experiment is shown in Fig. 3, where the X-ray reflectivity from the

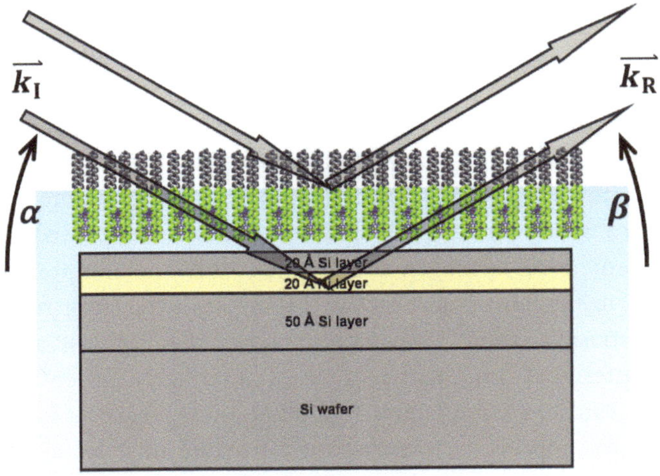

Fig. 3 Schematic representation of the interferometric approach to X-ray reflectivity to determine the xSLD profile structure of an ultrathin Langmuir film of an amphiphile at the liquid–gas interface. The close proximity of the multilayer reference structure without contacting the otherwise unperturbed Langmuir monolayer of the amphiphile at the liquid–gas interface allows for the critical interference of X-rays reflected by the underlying reference structure with those reflected by the amphiphile. For specular X-ray reflectivity, the angle of incidence α is equal to the angle of reflection β, with the plane containing the incident photon momentum vector $\overrightarrow{k_I}$ and reflected photon momentum vector $\overrightarrow{k_R}$ perpendicular to the plane of the interface. *Adapted with permission from Krishnan, V., Strzalka, J., Liu, J., Liu, C., Kuzmenko, I., Gog, T., et al. (2010). Interferometric enhancement of X-ray reflectivity from unperturbed Langmuir monolayers of amphiphiles at the liquid-gas interface.* Physical Review E, 81, *021604-1-10; Copyrighted by the American Physical Society.*

Langmuir monolayer of an amphiphilic four-helix bundle peptide was first measured in the absence of the interferometric effect with the inorganic Si–Ni–Si multilayer substrate immersed deeply in the aqueous subphase beneath the Langmuir monolayer (not shown) and subsequently in the presence of the interferometric effect (shown) with the multilayer substrate just beneath the otherwise unperturbed Langmuir monolayer of the peptide with an intervening thin layer of the aqueous medium. Using monochromatic X-rays, the reflectivity experiment involves recording the elastically reflected X-rays as a function of the angle of incidence (α) which is maintained equal to the angle of reflection (β) for the case of specular reflectivity, both angles (greatly exaggerated here) measured with respect to the plane of the water–helium interface. For the incident X-ray photon momentum vector $\overrightarrow{k_I}$ and reflected X-ray photon momentum vector $\overrightarrow{k_R}$, whose magnitudes are each $2\pi/\lambda$ for elastic scattering where λ is the photon wavelength, the photon momentum transfer vector for the scattering process is $\vec{Q}_z = \left(\vec{k}_R - \vec{k}_I\right)$, which is parallel to the normal to the interface $\vec{z}$, where $\left|\vec{Q}_z\right| \equiv Qz = (4\pi \sin\alpha)/\lambda$. Given the preceding definitions, the resulting X-ray reflectivity data for the three relevant cases are shown in

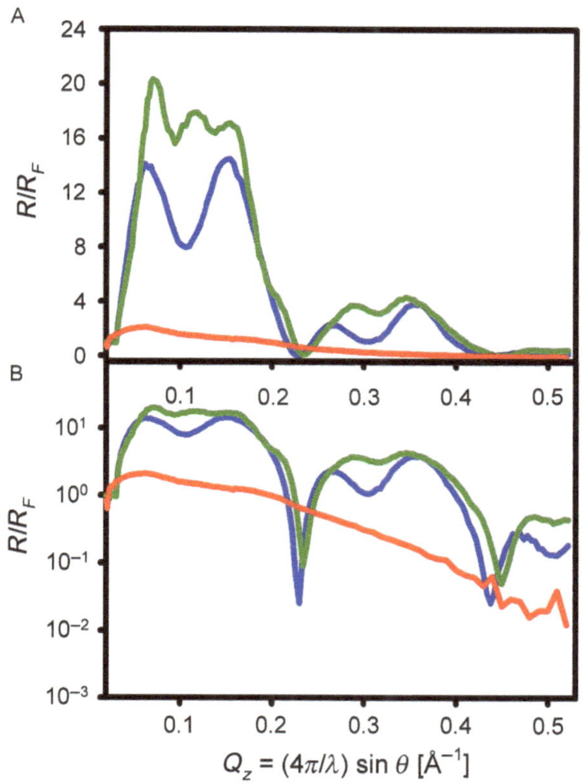

Fig. 4 Specular X-ray reflectivity (X-ray energy of 22.017 keV) from a Langmuir monolayer of the AP0-RuPZn amphiphile obtained using noninterferometric and interferometric approaches shown along with those for the Si–Ni–Si multilayer reference structure alone that possessed a macroscopically *hydrophobic* upper surface [*blue curves*: Si–Ni–Si reference structure in He, *green curves*: Si–Ni–Si reference structure with adjacent Langmuir monolayer (interferometric case), and *red curves*: Langmuir monolayer of the amphiphile only (noninterferometric case)]. (A) Fresnel normalized X-ray reflectivity data shown on a linear scale, and (B) Fresnel normalized X-ray reflectivity data shown on a semilogarithmic scale. *Adapted with permission from Krishnan, V., Strzalka, J., Liu, J., Liu, C., Kuzmenko, I., Gog, T., et al. (2010). Interferometric enhancement of X-ray reflectivity from unperturbed Langmuir monolayers of amphiphiles at the liquid-gas interface.* Physical Review E, 81, *021604-1-10; Copyrighted by the American Physical Society.*

Fig. 4, where the Fresnel-normalized reflectivity $R(Q_z)/R_F(Q_z)$ is defined as the specular reflectivity $R(Q_z)$ divided by that from an ideal (flat and smooth) interface $R_F(Q_z)$. As can be seen, especially from Fig. 4B on the semilogarithmic scale, the presence of the interference effect, manifest as the *difference* between the Fresnel-normalized reflectivity for the Langmuir monolayer of the amphiphilic peptide with the multilayer substrate adjacent the monolayer from below and that for the multilayer substrate alone, exhibits a readily measureable signal with minimal noise extending beyond $Q_z > 0.4\,\text{Å}^{-1}$. By contrast, the Fresnel-normalized reflectivity for the Langmuir, monolayer of the amphiphilic peptide alone in the absence of the interferometric effect has already fallen into the noise with no significant information content. Because the maximum value of momentum transfer for which statistically significant data is obtained in the experiment determines the spatial resolution (or level of detail) achieved in the membrane profile structure, the effect is hereby seen to extend the effective spatial resolution (Krishnan et al., 2010). The other effect is to provide the phase information critical to inverting the data in order to unambiguously derive the SLD profile for the membrane, in this case the xSLD or electron density profile for the Langmuir monolayer of the amphiphilic four-helix bundle peptide, as described in the subsequent paragraph.

The formalism describing the conventional X-ray or neutron reflectivity experiment was first developed for an amphiphilic macromolecule spread in a Langmuir monolayer at the liquid–gas (e.g., water–helium) interface (Als-Nielsen et al., 1994), although it applies to the solid–gas and solid–liquid interfaces as well. The Fresnel-normalized reflectivity $R(Q_z)/R_F(Q_z) = |F(Q_z')|^2$, where $F\left(Q_z'\right) \equiv \frac{1}{\rho_{\text{ave}}} \int \frac{d\rho(z)}{dz} e^{iQ_z' z} dz$ and $(Q_z')^2 = \{(Q_z)^2 - \text{sign}(\rho_{\text{ave}})(Q_z)^2_{\text{crit}}\}$ with $(Q_z)_{\text{crit}}$defined as Q_z at the critical angle for reflection. In this formalism developed in the Born approximation for elastic X-ray/neutron scattering, $F(Q_z')$ is proportional to the Fourier transform of the *gradient* of the SLD profile $d\rho(z)/dz$. In general, $F(Q_z') = |F(Q_z')| e^{i\phi(Q_z')}$ where $|F(Q_z')|$ is the modulus and $\phi(Q_z')$ is the phase of the Fourier transform, the latter being lost upon recording the Fresnel-normalized reflectivity data $R(Q_z)/R_F(Q_z)$. This missing phase information must be recovered in order to unambiguously derive the *gradient* of the SLD profile $d\rho(z)/dz$ from the data, and hence the SLD profile itself, $\rho(z)$, via integration. Note that in this formalism, with $F(Q_z')$ defined as the Fourier transform (FT) of the gradient SLD profile $d\rho(z)/dz$, then $d\rho(z)/dz$ is the inverse Fourier transform (FT^{-1}) of $F(Q_z')$, where both transforms are mathematically unique. Interferometry is a well-developed method for recovering this missing phase information which employs a so-called inorganic multilayer *reference* structure positioned sufficiently near the bioorganic ultrathin film, which would include immediately adjacent, for the interference effect to occur as described in the above paragraph. Its application for the neutron case, using the above formalism, has been thoroughly presented in numerous references (beginning with Kneller et al., 2001). Our approach utilizes a constrained refinement to determine the unknown phase information, which is particularly well suited to this formalism as described later. Briefly, $d\rho(z)/dz$ is of finite extent Δ, i.e., it is bounded, for an ultrathin bioorganic film with an adjacent inorganic multilayer at either the solid–liquid or the solid–gas interface because the SLD, $\rho(z)$, itself becomes constant for distances along the interface normal sufficiently far from the film. As a result, the autocorrelation of $d\rho(z)/dz$, namely, $\left\{\frac{d\rho(z)}{dz} * \frac{d\rho(-z)}{dz}\right\}$, which can be obtained from the inverse Fourier transform of the data $R(Q_z)/R_F(Q_z) = |F(Q_z')|^2$ without any phase information, is similarly of finite extent 2Δ, possessing no information content for $|z| > \Delta$, thereby providing a key constraint on the to-be-derived gradient SLD profile. Because the SLD, $\rho(z)$, of the inorganic multilayer reference structure is essentially known from its fabrication specifications, as can be confirmed by X-ray reflectivity in the absence of the adjacent bioorganic film, the phase of the Fourier transform of its gradient SLD profile $d\rho(z)/dz$ is also known by simple calculation. The constrained refinement is initiated, utilizing an inverse Fourier transform (FT^{-1}) of the experimental $|F(Q_z')|$ modulus data for the ultrathin bioorganic film adjacent the inorganic multilayer reference structure combined with the known phase of the reference structure's gradient SLD profile to calculate the first approximation to the gradient SLD profile for the ultrathin bioorganic film adjacent the inorganic multilayer reference structure. The key constraint is then applied by truncating this first approximation to the gradient SLD profile for $|z| > \Delta/2$, and a new phase function $\phi(Q_z')$ is created for the sotruncated gradient SLD profile via Fourier transformation (FT), which is then combined with the experimental $|F(Q_z')|$ modulus data for the ultrathin bioorganic film adjacent the inorganic multilayer reference structure to calculate the second approximation to the gradient SLD profile via inverse Fourier transformation (FT^{-1}). This process is then repeated until convergence is achieved, usually within only 10–20 iterations.

There is a further consideration with regard to the formalism developed by Als-Nielsen et al., which is important to the "pump–probe" neutron interferometry approach described later. The reflectivity itself, $R(Q_z)$, decays over six to seven orders of magnitude with increasing momentum transfer Q_z (Gupta et al., 2012). The Fresnel function $R_F(Q_z)$ is a precise analytic function, and hence division by the Fresnel function to result in the Fresnel-normalized reflectivity $R(Q_z)/R_F(Q_z)$ has no effect on the statistical accuracy inherent in the reflectivity data. However, the normalization does result in the Fresnel-normalized reflectivity being of comparable amplitude over the entire range of momentum transfer Q_z accessed in the experiment. Since the changes in the membrane nSLD profile $\rho(z)$ dependent on the applied transmembrane voltage are expected to be relatively small due to the nature of the profile projection, the Fresnel normalization of the reflectivity data makes it much easier to readily ascertain whether the voltage-induced changes are statistically significant.

When the neutron counting statistics in the Fresnel-normalized reflectivity $R(Q_z)/R_F(Q_z)$ data are minimal, i.e., just sufficient to establish the experimental significance of the voltage-dependence of the data, it can be helpful to apply an additional constraint in the constrained refinement approach, namely, that there must be no voltage-dependent changes within the profile structure of the inorganic multilayer substrate to which the single-membrane specimen is tethered. This approach was required for the case of the isolated VSD of KvAP, and one form of its implementation is fully described in (Tronin et al., 2014). Such an additional constraint was not required for the case of the complete KvAP channel because the statistics were better.

2.4 Design of the Neutron Interferometry "Pump–Probe" Experiment

In a classic "pump–probe" experiment, the membrane containing the vectorially oriented Kv-channel protein is subjected to either a periodic sequence of polarizing, nonpolarizing, and depolarizing pulses, or a periodic sequence of polarizing and depolarizing pulses with respect to the resting transmembrane voltage, each of constant amplitude and finite duration, and the membrane structure is synchronously probed with a small number of neutron pulses for each voltage pulse applied to the membrane. This approach has the absolute requirement that the activation (deactivation) of the VSDs induced by depolarizing (polarizing) voltage pulses, coupled with the subsequent opening (closing) of the channel pore for a complete Kv-channel, be completely reversible. Utilizing a spallation neutron source such at the Spallation Neutron Source at Oak Ridge National Laboratory, although the incident neutron pulses are of only ~10 μs duration, the

$\vec{Q}$-vector for each scattered neutron is determined by time of flight (TOF) which requires 16.5 ms. Hence, the number of neutron pulses employed for each voltage pulse depends on the duration of the latter, down to the lower limit of one neutron pulse for a voltage pulse of 16.5 ms duration.

2.5 Experimental Details

The basic features of the electrochemical cells used for the time-resolved neutron interferometry experiments of the complete KvAP channel were similar to those for the cells used previously for the isolated VSD of KvAP (Tronin et al., 2014). For the complete KvAP channel, instead of a Si–Ni–Si multilayer on silicon substrate as in the reference, the working electrode was provided by a Si–Ge–Si multilayer on silicon substrate, the latter being heavily doped with phosphorus to make it sufficiently conductive ($R < 1\ \Omega cm^2$), which was held at ground potential. For both the isolated VSD and the complete KvAP, the electric potential (voltage) was applied to the membrane on the surface of the inorganic multilayer substrate via a platinum counter electrode in the aqueous electrolyte comprised of 0.1 *M* KCl in 1 m*M* Tris buffer at pH 8. For either the isolated VSD or KvAP itself tethered to the substrate surface via a His-tag on the protein's C-terminus, this is exactly opposite to the electrophysiological convention where the extracellular side of the membrane is at ground potential and the electric potential (voltage) is applied to the cytoplasmic side. The isolated VSD of KvAP and the single-membrane specimens, containing the vectorially oriented VSD tethered to the surface of the Si–Ni–Si multilayer on silicon substrates, were prepared as described previously (Tronin et al., 2014). The complete KvAP protein was provided by Prof. Patrick Loll, Drexel University College of Medicine. KvAP protein was expressed in *E. coli* with a C-terminal hexa-His tag, purified by immobilized metal affinity chromatography in the detergent DM (*n*-decyl-β-D-maltopyranoside), and flash-frozen in aliquots. KvAP:DM was tethered to the surface of the Si–Ge–Si multilayer on silicon substrates, each silicon or germanium layer of ~20 Å thickness, fabricated by magnetron sputtering at the Advanced Photon Source, Argonne National Laboratory, and the detergent was subsequently exchanged for POPC (1-palmitoyl-2-oleoylphosphatidylcholine) using the self-assembly method as described (Gupta et al., 2011). Time-resolved neutron reflectivity data for both the VSD:POPC and KvAP:POPC single-membrane specimens were collected with the Magnetism Reflectometer on Beamline 4A at the Spallation Neutron Source, Oak Ridge National Laboratory. The incident neutron pulses are polychromatic, of ~10 μs duration at a repetition rate of 60 Hz. 16.5 ms is required to determine the momentum transfer vector $\vec{Q_z}$ of each reflected neutron via TOF. For the VSD:POPC membranes, the cyclic series of nonpolarizing pulses of 0 mV, polarizing voltage pulses of +100 mV and depolarizing pulses of −100 mV, were each of 20 s duration because the isolated VSD does not exhibit inactivation when subjected to long depolarizing voltage pulses, unlike the complete KvAP channel. For the complete KvAP channel exhibiting inactivation (Schmidt et al., 2009), the cyclic sequence of hyperpolarizing voltage pulses of +150 mV and depolarizing voltage pulses of −20 mV, each of 50 ms duration and constant amplitude, was synchronized with the incident neutron pulses so that there were three neutron pulses for each voltage pulse. In order to achieve acceptable reflected neutron counting statistics, reflectivity data were collected from each type of single-membrane specimen for 3–200 min, for each of the six angles of incidence employed to span the range of Q_z accessed using the polychromatic neutron pulses, the time required increasing with angle, and the reflected neutrons for each triplet of incident neutron pulses were collected separately and subsequently averaged over each of the hyperpolarizing or depolarizing voltage pulses. Given the incident neutron pulse rate, this implies that the reflectivity data for each transmembrane voltage were averaged over 10^3–10^6 cycles of the voltage pulse sequence. The time-averaged specular reflectivity data for each transmembrane voltage were then calculated, correcting for the incident beam footprint on the planar specimen depending on the angle of incidence, and for off-specular background scattering.

3. RESULTS

3.1 "Pump–Probe" Neutron Interferometry of a Reconstituted VSD:POPC Membrane

In Fig. 5, we compare the nSLD profile for a VSD:POPC membrane at 0 mV and the *difference* ΔnSLD profiles for polarizing *minus* nonpolarizing {p100mV-0mVave} and depolarizing *minus* nonpolarizing {m100mV-0mVave} transmembrane voltages with those from a specimen otherwise similar, but lacking only the VSD:POPC membrane. The nSLD profile for Si–Ni–Si multilayer substrate occurs for $z < 0$ Å (not shown here), while the nSLD profile of the VSD:POPC membrane for this specimen occurs within 25 Å $< z <$ 95 Å at negative contrast with respect to the average nSLD of the aqueous solvent. The *difference* ΔnSLD profiles were derived from the

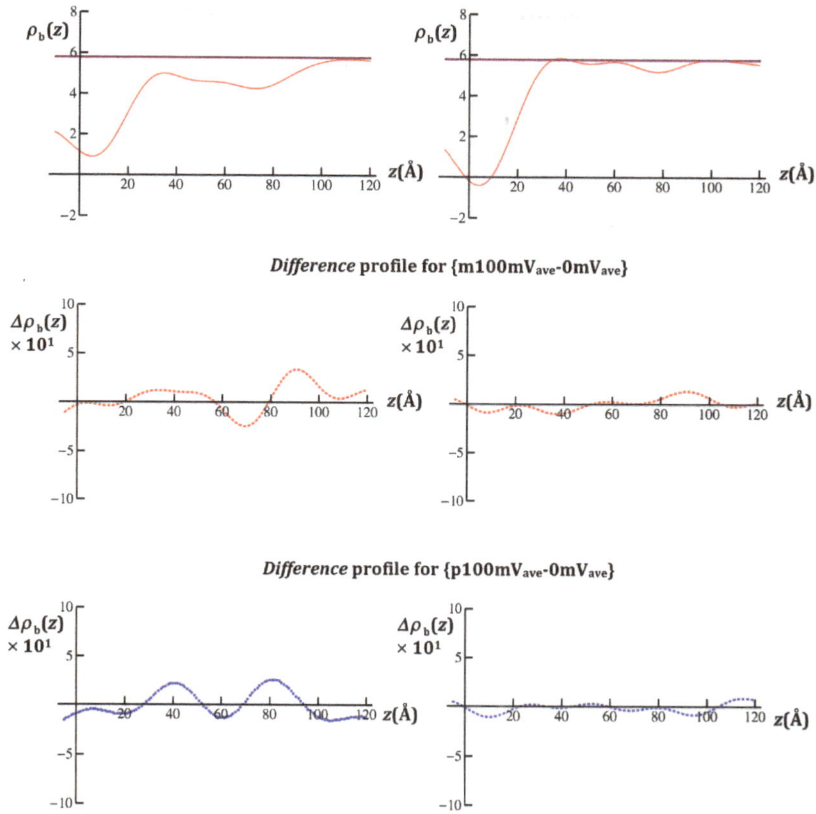

Fig. 5 Neutron scattering length density (nSLD) profiles, $\rho_b(z)$, calculated for the VSD: POPC membrane tethered to the surface of a Si–Ni–Si multilayer substrate (*upper left*) and for the organic self-assembled monolayer (SAM) chemisorbed onto the surface of a Si–Ni–Si multilayer substrate, but lacking the VSD:POPC membrane (*upper right*), each at a transmembrane voltage of 0 mV. The buffered KCl electrolyte was in 90% D_2O:10% H_2O. The respective *difference* nSLD profiles, $\Delta\rho_b(z)$ for each case, are shown in the panels below for the pairs of voltages indicated; here, "p100mV" represents +100 mV, "m100 mV" represents −100 mV. The nSLD profiles are in units of 10^{-6}/Å^2. *Adapted with permission from Tronin, A., Strzalka, J. W., Kuzmenko, I., Worcester, D., Lauter, V., Freites, J. A., et al. (2014). Direct evidence of conformational changes associated with voltage-gating in a voltage sensor protein by time-resolved X-ray/neutron interferometry.* Langmuir, 30*(16), 4784–4796; Copyrighted by the American Chemical Society.*

$R(Q_z)/R_F(Q_z)$ data using the constrained refinement approach described in Section 2.3, employing the additional constraint that no voltage-dependent changes could occur with the profile structure of the Si–Ni–Si multilayer substrate (Tronin et al., 2014). The *difference* ΔnSLD profiles exhibit substantial changes throughout the extent of VSD:POPC membrane profile, and are distinctly different depending on the particular voltage difference utilized. In particular, for the depolarizing voltage pair {m100mV-0mVave}, the major changes involve an increase in nSLD over the region $80\,\text{Å} < z < 100\,\text{Å}$, and corresponding decrease in nSLD over the adjacent region $55\,\text{Å} < z < 80\,\text{Å}$ of the membrane profile. For the polarizing voltage pair {p100mV-0mVave}, the major changes involve a comparable increase in nSLD over the two regions $30\,\text{Å} < z < 50\,\text{Å}$ and $70\,\text{Å} < z < 90\,\text{Å}$, with smaller decreases in nSLD for the region $z < 30\,\text{Å}$ and the intervening region $50\,\text{Å} < z < 70\,\text{Å}$ of the membrane profile. The corresponding *difference* profiles for the SAM are essentially featureless, providing an essential control.

3.2 "Pump–Probe" Neutron Interferometry of the Reconstituted KvAP:POPC Membrane

The voltage-dependent nSLD profiles for a KvAP:POPC single-membrane specimen at an aqueous solvent contrast provided by 90% D_2O/10% H_2O, as derived from the $R(Q_z)/R_F(Q_z)$ data using the constrained refinement approach described in Section 2.3, are shown in Fig. 6. The nSLD profile for the heavily doped Si–Ge–Si multilayer substrate occurs for $z < 0\,\text{Å}$ (not shown here), while the nSLD profile for the KvAP:POPC membrane occurs within $30\,\text{Å} < z < 110\,\text{Å}$ with the membrane center located at $z \sim 75\,\text{Å}$. The nSLD profiles for the membrane appear at negative contrast with respect to the average nSLD of the aqueous solvent. The changes in the membrane nSLD profiles for the hyperpolarizing vs the depolarizing voltage pulses are seen to occur within the region $65\,\text{Å} \pm 30\,\text{Å}$ for this contrast. They demonstrate a loss of nSLD within the extracellular half of the membrane for $z > 75\,\text{Å}$ and a gain of nSLD within the cytoplasmic half of the membrane for $z < 75\,\text{Å}$ for this contrast. The *difference* ΔnSLD profile, hyperpolarizing *minus* depolarizing, manifests the changes noted. While the changes are similar for solvent contrasts provided by both 90% D_2O/10% H_2O and 60% D_2O/40% H_2O, they are substantially larger for 60% D_2O/40% H_2O (not shown). This is due to the difference in the distribution of water, and water-accessible exchangeable hydrogen atoms, within the profile structure of the membrane for hyperpolarizing vs depolarizing transmembrane voltages, as noted in Section 2.2. By comparison, the *difference* ΔnSLD profile for a specimen otherwise similar, but lacking only the KvAP:POPC membrane, is nearly featureless on the same ordinate-scale, providing an essential control.

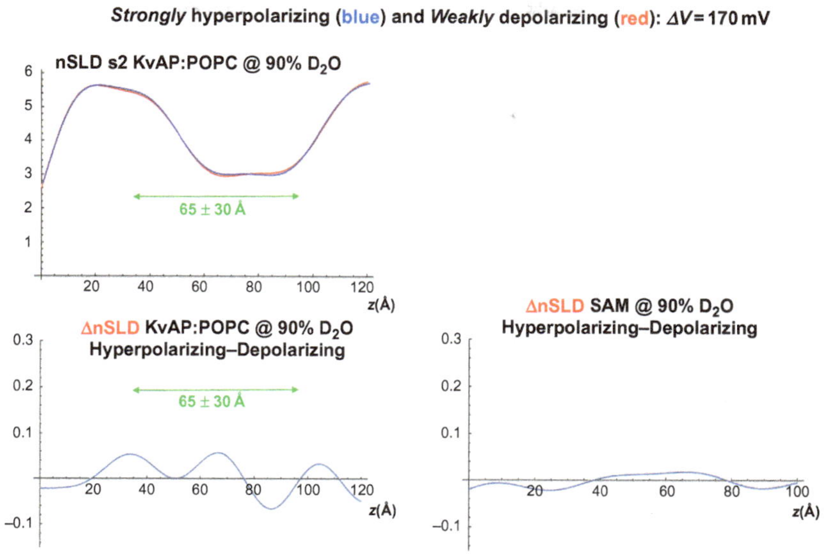

Fig. 6 *Upper left panel*: nSLD profile for a KvAP:POPC membrane at the aqueous solvent contrast noted. At this low spatial resolution of about 30 Å and the membrane center located at $z \sim 75$ Å, the extracellular half of the membrane occurs approximately within the region 75 Å $< z <$ 110 Å, the intracellular half of the membrane occurs approximately within 50 Å $< z <$ 75 Å, the cytoplasmic surface of the membrane occurs within 25 Å $< z <$ 50 Å, and the organic self-assembled monolayer (SAM) tethering the KvAP protein, and hence the membrane, to the surface of the inorganic Si–Ge–Si multilayer substrate's surface occurs within 0 Å $< z <$ 25 Å. The nSLD profile of the Si–Ge–Si multilayer itself occurs for $z <$ 0 Å and is not shown here. Changes in the nSLD profiles for the membrane, dependent on the applied transmembrane voltage, are seen to occur over the region 65 ± 30 Å (*horizontal green arrow*). *Lower left panel*: These changes are likewise manifest in the *difference* ΔnSLD profile, predominately over the same region 65 ± 30 Å, being substantially larger at an aqueous solvent contrast provided by 60% D_2O/H_2O (not shown) than for 90% D_2O/H_2O. *Lower right panel*: By comparison, the voltage-dependent changes in the *difference* ΔnSLD profile for only the organic SAM on the inorganic Si–Ge–Si multilayer substrate's surface, but otherwise lacking the tethered KvAP:POPC membrane, exhibits features of much smaller amplitude than for the KvAP:POPC membrane for either contrast on the same ordinate scale.

4. DISCUSSION

As described in Section 1, it is now possible for all-atom molecular dynamics simulations to address the conformational states responsible for voltage gating in a VGIC directly in response to the applied transmembrane voltage. For the isolated voltage sensor, the VSD investigated was that for the prokaryotic KvAP within a hydrated POPC bilayer membrane.

As described in Tronin et al. (2014), the *difference* ΔnSLD profiles calculated from the simulation for both the polarizing voltage pair and the depolarizing voltage pair were in good agreement with their counterparts from the time-resolved neutron interferometry experiments shown in Fig. 5. As a result, the time-resolved neutron interferometry results supported the voltage-dependent conformational changes in the 3-D structure of the VSD at the atomic level, despite the low spatial resolution of the nSLD profiles and the nature of the profile projection. In Tronin et al. (2014), we also showed that the *difference* ΔxSLD profiles calculated from the simulation for the same polarizing voltage pair and depolarizing voltage pair were also in good agreement with their counterparts from the time-resolved X-ray interferometry experiments. Because the origin of X-ray scattering and neutron scattering from atoms is physically different as described in Section 2.2, the results from the time-resolved X-ray interferometry experiments provided additional support for the simulations independent of that from the time-resolved neutron interferometry experiments.

For the complete KvAP channel, a time-resolved X-ray interferometry experiment could not be performed. Unlike the case for the VSD:POPC membrane, which exhibited an average xSLD greater than that of the aqueous medium hydrating the membrane, the average xSLD of the KvAP: POPC membrane was comparable to that of the aqueous medium hydrating the membrane, rendering the membrane relatively invisible to X-rays unless the average xSLD of the medium, was reduced to that for water-saturated helium gas. As described in Section 2.2, this is not a problem for neutrons. The complete Kv-channel investigated in the simulations (Jensen et al., 2012) was the eukaryotic Kv1.2/2.1 "paddle chimera," in which the voltage sensing S3b–S4 "paddle" within the four VSDs from Kv2.1 was transferred into Kv1.2, resulting in the highest resolution X-ray crystal structure available for a Kv-channel (Long et al., 2007). Each subunit of eukaryotic Kv-channels possesses a N-terminal cytoplasmic T1 domain, in addition to the transmembrane domain comprised of the four VSDs and the PD in the homotetramer. The transmembrane domain of eukaryotic Kv-channels is structurally homologous with that for prokaryotic Kv-channels like KvAP lacking the cytoplasmic T1 domain. In this work, the authors also investigated only the transmembrane domain of the Kv1.2/2.1 chimera, designated as Kv1.2/2.1-T1(−) embedded within a hydrated phospholipid bilayer as a function of the transmembrane voltage. The *difference* ΔnSLD profile calculated from the simulation for the transmembrane domain of the Kv1.2/2.1 chimera for hyperpolarizing *minus* depolarizing voltages was in good agreement with its counterpart from the time-resolved neutron interferometry experiment shown in Fig. 6. In addition, the *double difference* ΔΔnSLD profile for 90% D_2O/10% H_2O *minus* 60% D_2O/40% H_2O contrasts and hyperpolarizing *minus* depolarizing voltage pulses provided the ΔnSLD profile for water only within the KvAP channel (not shown here), which was also in good agreement with its counterpart calculated from the simulation. As a result, the time-resolved neutron interferometry results provided two independent tests in support of the voltage-dependent conformational changes in the 3-D structure of the complete Kv1.2/2.1channel at the atomic level, again despite the low spatial resolution of the nSLD profiles and the nature of the profile projection.

5. CONCLUSIONS AND FUTURE PROSPECTS

The time-resolved "pump–probe" neutron interferometry approach described herein has been successful in providing the nSLD profiles for both the activated and the deactivated state of the isolated VSD from KvAP, and for both the activated, open state and the deactivated, closed state of the complete KvAP homotetramer, each protein vectorially oriented within a hydrated POPC bilayer membrane. Given that collection of the neutron reflectivity data utilized 10^3–10^6 cycles (repetitions) of the transmembrane voltage pulses, the activation/deactivation of the VSDs within the VSD: POPC membrane, and the activation/deactivation of the four VSDs coupled with opening/closing the PD within the KvAP:POPC membrane, were fully reversible and achieved without any detectable inactivation of the complete channel.

The level of agreement achieved between the experimental *difference* ΔnSLD profiles for both the VSD:POPC membrane and the KvAP:POPC membrane with their respective counterparts from the simulations provides strong and direct experimental evidence for the correctness of the simulations, despite the assumptions required for the simulations as noted. However, these direct experimental tests must still be considered only as a "necessary" condition, given the relatively low spatial resolution and the nature of the profile projections employed. They succeeded because the changes in the membrane profile structure depending upon the applied transmembrane voltages extended over a sufficiently large region of the profile along the membrane normal. Despite, the inherent nature of the profile projection, the spatial resolution could be enhanced substantially through the deuteration of selected components of the VSDs within the

Kv-channel, at the level of the individual helices S1–S6 including the S4–S5 linker, or perhaps even just the residues at the ends of these helices, deuterating only one helix per experiment. Such site-selective deuteration, which is generally considered to be isomorphous in structural biology, can be now achieved via protein semisynthesis, which has been accomplished more recently for KvAP itself, although for a different purpose (Devaraneni et al., 2013). Clearly, such site-selective deuteration would make such time-resolved neutron interferometry experiments as described herein even more definitive, despite still not providing 3-D structural information at the atomic level.

In Section 1, it was stated that *in principle*, it should be straightforward to extend our "simulation" vs "experiment" approach to include investigations into the effects of anesthetics on the mechanism of electromechanical coupling for both Kv- and Nav-channels. With respect to the former Kv-channels, an excellent example is provided from work by the Covarrubias group demonstrating how the otherwise anesthetic-insensitive Kv1.2 can be transformed into an exquisitely anesthetic-sensitive Kv-channel by mutation of either only five residues within the S4–S5 linker helix, Kv1.2-FRAKT, or only a single residue within the linker, Kv1.2-G329T (Liang et al., 2015). However, the level of difficulty involved in comparing time-resolved neutron interferometry results from these two channels, the insensitive Kv1.2 "control" with the sensitive Kv1.2-FRAKT or Kv1.2-G329T "experimental," each in the presence and absence of one anesthetic (e.g., sevoflurane), is already increased by 4 ×. In addition, only those scientists working at Shaw Research, where the Anton supercomputers and associated software for such simulations were originally (and continue to be) developed, currently have access to sufficient Anton-level resources necessary to carry out such long simulations as required to elicit a response to the applied transmembrane voltage, e.g., hundreds of microseconds. Thus, without greater access to both suitable neutron reflectometer instruments at the Spallation Neutron Source and Anton-level supercomputers at the National Computer Facilities, this next level may be even more difficult to attain, at least for scientists in the United States.

ACKNOWLEDGMENTS

The work described was supported by the National Institutes of Health Grants P01-GM86685 (isolated VSD of KvAP) and P01-GM55876 (complete KvAP channel). The Spallation Neutron Source, Oak Ridge National Laboratory, is a user facility sponsored by the Scientific User Facilities Division, Office of Basic Energy Sciences, U.S. Department of Energy.

REFERENCES

Als-Nielsen, J., Jacquemain, D., Kjaer, K., Leveiller, F., Lahav, M., & Leiserowitz, L. (1994). Principles and applications of grazing incidence X-ray and neutron scattering from ordered molecular monolayers at the air-water interface. *Physics Reports*, *246*, 251–313.

Bacon, G. E. (1975). *Neutron diffraction* (3rd ed.). Oxford: Oxford University Press.

Devaraneni, P. K., Komarov, A. G., Costantino, C. A., Devereaux, J. J., Matulef, K., & Valiyaveetil, F. I. (2013). Semisynthetic K^+ channels show that the constricted conformation of the selectivity filter is not the C-type inactivated state. *Proceedings of the National Academy of Sciences of the United States of America*, *110*(39), 15698–15703.

Freites, J. A., Schow, E. V., White, S. H., & Tobias, D. J. (2012). Microscopic origin of gating current fluctuations in a potassium channel voltage sensor. *Biophysical Journal*, *102*, L44–L46.

Grizel, A. V., Glukov, G. S., & Sokolova, O. S. (2014). Mechanisms of activation of voltage-gated potassium channels. *Acta Naturae*, *6*(4), 10–26.

Gupta, S., Dura, J., Freites, A., Tobias, D., & Blasie, J. K. (2012). Structural characterization of the voltage sensor domain and voltage-gated K^+-channel proteins vectorially-oriented within a single bilayer membrane at the solid/vapor and solid/liquid interfaces via neutron interferometry. *Langmuir*, *28*(28), 10504–10520.

Gupta, S., Liu, J., Strzalka, J., & Blasie, J. K. (2011). Profile structures of the VSD and KvAP channel vectorially-oriented in single membranes at solid-vapor or solid-liquid interfaces via X-ray reflectivity. *Physical Review E*, *84*(3), 031911-1-15.

Hemmings, H. C., Jr. (2009). Sodium channels and the synaptic mechanisms of inhaled anesthetics. *British Journal of Anaesthesia*, *103*(1), 61–69.

Henrion, U., Renhorn, J., Börjesson, S. I., Nelson, E. M., Schwaiger, C. S., Bjelkmar, P., et al. (2012). Tracking a complete voltage-sensor cycle with metal-ion bridges. *Proceedings of the National Academy of Sciences of the United States of America*, *109*(22), 8552–8557.

Hodgkin, A. L. (1951). The ionic basis of electrical activity in nerve and muscle. *Biological Reviews*, *26*, 339–409.

Hodgkin, A. L., & Huxley, A. F. (1952). A quantitative description of membrane current and its application to conduction and excitation in nerve. *Journal of Physiology*, *117*, 500–544.

Jensen, M. Ø., Jogini, V., Borhani, D. W., Leffler, A. E., Dror, R. O., & Shaw, D. E. (2012). Mechanism of voltage gating in potassium channels. *Science*, *336*, 229–233.

Jiang, Y., Lee, A., Chen, J., Ruta, V., Cadene, M., Chait, B. T., et al. (2003). X-ray structure of a voltage-dependent K^+ channel. *Nature*, *423*, 33–41.

Kneller, L. R., Edwards, A. M., Majkrzak, C. F., Berk, N. F., Krueger, S., & Blasie, J. K. (2001). Hydration state of a single cytochrome c monolayer vectorially-oriented at a soft interface investigated via neutron interferometry. *Biophysical Journal*, *80*(5), 2248–2261.

Krepkiy, D., Mihailescu, M., Freites, J. A., Schow, E. V., Worcester, D. L., Gawrisch, K., et al. (2009). Structure and hydration of membranes embedded with voltage-sensing domains. *Nature*, *462*, 473–479.

Krishnan, V., Strzalka, J., Liu, J., Liu, C., Kuzmenko, I., Gog, T., et al. (2010). Interferometric enhancement of X-ray reflectivity from unperturbed Langmuir monolayers of amphiphiles at the liquid-gas interface. *Physical Review E*, *81*, 021604-1-10.

Liang, Q., Anderson, W. D., Jones, S. T., Souza, C. S., Hosoume, J. M., Treptow, W., et al. (2015). Positive allosteric modulation of Kv channels by sevoflurane: Insights into the structural basis of inhaled anesthetic action. *PLoS One*, *10*, e0143363.

Long, S. B., Campbell, E. B., & MacKinnon, R. (2005). Crystal structure of a mammalian voltage-dependent *Shaker* family K^+ channel. *Science*, *309*, 897–902.

Long, S. B., Tao, X., Campbell, E. B., & MacKinnon, R. (2007). Atomic structure of a voltage-dependent K^+ channel in a lipid membrane-like environment. *Nature*, *450*, 376–382.

Payandeh, J., El-Din, T. M. G., Scheurer, T., Zheng, N., & Catterall, W. A. (2012). Crystal structure of a voltage-gated sodium channel in two potentially inactivated states. *Nature*, *486*, 135–139.

Payandeh, J., Scheurer, T., Zheng, N., & Catterall, W. A. (2011). The crystal structure of a voltage-gated sodium channel. *Nature*, *475*, 353–358.

Ruta, V., Chen, J., & MacKinnon, R. (2005). Calibrated measurement of gating-charge arginine displacement in the KvAP voltage-dependent K^+ channel. *Cell*, *123*, 463–475.

Schmidt, D., Cross, S. R., & MacKinnon, R. (2009). A gating model for the archeal voltage-dependent K^+ channel in DPhPC and POPE:POPG decane lipid bilayers. *Journal of Molecular Biology*, *390*(5), 902–912.

Schow, E. V., Freites, J. A., Gogna, K., White, S. H., & Tobias, D. J. (2010). Down-state model of the voltage-sensing domain of a potassium channel. *Biophysical Journal*, *98*, 2857–2866.

Schow, E. V., Freites, J. A., Nizkorodov, A., White, S. H., & Tobias, D. J. (2012). Coupling between the voltage-sensing and pore domains in a voltage-gated potassium channel. *Biochimica et Biophysica Acta*, *1818*, 1726–1736.

Tronin, A.Y., Maciunas, L., Grasty, K.C., Loll, P.J., Ambaye, H., Parizzi, A. et al. (2018). Voltage dependent profile structures of a voltage-gated K^+ channel via time-resolved neutron interferometry, Biophysical Journal, under review.

Tronin, A., Strzalka, J. W., Kuzmenko, I., Worcester, D., Lauter, V., Freites, J. A., et al. (2014). Direct evidence of conformational changes associated with voltage-gating in a voltage sensor protein by time-resolved X-ray/neutron interferometry. *Langmuir*, *30*(16), 4784–4796.

Vargas, E., Yarov-Yarovoy, V., Khalili-Araghi, F., Catterall, W. A., Klein, M. L., Tarek, M., et al. (2012). An emerging consensus on voltage-gating from computational modeling and molecular dynamics simulations. *Journal of General Physiology*, *140*(6), 587–594.

SECTION II

Evolving Biophysical Technologies

CHAPTER FIVE

Fluorescent Anesthetics

Daniel J. Emerson, Ivan J. Dmochowski[1]
University of Pennsylvania, Philadelphia, PA, United States
[1]Corresponding author: e-mail address: ivandmo@sas.upenn.edu

Contents

Abstract

Methods for using exogenous fluorophore and general anesthetic 1-aminoanthracene (1-AMA) and its photoactive derivative 1-azidoanthracene (1-AZA) are provided. 1-AMA potentiates $GABA_A$ chloride currents and immobilizes *Xenopus laevis* tadpoles. Cellular and tissue anesthetic distribution can be imaged for quantifying "on-pathway" and "off-pathway" targets. 1-AZA shares targets with 1-AMA and offers further optoanesthetic spatial and temporal control upon near-UV laser irradiation. Furthermore, 1-AZA adduction provides screening of possible relevant anesthetic protein targets and binding site characterization. We highlight several useful imaging and binding assays to demonstrate utility of 1-AMA and its derivative 1-AZA.

ABBREVIATIONS

1-AMA 1-aminoanthracene
1-AZA 1-azidoanthracene
epoD epothilone D
$GABA_A$ gaba aminobutyric acid receptor
HSAF horse spleen apoferritin
SDS-PAGE sodium dodecyl sulfate polyacrylamide gel electrophoresis

Methods in Enzymology, Volume 603
ISSN 0076-6879
https://doi.org/10.1016/bs.mie.2018.02.006

1. INTRODUCTION

Efforts to identify safer, more efficacious general anesthetic candidates have been hindered by poor understanding of relevant molecular targets and in vivo distribution of clinical anesthetics (Butts et al., 2009). Fluorescent probe development can be one approach for tackling the problem while uncovering "on-pathway" and "off-pathway" sites of action. Previously, fluorophore and fluorophore tethering for probing local and general anesthetic binding sites have been implemented with limited utility for elucidating anesthetic mechanisms. General anesthetics are small molecules, thus perturbing their size and shape with fluorophore attachment affects potency, likely due to steric constraints of binding sites (Butts et al., 2009; Krasowski et al., 2001).

Very broadly, fluorescence has been sparingly used for identifying or characterizing local anesthetic targets. The fluorophore quinacrine has been shown to mimic local anesthetic action through noncompetitive inhibition of electrophysiological response to acetylcholine in electroplaque preparations. Furthermore, binding to acetylcholine receptor ionophore (Greenberg & Tsung, 1982) was confirmed through displacement by clinical local anesthetics in membrane homogenates. Additionally, amino benzoates have been used to deduce binding interaction in phosphatidylcholine bilayers based on emission energy, fluorescence quantum yield, and lifetime in a variety of solvents (Birmachu & Reed, 1988).

For general anesthetic behavior, drug–fluorophore couplings have been devised for probing function and physiology of ligand-gated ion channels such as 5-HT3, gaba aminobutyric acid (GABA), and glycine receptors. In vivo imaging for cell labeling and distribution of 5-HT3 with deep tissue penetration has been performed with granisetron–fluorophore conjugates (Simonin et al., 2012). Furthermore, photochromic potentiators of $GABA_A$ receptors have been developed to induce GABA currents via light using azobenzene derivatives of propofol wherein an aryldiazene unit is directly coupled to the pharmacophore, termed azo-propofols (Chen et al., 2012). One of the azo-propofol molecules, AP2, can function as an UV-light-dependent photoswitchable anesthetic altering between *cis* and *trans* states to modulate GABA currents and resultant anesthesia as demonstrated in tadpoles. Additionally, fluorophore 1-anilino-8-naphtalenesulphonate (ANS) has been used to assess changes in the conformational state of protein components of biological membranes. ANS fluorescence changes coincide with depletion of accessible binding sites due to disruption of protein hydrophobic regions upon anesthetic binding (Augustin & Hasselbach, 1973; Gasymov & Glasgow, 2007).

These previous probes for screening targets either lack fluorescence imaging capabilities or are fluorophores or fluorophore couplings that lack anesthetic properties themselves. Circumventing these limitations, an endogenous fluorophore that behaves as a general anesthetic itself was discovered with 1-aminoanthracene (1-AMA). A photoactive derivative, 1-azidoanthracene (1-AZA), covalently labels proteins, which alters its biological properties and facilitates labeling studies.

2. VALUE OF 1-AMA FOR STUDYING GENERAL ANESTHETIC MECHANISMS

A known fluorophore, 1-AMA, was initially discovered to bind to the general anesthetic site in horse spleen apoferritin (HSAF), which consists of 4-helix bundle motifs that mimic the $GABA_A$ receptor 4-helix pentamer bundle (Butts et al., 2009). 1-AMA was subsequently found to immobilize model organisms and potentiate GABAergic transmission. 1-AMA has physicochemical properties similar to propofol, with similar molecular mass (193 g/mol for 1-AMA vs 178 g/mol for propofol), hydrophobicity based on octanol/water coefficient, and polar surface area (26 cm^2 for 1-AMA vs 20 cm^2 for propofol) (Butts et al., 2009). 1-AMA overcomes the severe limitation of conjugating a fluorophore that greatly reduces anesthetic potency. Notably, anesthetic potency decreases dramatically when molecular mass increases above 180 g/mol, corresponding to dodecanol in the *n*-alkanol series (Alifimoff, Firestone, & Miller, 1989; Butts et al., 2009).

1-AMA can be utilized in competition assays via fluorometry, a useful technique given 1-AMA's binding affinity to sites in model HSAF or $GABA_A$ receptor, with greater speed and sensitivity than isothermal titration calorimetry. Additionally, in vivo studies of temporal anesthetic distribution in *Xenopus laevis* tadpoles can be performed (see Fig. 1) (Emerson, Liao, Eckenhoff, & Dmochowski, 2012). Its solubility in pond solutions with minimal ethanol concentration allows for easy ingestion with minimal solvent toxicity. Histological fluorescence studies in mouse brain of targeted sites via competition assays can also be performed. Its fluorescence can be exploited at the cellular level with live cell fluorescence colocalization studies with other markers (e.g., MitoTracker to study targets in mitochondria). These cellular studies are aided by its small fluorophore size allowing for

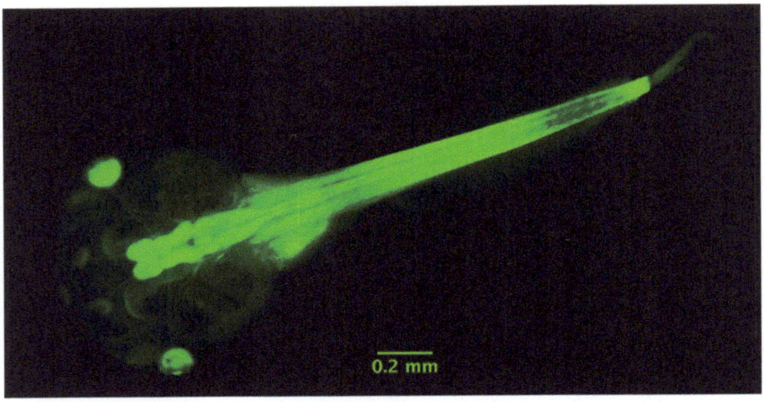

Fig. 1 Fluorescence micrograph of 1-AMA distribution (15 μ*M*) in *Xenopus laevis* tadpole after incubation for 30 min in pond water. *Reprinted with permission from Emerson, D. J., Liao, Z., Eckenhoff, R. G., Dmochowski, I. J. (2012). A novel fluorescent general anesthetic enables imaging of sites of action in vivo.* Anesthesiology, 116*(6), 1363.*

maintenance of labile general anesthetic binding sites avoiding tagging constructs with variation in cellular compartment labeling depending on tag design. 1-AMA provides reasonably strong fluorescence with quantum yield of 0.61 in cyclohexane, comparable to many GFP variants (Patterson, Day, & Piston, 2001). Absorbance in the 350–450 nm range in pond water ethanol solution with fluorescence emission in 460–600 nm range permits dyes in longer wavelength 600–700 nm emission/excitation range for minimal cross talk. Given its low toxicity in tadpoles (full recovery after incubation for half hour), 1-AMA can be used for neuronal tissue studies. Electrophysiological recordings with patch clamp current measurements in cultured neurons in combination with imaging of spatiotemporal distribution allows for in tandem measurements of both its fluorophore and anesthetic properties.

3. VALUE OF 1-AZA FOR STUDYING GENERAL ANESTHETIC MECHANISMS

1-AMA can be readily converted into a photoactive analog 1-AZA by replacing the amine group with an azide in a simple one-step reaction (Emerson et al., 2013). 1-AZA enables in vivo spatial photolabeling at the same conserved anesthetic sites occupied by 1-AMA. Upon covalent binding to relevant ligand binding sites, a light-induced optoanesthetic state occurs, which persists for tens of minutes, even after animal is moved to clean pond water. Using confocal microscopy with *X. laevis* tadpoles, 1-AZA can be irradiated with near-UV laser in particular neuronal regions such as the forebrain (see Fig. 2) for site-selective adduction. Site-specific nitrene formation provides spatial control over fluorescence and immobilization. For animals treated with 1-AZA and irradiated in all regions of the central nervous system, a similar fluorescence distribution profile is achieved as found for 1-AMA with nonselective irradiation. The nitrene photoproduct can be quantified by UV–vis spectroscopy in which three intense absorbance peaks form between 350 and 400 nm after irradiation. Prior to irradiation, the azide moiety quenches anthracene fluorescence by photo-induced electron transfer. After irradiation, "turn-on" of

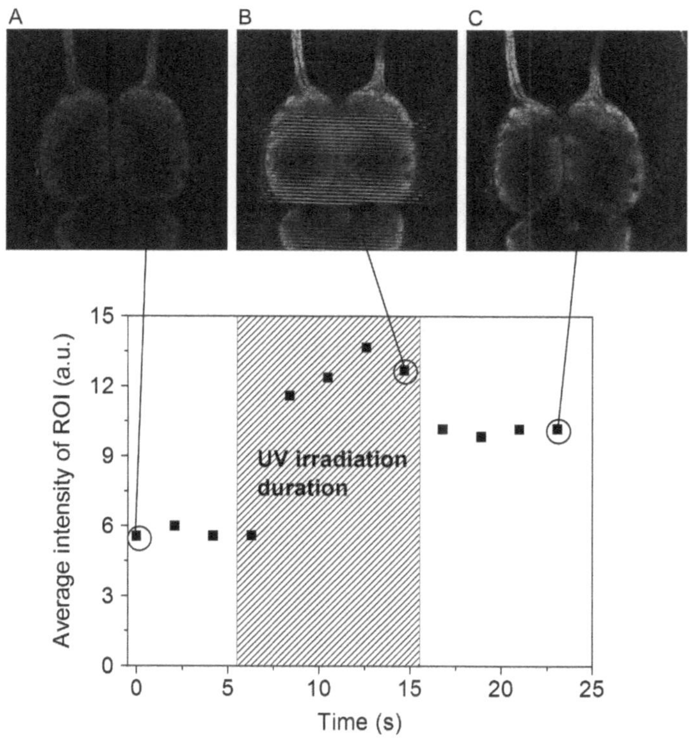

Fig. 2 Photoactivation of 1-AZA in vivo. Representative confocal fluorescence images of 1-AZA in a tadpole brain: before (A), during (B), and after (C) near-UV laser photolysis of a tadpole forebrain. The forebrain area was irradiated for 10 s by 351 and 364 nm lasers (100% power). *Bottom plot* shows the average fluorescent pixel intensity over time under 488 nm laser excitation. *Reprinted with permission from Emerson, D. J., Weiser, B. P., Psonis, J., Liao, Z., Taratula, O., Fiamengo, A., et al. (2013). Direct modulation of microtubule stability contributes to anthracene general anesthesia. Journal of the American Chemical Society, 135(14), 5389–5398. Copyright © 2013. American Chemical Society.*

anthracene fluorophore coincides with optoanesthesia. Thus, 1-AZA combines selective anesthesia with selective fluorescence imaging to offer visual image of active anesthetic in fluorescent regions.

Beyond providing spatiotemporal control over anesthetic activity, a covalently linked AZA-photolabel provides an identifiable tag for mass spectrometry for protein target analysis. Conversion of azides to nitrenes occurs within minutes via low-flux near-UV exposure, which allows for rapid protein adduction. High-throughput protein target screening assays can be performed via neuronal tissue photolabeling, followed by tissue dissection and isolation with sucrose isolation buffer, isoelectric focusing protein solubilization, SDS-PAGE, and mass spectrometric analysis. Precise residue locus can be uncovered through adduction.

Using in vitro tadpole 1-AZA photolabeling, tubulin was identified as a prominent protein target. In biochemical assays, 1-AMA was subsequently found to destabilize microtubules and inhibit tubulin polymerization. Tadpoles preincubated with a microtubule stabilizing agent epothilone D (epoD) that crosses the blood–brain barrier showed protection against 1-AMA: In epoD-protected tadpoles, higher 1-AMA concentrations were required for immobilization (Emerson et al., 2013). (In a previous study, epoD was synthesized in laboratory of Dr. Amos Smith at the University of Pennsylvania, but can be purchased commercially; Emerson et al., 2013). These assays help to confirm that microtubule destabilization mediated by 1-AMA/1-AZA can contribute toward general anesthesia. We hypothesize that microtubule-associated proteins required for animal movement (including ligand-gated ion channels) may be less present and/or less active at the cell surface in organisms with destabilized microtubules. While this route to general anesthesia would have significant drawbacks clinically, these molecular probes provide a target-based approach for studying the anesthetic state. Fluorescence labeling data visualized by SDS-PAGE indicate that 1-AZA binds to additional protein targets, and it remains to determine whether these sites are relevant to general anesthesia, and either "on-target" or "off-target."

4. EXPERIMENTS

4.1 Tadpole Immobilization, Imaging, and Recovery

1. Synthesize 1-AZA from 1-AMA following published protocol (Emerson et al., 2013) and measure concentration of 1-AZA in ethanol stock with UV–Vis spectrometer ($\varepsilon_{372} = 9090\ M^{-1}\ \mathrm{cm}^{-1}$).

2. Make pond solution (3 m*M* NaCl, 30 μ*M* $CaCl_2$, 6 μ*M* $NaHCO_3$ in deionized water) with 0.5% ethanol in Petri dish.
3. Measure 1-AMA (Sigma-Aldrich, St. Louis, MO) concentration in ethanol stock with UV–Vis spectrometer ($\varepsilon_{368} = 4240\,M^{-1}\,cm^{-1}$).
4. Incubate *X. laevis* tadpoles (Nasco, Fort Atkinson, WI) for 0.5 h with 15 μ*M* 1-AMA or 20 μ*M* 1-AZA.
5. For reduced tadpole movement during confocal imaging, make tapered channel lined by low-melt agarose (1% in deionized water) on imaging dish. Imaging dish should have cover glass on bottom to allow passage of excitation and emission light.
6. Transfer tadpoles to dish chamber after 0.5 h incubation.
7. Photolyze tadpole forebrain for 10 s by rastering with near-UV laser (351 and/or 364 nm).
8. Collect fluorescence emission in 500–600 nm range under excitation with 488 nm Ar ion laser, scanning speed of 2.1 s/frame.
9. Transfer tadpoles to fresh pond water.
10. Measure immobilization recovery by tapping with a sterile cotton swab.

4.2 In Vitro Tadpole Photolabeling

1. Anesthetize tadpoles in tricaine methanesulfonate.
2. Decapitate and remove brain and spinal cord with forceps under dissecting microscope and homogenize in isolation buffer (0.32 *M* sucrose, 5 m*M* Tris buffer, pH 7.4, supplemented with protease inhibitors).
3. Centrifuge homogenates at 100,000 × *g* for 10 min, wash with isolation buffer, and recentrifuge.
4. Suspend in isolation buffer and dilute to 1 mg/mL in a microcentrifuge tube and vortex briefly.
5. Add saturated solution of 1-AZA (~200 μ*M*), incubate for 5 min in the dark, and photolabel for 1 min in a quartz cuvette with an ultraviolet transilluminator (around 365 nm peak intensity)
6. Centrifuge at 100,000 × *g*, homogenize in 5 m*M* Tris buffer, recentrifuge at 100,000 × *g*, wash, and store at −80°C in 2 m*M* Tris.
7. Thaw and centrifuge 100 μg of neuronal membrane protein for 15 min at 15,000 × *g*.
8. Dissolve pellet in 125 μL of 7 *M* urea, 2 *M* thiourea, 20 m*M* dithiothreitol, and 0.2% carrier ampholytes.

9. Perform isoelectric focusing and SDS-PAGE with nonlinear pH gradient strip and 4%–15% polyacrylamide gel.
10. After separation, wash gel with water and scan with 400 nm excitation and 535 nm emission filters.
11. Stain with Coomassie G-250 and reimage.
12. Excise spots and perform trypsin digest.
13. Perform LC-MS/MS.
14. Perform b and y ion search with 1.5 amu parent ion mass tolerance, 1 amu fragment ion mass tolerance, and 1 missed cleavage with 99.9% protein identification confidence.

4.3 Tubulin Polymerization Dynamics via Turbidity

1. Resuspend tubulin in 1 × BRB80 buffer (1 m*M* $MgCl_2$, 1 m*M* EGTA, 80 m*M* PIPES, pH 6.9) at 6 mg/mL. Bovine tubulin (>99% purity) purchase from Cytoskeleton, Inc., Denver, CO.
2. Snap freeze aliquots in liquid nitrogen.
3. Thaw aliquot on ice.
4. Add glycerol, then test compound (1-AMA, 1-AZA, isoflurane, propofol).
5. Add 1 × BRB80 buffer to aliquot to dilute to near final concentration (5% glycerol, 10 μ*M* tubulin, 14 μ*M* test compound).
6. Add GTP last (2 m*M*).
7. Insert into quartz cuvette and take UV–Vis spectrometer measurements at 37°C with absorbance recorded at 450 nm.

4.4 Competition Assay With Tubulin

1. Minimize tubulin polymerization and effects on binding site by keeping reagents on ice and fluorimeter settings at 4°C.
2. Incubate 20 μ*M* colchicine (colchicine stock in DMSO) or other test compound with 15 μ*M* tubulin in 100 μL buffer for 10 min (2 m*M* $MgCl_2$, 80 m*M* KPIPES, 0.5 m*M* EGTA in deionized water at pH 6.9).
3. Add 8 μ*M* 1-AZA (in ethanol stock) and photolyze for 1.5 min at 350 nm with desktop transilluminator.
4. Set fluorimeter excitation wavelength to 400 nm.
5. Incubate 1-AZA with tubulin alone under same conditions and concentrations and photolyze for 1.5 min at 350 nm.
6. Compare fluorescence emission peak and wavelength with and without colchicine or test compound.

REFERENCES

Alifimoff, J. K., Firestone, L. L., & Miller, K. W. (1989). Anaesthetic potencies of primary alkanols: Implications for the molecular dimensions of the anaesthetic site. *British Journal of Pharmacology*, *96*(1), 9–16. https://doi.org/10.1111/j.1476-5381.1989.tb11777.x.

Augustin, J., & Hasselbach, W. (1973). Studies on the fluorescence of 1-anilino-8-naphthalenesulfonate by the membranes of the sarcoplasmic reticulum. *European Journal of Biochemistry*, *35*(1), 114–121. https://doi.org/10.1111/j.1432-1033.1973.tb02816.x.

Birmachu, W., & Reed, J. K. (1988). The fluorescence properties of ortho aminobenzoate anesthetics in defined solvents and phospholipid vesicles. *Photochemistry and Photobiology*, *47*(5), 675–683. https://doi.org/10.1111/j.1751-1097.1988.tb02764.x.

Butts, C. A., Xi, J., Brannigan, G., Saad, A. A., Venkatachalan, S. P., Pearce, R. A., et al. (2009). Identification of a fluorescent general anesthetic, 1-aminoanthracene. *Proceedings of the National Academy of Sciences of the United States of America*, *106*(16), 6501–6506. https://doi.org/10.1073/pnas.0810590106.

Chen, Z. W., Chen, L. H., Akentieva, N., Lichti, C. F., Darbandi, R., Hastings, R., et al. (2012). A neurosteroid analogue photolabeling reagent labels the colchicine-binding site on tubulin: A mass spectrometric analysis. *Electrophoresis*, *33*(4), 666–674. https://doi.org/10.1002/elps.201100434.

Emerson, D. J., Liao, Z., Eckenhoff, R. G., & Dmochowski, I. J. (2012). A novel fluorescent general anesthetic enables imaging of sites of action in vivo. *Anesthesiology*, *116*(6), 1363. https://doi.org/10.1097/ALN.0b013e31824cb4b1.

Emerson, D. J., Weiser, B. P., Psonis, J., Liao, Z., Taratula, O., Fiamengo, A., et al. (2013). Direct modulation of microtubule stability contributes to anthracene general anesthesia. *Journal of the American Chemical Society*, *135*(14), 5389–5398. https://doi.org/10.1021/ja311171u.

Gasymov, O. K., & Glasgow, B. J. (2007). ANS fluorescence: Potential to augment the identification of the external binding sites of proteins. *Biochimica et Biophysica Acta*, *1774*(3), 403–411. https://doi.org/10.1016/j.bbapap.2007.01.002.

Greenberg, M., & Tsung, T. Y. (1982). Binding of quinacrine, a fluorescent local anesthetic probe, to mammalian axonal membranes. *The Journal of Biological Chemistry*, *257*(15), 7.

Krasowski, M. D., Jenkins, A., Flood, P., Kung, A. Y., Hopfinger, A. J., & Harrison, N. L. (2001). General anesthetic potencies of a series of propofol analogs correlate with potency for potentiation of gamma-aminobutyric acid (GABA) current at the GABA(A) receptor but not with lipid solubility. *The Journal of Pharmacology and Experimental Therapeutics*, *297*(1), 338–351.

Patterson, G., Day, R. N., & Piston, D. (2001). Fluorescent protein spectra. *Journal of Cell Science*, *114*(Pt. 5), 837–838.

Simonin, J., Vernekar, S. K. V., Thompson, A. J., Hothersall, J. D., Connolly, C. N., Lummis, S. C. R., et al. (2012). High-affinity fluorescent ligands for the 5-HT(3) receptor. *Bioorganic & Medicinal Chemistry Letters*, *22*(2), 1151–1155. https://doi.org/10.1016/j.bmcl.2011.11.097.

CHAPTER SIX

Investigation of Anesthetic–Protein Interactions by a Thermodynamic Approach

Hasan Babazada, Renyu Liu[1]
Perelman School of Medicine at the University of Pennsylvania, Philadelphia, PA, United States
[1]Corresponding author: e-mail address: liur@uphs.upenn.edu

Contents

Abstract

Anesthetics can interact with a wide variety of proteins in the body, including ion channels and alter their activity, but little is known about the molecular mechanisms of the interactions responsible for the functional activity. Characterization of the nature of anesthetic–protein interactions therefore is important and requires the complete analysis of the binding energetics. Isothermal titration calorimetry (ITC) is the only technique that allows quantitative determination of all thermodynamic parameters, including the equilibrium binding constant (K_B), the standard Gibbs free energy change (ΔG), the enthalpy change (ΔH), the entropy change (ΔS), heat capacity change (ΔC_p), and stoichiometry (n) of the reaction. ITC does not require any labeling or modification of the interacting partners analyzed and can be performed in solution with small amounts of reagents. In this chapter we describe the general properties of the ITC method, highlighting some critical aspects of experimental planning and data analysis, with practical application to anesthetic–protein interactions.

Methods in Enzymology, Volume 603
ISSN 0076-6879
https://doi.org/10.1016/bs.mie.2018.02.023

1. INTRODUCTION

Despite clinical use of anesthetics for well over a century, mechanisms underlying general anesthesia are still poorly understood and under active investigation (Franks & Lieb, 1994; Mashour, Forman, & Campagna, 2005; Urban & Friederich, 1998; Woll, Dailey, Brannigan, & Eckenhoff, 2016). Anesthetics have direct interaction with and alter the activity of ligand-gated ion channels such as γ-aminobutyric acid type A, glycine, nicotinic acetylcholine, and 5-hydroxytryptamine3 receptors, as well as α-amino-3-hydroxy-5-methyl-4-isoxazole propionic acid-, kainic acid-, and *N*-methyl-D-aspartate-selective glutamate receptors to produce anesthetic effects (Covarrubias, Barber, Carnevale, Treptow, & Eckenhoff, 2015; Olsen et al., 2014; Yamakura, Bertaccini, Trudell, & Harris, 2001). Understanding anesthetic–protein interactions is crucial to understanding biological mechanisms at the molecular level and to facilitate drug discovery. A significant amount of data has been accumulated using well-defined off-target protein models previously crystallized in complex with anesthetics, such as albumin (Bhattacharya, Curry, & Franks, 2000; Eckenhoff, Petersen, Ha, & Bhagavan, 2000), firefly luciferase (Franks, Jenkins, Conti, Lieb, & Brick, 1998), and apoferritin (Liu, Loll, & Eckenhoff, 2005). Techniques that allow the direct measurement of anesthetics binding to protein targets rely on methods such as NMR spectroscopy (Dubois, Cherian, & Evers, 1993; Dubois & Evers, 1992), photoaffinity labeling (Eckenhoff & Shuman, 1993), and fluorescence spectroscopy (Johansson, Eckenhoff, & Dutton, 1995; Manderson & Johansson, 2002). However, it is crucial to understand how the information about interacting structures acquired from these methods correlates with the changes in macroscopic level thermodynamic and kinetic parameters. Therefore, complete quantification of energetics and kinetics of molecular recognition of anesthetics is required. Previously, we and others described traditional thermodynamic approaches using isothermal titration calorimetry (ITC) to investigate the entropic and enthalpic contributions to the binding of volatile and intravenous anesthetics to human and bovine serum albumins (Banerjee & Kishore, 2005; Liu et al., 2004; Sawas, Pentyala, & Rebecchi, 2004; Ueda & Yamanaka, 1997). In this chapter, we provide a practical guide to ITC due to its capacity to provide complete information on the binding thermodynamics of anesthetics–protein interactions. We examine the steps required to obtain thermodynamic information following each anesthetic ligand injection in the ITC experiment.

2. ITC MEASUREMENTS

ITC enables direct measurement of the heat change during the molecular interactions held at constant temperature. Data analysis from a single experiment simultaneously determines the equilibrium binding constant (K_B), the standard Gibbs free energy change (ΔG), the enthalpy change (ΔH), the entropy change (ΔS), and stoichiometry (n) of the interaction. Moreover, experiments conducted at different temperatures allow obtaining the heat capacity change (ΔC_p) of the binding reaction. Detailed theoretical aspects of ITC are described elsewhere (Fisher & Singh, 1995; Indyk & Fisher, 1998; Perozzo, Folkers, & Scapozza, 2004).

2.1 Instrumentation

Substantial developments in ITC instruments and software for data analysis have occurred over the last decade, and several easy-to-use instruments are now commercially available. There are two types—single- and dual-injection ITC instruments available (Freire, Mayorga, & Straume, 1990). The advantage of a dual-injection instrument is that the reactant is injected simultaneously to the cells containing both, macromolecule and buffer, eliminating the need for separate control experiment to measure the heat effect arising from ligand dilution. For this methodology, we describe the use of the computer-controlled VP-ITC titration calorimeter manufactured by MicroCal (Malvern Instruments Ltd.) operating in a large temperature range of 2–80°C. The MicroCal VP-ITC has a cylindrical adiabatic shield, in which two coin-shaped cells are enclosed. Macromolecule (e.g., protein) solution in the sample cell is titrated with ligand solution from a computer-driven syringe. The differences in temperatures between two cells (ΔT_1), and between the sample cell and the jacket (ΔT_2) are measured. Heat is released (exothermic) or absorbed (endothermic), as molecular interactions occur in the sample cell. While exothermic reactions result in decrease, endothermic reactions result in increase in feedback power. ΔT_1 is constantly held at zero by the increase of heat to the sample or reference cell using the cell feedback network system. The integral of the power necessary to keep $\Delta T_1 = 0$ over time is a differential measure of total heat resulting from the studied process (Sobhany & Negishi, 2006). The heat per unit time provided to the sample cell is the observed signal in an ITC procedure and a direct measure of the heat generated from the ligand–macromolecule binding.

2.2 Experimental Design and Procedure

A standard ITC experiment contains following steps: (1) a ligand is titrated into the solution containing macromolecule of interest; (2) released or absorbed heat associated with interaction process is assessed; and (3) received ITC data are processed and fitted to determine energetic parameters governing molecular interactions, such as K_B, ΔG, ΔH, ΔS, n, and their temperature dependence through ΔC_p. In the following section, we describe initial considerations for the experiment including concentrations of ligand and macromolecule, buffer choice, sample preparation, control experiments, and data analysis.

2.3 Initial Considerations

2.3.1 Concentrations

Determining accurate concentrations of ligand and macromolecule is critical, particularly when one or both interacting reagents are difficult to obtain in large quantities or they have low solubility. The choice of macromolecule concentration will influence the binding isotherm. The experimental binding isotherm can be characterized by the unitless c parameter, which is the product of the association equilibrium constant, K_a, total concentration of macromolecule, [M], and the stoichiometry of the reaction, n ($c = K_a \times [M] \times n$). For an accurate determination of the association constant, a c value in the range of 10–1000 is recommended (Wiseman, Williston, Brandts, & Lin, 1989). However, both K_a and n are unknown a priori in most cases of anesthetic–protein interactions. Unlike many enzymes, which are known to bind substrates with one-to-one stoichiometry, anesthetics can bind multiple sites on proteins (Bhattacharya et al., 2000; Franks et al., 1998; Tilton, Kuntz, & Petsko, 1984). A range of preliminary experiments is often necessary to determine the optimal experimental conditions. We suggest minimum of 15 μM protein solution, 5–10 mM solution of inhaled and 0.5 mM of injectable anesthetics as a good starting point. If early saturation is obtained, either increase of protein concentration and/or decrease of the ligand concentration is necessary. If full saturation cannot be reached after the injection of all the ligand, increase of the ligand concentration, decrease of protein concentration (if the heat signal is large enough), and 2–3 sequential portions of titrations should be attempted. However, optimal conditions cannot be met sometimes, especially for low-affinity bindings exemplified by most anesthetic–protein interactions. Low-affinity binding measurements in ITC require a high concentration of protein and therefore high ligand concentration. For soluble proteins, it is possible to work at high concentrations if the material is available. However, protein oligomerization, which should be avoided, may occur at high concentrations. The availability and the solubility are often limiting factors for membrane and other hydrophobic proteins to obtain optimal concentrations. All the inhaled anesthetics and some injectable anesthetics such as propofol are hydrophobic. For example, the solubility limit is 9.1 m*M* for methoxyflurane, 18.0 m*M* for halothane, and 13.5 m*M* for isoflurane in water at 25°C as determined by nuclear magnetic resonance spectroscopy (Seto, Mashimo, Yoshiya, Kanashiro, & Taniguchi, 1992). Thus, the solubility of anesthetics is a significant limiting factor. Because of the solubility limitation of anesthetics, full saturation of the binding sites in the macromolecule in the sample cell may not be reached in the case of weak interactions. Solutions with maximal or nearly maximal concentrations of anesthetics may be prepared, and sequential titrations (2 or 3 experiments) can be designed by repeatedly loading and titrating with the same ligand solution without removing the contents of the sample cell, until the heat signal decreases and is detected essentially from only blank injections. Another option is obviously to increase the solubility of anesthetics by using organic solvents (i.e., dimethyl sulfoxide) and/or changing the temperature. However, dilution of organic solvents into aqueous buffers in many cases leads to consequential heat effects, which may cause signal-to-noise ratio issues. Therefore, solvent concentrations in both protein and ligand solutions should be carefully matched. Organic solvents such as dimethyl sulfoxide (DMSO) at the concentrations of less than 2% (v/v) can be safely added to many proteins. However, careful data interpretation is necessary because the change in hydrophobicity of the buffer may affect the parameter values compared to those in purely aqueous buffers (Holdgate, 2009). To reduce the need for high ligand concentrations, a reverse titration can be tried, where ligand is loaded into the sample cell in the calorimeter and the macromolecule is loaded in the syringe. However, it should be noted that high concentrations of macromolecule are needed in this case and some proteins aggregate at the high concentrations in the injection syringe. Thus, most often protein is loaded into the sample cell.

2.3.2 Buffers

The choice of buffer is also critical when planning the ITC experiment. To reflect physiological conditions, buffer pH around 7.0–7.4 may be preferred although the sample cell of VP-ITC can tolerate broad range of pH of 2–12. It might be wise to use the same buffering system that has been tested in other studies. Experimental data suggest that even 1 m*M* dithiothreitol could cause a large artifact in the ITC baseline, so this should be avoided in solutions for ITC experiments. In a carefully designed experiment, buffer for protein should be matched (pH, salt concentration, etc.) with that for ligand to reduce the heat signal arising from mismatched buffer constituents, i.e., because of buffer protonation effects. Therefore, it is recommended to exhaustively dialyze both ligand and macromolecule in buffer. The final dialysis buffer should be used for further concentration adjustments of the macromolecule or ligand solutions. As a first choice, we recommend using buffers with a low ionization enthalpy such as sodium phosphate. If the selected buffer has a large enthalpy of ionization (e.g., Tris–HCl), the measured enthalpy will reflect both buffer ionization and binding complex formation (Pierce, Raman, & Nall, 1999). Control experiments are needed to correct unwanted heat released from both dilution and mixing effects. The heat change from injecting buffer into protein solution should be subtracted from that of ligand into protein, especially when high concentrations of protein are used, which is usually the case of investigating weak anesthetic–protein interactions. If the buffer conditions were carefully optimized, buffer into buffer injection should only generate very small blank injection heat, which is the thermal energy associated with the force of the injection, mixing and any temperature differences between the cell and the injectant. Furthermore, in many instances it is apparent that the heat of dilution of the anesthetics into the buffer is negligible compared to the heat from the anesthetic–protein interactions, so subtracting it would not improve the data quality but merely introduce noise.

2.3.3 Sample Preparation

Prior to injection loading, macromolecule and ligand solutions should be filtered or centrifuged to remove any artifacts. Finally, solutions must be degassed before loading to remove residual air bubbles, which can potentially interfere with the feedback circuit and result in poor baseline. This can be facilitated by the ThermoVac accessory available from MicroCal. Degassing can be achieved in vacuum for about 5 min with stirring or 20 min without stirring. Solutions of volatile anesthetics should be prepared with predegassed buffer and kept gastight until loading. Water or buffer for the reference cell should also be degassed before introduction to the system, although this does not need to be changed in each experiment. Some researchers recommend weekly based change with very low concentration of bactericidal agent (i.e., 0.05% sodium azide).

2.4 ITC Measurements

Here, we present a typical experiment using ITC. ITC was performed using a VP-ITC titration microcalorimeter (MicroCal, Inc., Northampton, MA) at 20°C. Human serum albumin (HSA) at concentration of 0.21 m*M* in 20 m*M* sodium phosphate buffer, containing 130 m*M* NaCl, pH 7.0, was placed into the 1.4-mL calorimeter cell (less material would be required in the newer versions of the ITC equipment such as MicroCal PEAQ-ITC), and 12 m*M* isoflurane was added sequentially in 12 μL aliquots at 5-min intervals. The reference cell was filled with distilled water. For most buffers with low ionization enthalpy, distilled water is suitable to use as the reference solution. In order to ensure full occupancy of the binding sites sequential titrations were conducted by loading and titrating using the same ligand, while keeping the samples within the cell, until constant titration signal is achieved. If multiple separate titrations are needed to ensure full occupancy, the titrations can be linked together for data analysis using ConCat32 software (MicroCal). In this case, two separate sequential titrations are linked together as shown in Fig. 1. Origin software can be utilized to assess the heat of the reaction per injection (μcal/s) by integrating the areas under the peaks. This calculation provides the best-fit values of the thermodynamic parameters, ΔH, n, and the association constant (K_a) from profiles of the heat evolved from injected isoflurane (kcal/mol) vs the isoflurane/HSA molar ratio. Control experiments including buffer to buffer, ligand to buffer, buffer to macromolecule need to be carried out in parallel (using exactly the same setup for ligand into macromolecule) to determine the heats of dilution. These heats of dilution are subtracted from the corresponding anesthetic–protein binding experiments before curve fitting.

2.5 Data Analysis

Representative calorimetric titration at pH 7.0 of isoflurane with HSA is shown in Fig. 1. The negative deflections from the baseline upon addition of isoflurane indicate that heat was evolved, i.e., an exothermic reaction. Total heat released upon the isoflurane injection into the cell corresponds to plot of the area under the signal vs time (Fig. 1, upper panel). Ligand dilution control data should be carefully subtracted from the raw data. Fitting of the data can be performed using data fitting programs supplied by the

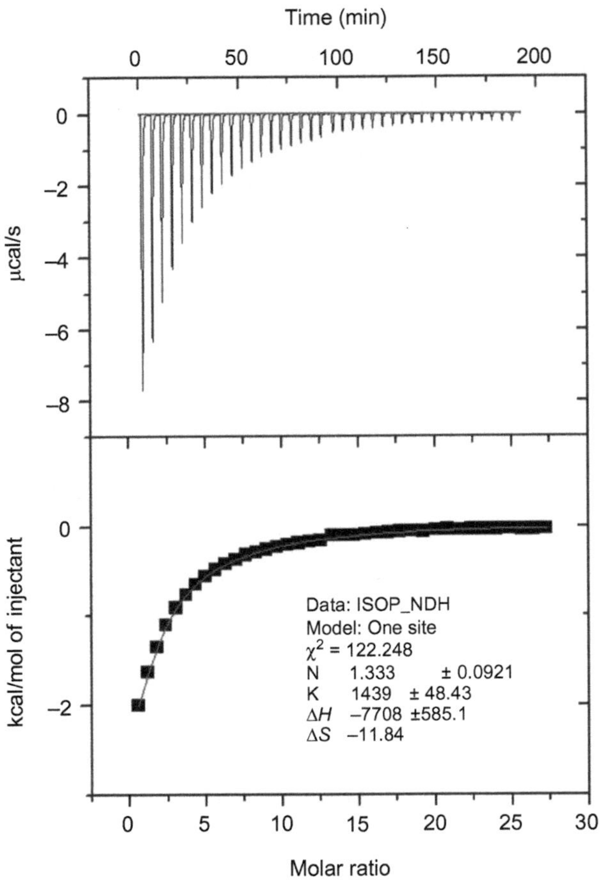

Fig. 1 Binding of isoflurane to HSA observed by ITC. The titration of 12 m*M* isoflurane into 0.21 m*M* HSA at 20°C in buffer containing 130 m*M* NaCl and 20 m*M* $NaHPO_4$, pH 7.0, is shown. Two portions of sequential titration were performed. Programmed injection was performed in the first portion of titration to obtain the isothermal curve at a lower molar ratio. The interaction of isoflurane with HSA is an exothermic process.

manufacturers such as Origin software from MicroCal. Nonlinear regression procedure is used to fit a model to the data to analyze the thermogram. The data can be fitted with initial assumptions for a binding model, e.g., one-site (or *n* identical sites), two independent sites or sequential binding, and thermodynamic parameters—stoichiometry, enthalpy, and binding affinity. The goal is to fit the experimental data within the reasonable experimental error, using the simplest model possible, which also should make sense in view of available information about the system (e.g., stoichiometry). However, investigators may be required to develop their own analysis for more complicated models such as having three independent sites (for review, see Freyer & Lewis, 2008). Isoflurane data were fitted to a model for one binding site. The enthalpy change associated with each injection of isoflurane was plotted vs the isoflurane/HSA molar ratio (Fig. 1, lower panel). The ΔH (cal/mol), K_a (1/M), n, and ΔS (cal/mol K) were determined as -7708 ± 585.1, 1439 ± 48.4, 1.333 ± 0.092 and -11.84, respectively.

3. CONCLUSION

The techniques used to probe anesthetic–protein interactions have improved dramatically over recent years due to optimization of the protocols, strategies, and instrumentation. In this chapter we described ITC, which can be used as a powerful technique for full thermodynamic characterization of anesthetic–macromolecule interactions, using specific examples. Binding data obtained with HSA and isoflurane show that ITC can be used to determine complete sets of thermodynamic binding parameters. We demonstrated that isoflurane–HSA interaction is an exothermic and enthalpy-driven process. However, care must be taken to properly consider the initial conditions of experiments including choice of both ligand (anesthetic) and macromolecule concentrations, buffer, and sample preparation. When properly designed and carefully conducted, ITC will certainly continue to shed light on the molecular basis of anesthetic–protein interactions.

REFERENCES

Banerjee, T., & Kishore, N. (2005). Does the anesthetic 2,2,2-trifluoroethanol interact with bovine serum albumin by direct binding or by solvent-mediated effects? A calorimetric and spectroscopic investigation. *Biopolymers*, *78*(2), 78–86.

Bhattacharya, A. A., Curry, S., & Franks, N. P. (2000). Binding of the general anesthetics propofol and halothane to human serum albumin. High resolution crystal structures. *The Journal of Biological Chemistry*, *275*(49), 38731–38738.

Covarrubias, M., Barber, A. F., Carnevale, V., Treptow, W., & Eckenhoff, R. G. (2015). Mechanistic insights into the modulation of voltage-gated ion channels by inhalational anesthetics. *Biophysical Journal*, *109*(10), 2003–2011.

Dubois, B. W., Cherian, S. F., & Evers, A. S. (1993). Volatile anesthetics compete for common binding sites on bovine serum albumin: A 19F-NMR study. *Proceedings of the National Academy of Sciences of the United States of America*, *90*(14), 6478–6482.

Dubois, B. W., & Evers, A. S. (1992). Fluorine-19 NMR spin-spin relaxation (T2) method for characterizing volatile anesthetic binding to proteins. Analysis of isoflurane binding to serum albumin. *Biochemistry*, *31*, 7069–7076.

Eckenhoff, R. G., Petersen, C. E., Ha, C. E., & Bhagavan, N. V. (2000). Inhaled anesthetic binding sites in human serum albumin. *The Journal of Biological Chemistry*, *275*(39), 30439–30444.

Eckenhoff, R. G., & Shuman, H. (1993). Halothane binding to soluble proteins determined by photoaffinity labeling. *Anesthesiology*, *79*(1), 96–106.

Fisher, H. F., & Singh, N. (1995). Calorimetric methods for interpreting protein-ligand interactions. *Methods in Enzymology*, *259*, 194–221.

Franks, N. P., Jenkins, A., Conti, E., Lieb, W. R., & Brick, P. (1998). Structural basis for the inhibition of firefly luciferase by a general anesthetic. *Biophysical Journal*, *75*(5), 2205–2211.

Franks, N. P., & Lieb, W. R. (1994). Molecular and cellular mechanisms of general anaesthesia. *Nature*, *367*(6464), 607–614.

Freire, E., Mayorga, O. L., & Straume, M. (1990). Isothermal titration calorimetry. *Analytical Chemistry*, *62*(18), 950A–959A.

Freyer, M. W., & Lewis, E. A. (2008). Isothermal titration calorimetry: Experimental design, data analysis, and probing macromolecule/ligand binding and kinetic interactions. *Methods in Cell Biology*, *84*, 79–113.

Holdgate, G. (2009). Isothermal titration calorimetry and differential scanning calorimetry. *Methods in Molecular Biology (Clifton, N.J.)*, *572*, 101–133.

Indyk, L., & Fisher, H. F. (1998). Theoretical aspects of isothermal titration calorimetry. *Methods in Enzymology*, *295*, 350–364.

Johansson, J. S., Eckenhoff, R. G., & Dutton, P. L. (1995). Binding of halothane to serum albumin demonstrated using tryptophan fluorescence. *Anesthesiology*, *83*(2), 316–324.

Liu, R., Loll, P. J., & Eckenhoff, R. G. (2005). Structural basis for high-affinity volatile anesthetic binding in a natural 4-helix bundle protein. *FASEB Journal: Official Publication of the Federation of American Societies for Experimental Biology*, *19*(6), 567–576.

Liu, R., Meng, Q., Xi, J., Yang, J., Ha, C.-E., Bhagavan, N. V., et al. (2004). Comparative binding character of two general anaesthetics for sites on human serum albumin. *The Biochemical Journal*, *380*(1), 147–152.

Manderson, G. A., & Johansson, J. S. (2002). Role of aromatic side chains in the binding of volatile general anesthetics to a four-alpha-helix bundle. *Biochemistry*, *41*(12), 4080–4087.

Mashour, G. A., Forman, S. A., & Campagna, J. A. (2005). Mechanisms of general anesthesia: From molecules to mind. *Best Practice & Research. Clinical Anaesthesiology*, *19*(3), 349–364.

Olsen, R. W., Li, G.-D., Wallner, M., Trudell, J. R., Bertaccini, E. J., Lindahl, E., et al. (2014). Structural models of ligand-gated ion channels: Sites of action for anesthetics and ethanol. *Alcoholism: Clinical and Experimental Research*, *38*(3), 595–603.

Perozzo, R., Folkers, G., & Scapozza, L. (2004). Thermodynamics of protein-ligand interactions: History, presence, and future aspects. *Journal of Receptor and Signal Transduction Research*, *24*(1–2), 1–52.

Pierce, M. M., Raman, C. S., & Nall, B. T. (1999). Isothermal titration calorimetry of protein–protein interactions. *Methods*, *19*(2), 213–221.

Sawas, A. H., Pentyala, S. N., & Rebecchi, M. J. (2004). Binding of volatile anesthetics to serum albumin: Measurements of enthalpy and solvent contributions. *Biopolymers*, *43*(39), 12675–12685.

Seto, T., Mashimo, T., Yoshiya, I., Kanashiro, M., & Taniguchi, Y. (1992). The solubility of volatile anaesthetics in water at 25.0 degrees C using 19F NMR spectroscopy. *Journal of Pharmaceutical and Biomedical Analysis*, *10*(1), 1–7.

Sobhany, M., & Negishi, M. (2006). Characterization of specific donor binding to alpha1,4-N-acetylhexosaminyltransferase EXTL2 using isothermal titration calorimetry. *Methods in Enzymology*, *416*, 3–12. https://doi.org/10.1016/S0076-6879(06)16001-2.

Tilton, R. F., Kuntz, I. D., & Petsko, G. A. (1984). Cavities in proteins: Structure of a metmyoglobin-xenon complex solved to 1.9 Å. *Biochemistry*, *23*(13), 2849–2857.

Ueda, I., & Yamanaka, M. (1997). Titration calorimetry of anesthetic-protein interaction: Negative enthalpy of binding and anesthetic potency. *Biophysical Journal*, *72*(4), 1812–1817.

Urban, B. W., & Friederich, P. (1998). Anesthetic mechanisms in-vitro and in general anesthesia. *Toxicology Letters*, *100-101*, 9–16.

Wiseman, T., Williston, S., Brandts, J. F., & Lin, L. N. (1989). Rapid measurement of binding constants and heats of binding using a new titration calorimeter. *Analytical Biochemistry*, *179*(1), 131–137.

Woll, K. A., Dailey, W. P., Brannigan, G., & Eckenhoff, R. G. (2016). Shedding light on anesthetic mechanisms: Application of photoaffinity ligands. *Anesthesia and Analgesia*, *123*(5), 1253–1262. https://doi.org/10.1213/ANE.0000000000001365.

Yamakura, T., Bertaccini, E., Trudell, J. R., & Harris, R. A. (2001). Anesthetics and ion channels: Molecular models and sites of action. *Annual Review of Pharmacology and Toxicology*, *41*(1), 23–51.

CHAPTER SEVEN

Electron Spin Resonance (EPR) in *Drosophila* and General Anesthesia

Luca Turin[1], Efthimios M.C. Skoulakis
Biomedical Science Research Center "Alexander Fleming", Vari, Greece
[1]Corresponding author: e-mail address: lucaturin@me.com

Contents

Abstract

Changes in electron spin content can be detected by X-band continuous-wave electron spin resonance (ESR, EPR) in Drosophila fruit flies without requiring the use of spin traps. The spin changes are related to cellular respiration and behave differently in anesthesia-resistant fly strains. We describe the method used in these measurements and its possible applications to the problem of the mechanism of general anesthesia.

Methods in Enzymology, Volume 603
ISSN 0076-6879
https://doi.org/10.1016/bs.mie.2018.01.020

1. INTRODUCTION

Almost the only thing we know for sure about consciousness is that it is, so to speak, soluble in ether, chloroform, and a variety of other solvents. Consequently general anesthesia, whose chief attribute is to make consciousness go away reversibly, is probably the medical discovery we most need and least understand (Eger, 2014). The word general in general anesthesia (GA) was originally intended to denote a contrast to local anesthesia, but GA is also general in the sense that the same set of compounds acts on all animals, though with various patterns of potency. While it is not known to what extent fruit flies are conscious, they are most definitely unconscious when exposed to chloroform or ether. This, in our view, makes it unlikely that an emergent "systems" property (Alkire, Hudetz, & Tononi, 2008) peculiar to the central nervous systems of higher animals is responsible. There may or may not be a unitary mechanism of GA in the old sense of a single target membrane, enzyme, or receptor (Perouansky, 2012), but we think likely that the target, however, extended or complex, is shared by all animal cells (and possibly some plants) (Grémiaux, Yokawa, Mancuso, & Baluška, 2014; Yokawa et al., 2017). If systems are involved, they seem more likely to be subcellular. Our current ignorance of what is going on therefore has at least one salient advantage: it probably does not matter what organism we study GA in. Simplest, fastest, and cheapest is best. Drosophila is a strong contender for these reasons and two additional ones: (1) the availability of RNAi strains for almost every one of the ≈14,000 genes of Drosophila (http://www.vbcf.ac.at/) and (2) the existence of anesthetic-resistant strains, which are either known or suspected single-point mutations (Madhavan, Kumar, & Krishnan, 2000).

The discovery of a connection between electron spin and general anesthetic action in fruit flies was largely accidental. As part of a study of possible electron current flow through olfactory receptors (Turin, 1996), we measured electron spin by resonance spectroscopy (ESR) in fruit flies exposed to various odorants. The results were unconvincing, but there did seem to be some spin changes in response to octanol and acetophenone, both of which are known general anesthetics. A closer look at spin changes induced by more potent anesthetics revealed interesting signals (Turin, Skoulakis, & Horsfield, 2014), some of which later turned out to be due to interference between anesthetics and traces of oxygen rather than the anesthetics themselves. It is now clear that chloroform, the most powerful anesthetic tried, does cause reproducible spin changes in Drosophila and that these spin changes are different in anesthetic-resistant mutants. This chapter will describe the methods—and pitfalls—of ESR measurements in Drosophila in the context of GA mechanism.

2. ELECTRON SPIN RESONANCE PRIMER

Only the most basic account of continuous-wave (CW) ESR will be given here, and the reader is referred to classical (Poole, 1967) and modern (Eaton, Eaton, Barr, & Weber, 2010) texts for further information. Pulsed ESR techniques require a deeper understanding of quantum mechanics and physical chemistry, but basic CW ESR is akin to simple electrophysiological recordings and can in principle be practiced by beginners. Our intent is to give confidence to nonspecialists in learning the fascinating technique of CW ESR, much as Nick Franks and Bill Lieb learned electrophysiology to perform their pioneering studies (Franks, 2014). Only one earlier publication reported spin measurements in live Drosophila (Trapp, Waters, Lebendiger, & Perkins, 1983), and the two recent ones are about anesthesia. There is clearly plenty of scope for more ESR studies on anesthesia in this remarkable organism.

ESR measures the resonant absorption of microwaves by a population of unpaired electron spins in a magnetic field. The vast majority of molecular orbitals in the molecules making up a living organism contain two paired electrons of opposite spin adding up to a zero magnetic moment, and are therefore undetectable by ESR. These are known as closed-shell molecules. Exceptions to this rule are metal ions, some of which have partially filled orbitals, and free radicals (Hallwell & Gutteridge, 1985; Pryor, 2012). The majority of free radicals (with the signal exception of nitroxides) in biological systems are carbon centered, in the sense that the unpaired electron resides mainly on carbon atoms. Dioxygen is a ground-state triplet with two unpaired electrons, but its ESR signal is so broad that it is undetectable.

Unpaired electrons switch between up and down spin states under the influence of thermal perturbations. The difference in energy between spin states is zero in the absence of a magnetic field, and increases linearly with field according to the equation.

$$E = m_s g_e \mu_B B_0 \qquad (1)$$

where m_s is $\pm 1/2$, g_e is the Landé g-factor, equal to 2.0023 for the free electron, μ_B is the Bohr magneton, and B_0 is the magnetic field.

For carbon-centered radicals the value of g is very close to 2.0, and resonance will occur and be detected by ESR at a frequency close to 28 GHz/T (1 T = 10,000 gauss). Typically in CW ESR, the frequency of the microwaves is held fixed with high accuracy and the magnetic field is varied to detect a resonance. The instrument used in our study was an Adani 8400 ESR spectrometer working in the microwave X-band, i.e., 9.4 GHz. Resonance for $g = 2.0$, therefore, occurs at 3350 gauss. The basic RF absorption signal is a bell-shaped curve of a width determined by the environment of the spins, typically of the order of 2–10 gauss. The exact shape of the curve usually lies between a Lorentzian and a Gaussian. In practice the derivative signal is what is usually measured, by modulating the magnetic field at $\approx$100 kHz and using a lock-in amplifier to measure the resultant fluctuations in RF absorption. The resulting curve is shown in Fig. 3. Modulation width should be as large as possible to maximize signal amplitude without distorting the signal by "over modulation." Typically a modulation width equal or smaller than the resonance width is adequate.

The derivative measurement method gives good signal to noise at the expense of one drawback: the slope at the peak of the bell-shaped curve is zero and changes sign on either side. The derivative is therefore zero at the peak of the absorption. If one is interested in measuring changes in spin content of a preparation, derivative measurement therefore compels one to measure the signal at the position of maximum *slope* of the underlying Gaussian. Unless gross changes in shape of the absorption bell curve occur, the maximum value of derivative and amplitude of the underlying absorption will be linearly related.

In this chapter we are concerned mainly with changes in quantity of spin species of constant or near-constant g-factor as a function of time. The recording method typically involves two stages: (1) determination of the full derivative curve by a sweep in magnetic field and (2) setting the instrument to a continuous recording at a fixed value of magnetic field corresponding to the positive derivative signal peak. Typically, simple ESR instruments collect a fixed number of samples during a "sweep" of magnetic field of variable duration. The arrangement of the lock-in within the ESR spectrometer is such that its output filter typically has a time constant of few tens of milliseconds. There is therefore nothing to be gained by sampling at more than the Nyquist frequency. For true continuous recordings, some instruments output detector signal and magnetic field oscillator and allow the use of an external lock-in.

3. QUANTITATION

Classical CW ESR is a relatively insensitive and unquantitative technique. A figure of 10^{11}–10^{12} spins is often given as the limit of detection. The insensitivity is ultimately due to the fact that the energy of the absorbed RF photon (<40 μeV for X-band) is very small compared to room temperature energy (25 meV). This means that the two spin states will be almost equally populated at room temperature, and therefore that promotion from lower to upper level by absorption of an RF photon will be a rare event. The difficulty in quantitation arises from the fact that RF absorption takes place in a tuned resonant RF cavity which, by its very nature, will contain one or more nodes and bellies of the electromagnetic field. An extended sample placed in the cavity will therefore experience an inhomogeneous magnetic field, depending on its position in the chamber. Nevertheless, calibration to less than one order of magnitude is achievable using a dilute sample of a stable spin species with a geometry similar to the preparation to be measured. Many stable free radicals are available for calibration. At nominal purity, they carry one unpaired electron per molecule and are typically soluble in organic solvents. To simulate a 20-mm section of Teflon tubing packed with fruit flies we soaked a comparable length of tissue paper in a dilute solution of DPPH (diphenyl picrylhydrazyl, di(phenyl)-(2,4,6-trinitrophenyl) iminoazanium), which gives a narrow (<5 gauss) line at $g = 2.0036$.

4. SAMPLE PREPARATION

Teflon (PTFE) is almost completely transparent to X-band microwaves and is the ideal material as a sample holder. We used a length 4 mm i.d. tubing inserted into the chuck normally intended to hold quartz ESR tubing. The Adani ESR is fitted with a temperature control device which takes gas (typically N_2) from a cylinder, passes it through a Dewar flask filled with liquid N_2, and feeds it to a quartz tube friction-fit into the RF resonator. At the bottom of the quartz tube is an arrangement of heating coils and temperature sensors connected to a heating control to maintain gas temperature constant. The gas input leaks around the chuck and maintains a modest positive pressure differential above atmospheric. The fly sample tube is stoppered at the bottom end with a shortened pipette tip to allow gas entry. At the other end another pipette tip is fitted, connected to a plastic cannula

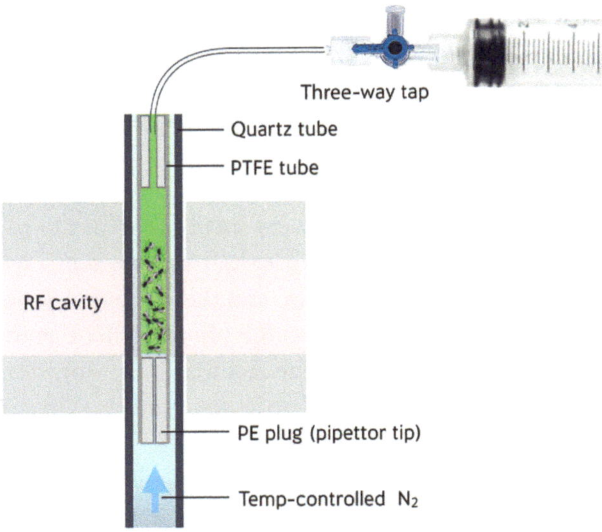

Fig. 1 Diagram of the experimental change inside the ESR spectrometer. Temperature-controlled nitrogen gas (*blue*) flows around the Teflon tube filled with ≈20 flies, stoppered at both ends with pipettor tips. The nitrogen flows around the Teflon tube, which can be filled with the desired anesthetic gas mixture (*green*) through a cannula connected to a syringe via a three-way tap.

tube and a syringe with a three-way tap. This simple arrangement operates in three modes: (1) three-way tap vented to outside air: the sample tube rapidly fills from below with nitrogen at the set temperature, (2) three-way tap connected to syringe: a rapid change from nitrogen to the contents of the 10 to 30-mL syringe can be effected. Thermal equilibration follows within seconds, and (3) three-way tap closed: whatever is in the tube remains there and only exchanges very slowly (tens of minutes) with the nitrogen flowing past it. This simple arrangement allows fast gas changes as well as steady exposures (Fig. 1).

5. MEASUREMENTS

When fruit flies are put in the tube and the nitrogen is allowed to flow, the ESR mechanically tunes the cavity for maximum quality factor, i.e., roughly speaking, for the maximum number of passes of the RF power back and forth inside the sample. The equivalent length of the sample, and therefore, its absorption is thus increased as it would be in a resonant cavity in the beam of an optical spectrometer. If, however, attenuation in the sample is too high due to excess water the spectrometer fails to tune the cavity. Under the conditions and with the instrument described earlier, a maximum of about 20–25 flies could be placed in the cavity and tuning achieved. A magnetic field sweep about 100 gauss wide centered on ≈3350 gauss, i.e., *the magnetic field value* resonant with the microwave frequency typical of X-band spectrometers is then performed to measure total spin. The magnetic field is then set to a constant value (sweep range set to 0) at the positive peak and the sweep started again. The *x*-axis then becomes the time axis.

5.1 Movement Artifacts

Fully awake flies generate very large artifacts in the ESR trace (Fig. 2). Fortunately, a brief exposure to nitrogen followed by a refill of the experimental tube with ambient air rapidly and permanently quietens them. This is not due to death of the flies by anoxia. Flies can be kept in O_2-free nitrogen for hours and promptly revive when exposed to air. Another, smaller movement artifact is the settling of the flies in the tube. This can be accelerated by doing a vigorous (≈20 mL/s) gas change in the tube after the flies have become motionless, which will gently tamp the flies down. Thereafter, recordings can be very stable aside from instrumental drift.

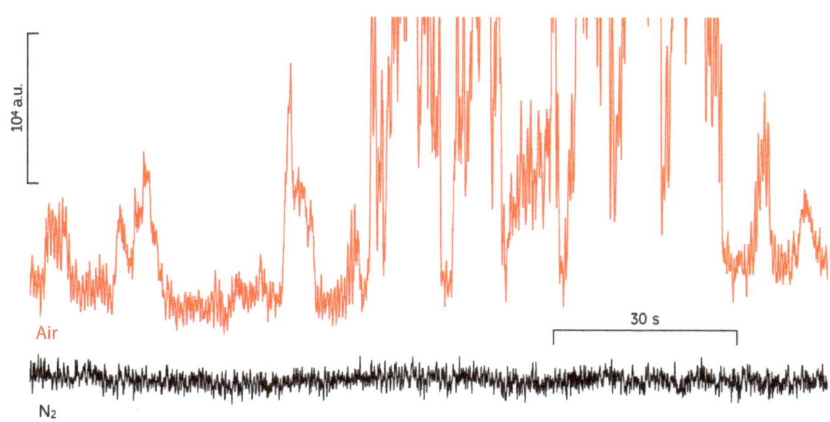

Fig. 2 Continuous ESR records at 3345.6 gauss from approximately 30 flies at 25°C. *Top*: immediately after transfer into PTFE recording chamber, in air. *Bottom*: 6 min after switching to N_2. The flies recover fully from anoxia even after N_2 exposure for more than 60 min.

5.2 Gas Artifacts

We have found it impossible to perform ESR recordings in the X-band in ether vapor, and more generally in the vapor of other halogenated anesthetic ethers such as isoflurane, insofar as they appear to interfere with the RF signal, likely by having strong absorptions in the microwave range.

6. ORIGIN OF THE SPIN SIGNAL IN *DROSOPHILA*

6.1 Melanin

Most of the spin signal in Drosophila comes from melanin, a pigment material whose properties are so remarkable they deserve a brief summary. Melanin comes in three forms (Meredith & Sarna, 2006): eumelanin, a blue-black pigment, largely formed from catechol monomers; pheomelanin, a reddish-brown pigment formed from a complex mixture of monomers, among which some sulfur-containing molecules; and neuromelanin—the gray in "gray matter"—about whose biosynthesis relatively little is known (Zecca, Youdim, Riederer, Connor, & Crichton, 2004). All three types of melanins can form spontaneously without enzyme action, and all are extraordinarily stable to chemical and enzymatic degradation, so much so that melanin is found in fossils embedded in rock hundreds of millions of years ago (Colleary et al., 2015). Melanin is also the largest biological free radical and contains approximately 10^{18} spins per gram. The stability of melanin as a radical is due to its extensively delocalized electronic structure. This property is also responsible for its broad-band light absorption, and for its remarkable electrical properties including conduction and switching (McGinness, Corry, & Proctor, 1974).

Wild-type Drosophila contain all three types of melanin. The "yellow" mutant strain of Drosophila does not contain eumelanin, and is markedly lighter in pigmentation and yellow reddish rather than brownish black. Neuromelanin is present in both. Remarkably, of the more than 76,000 fly strains kept in the Bloomington central repository, none are reported to be completely unpigmented. Neuromelanin, it should be noted, is both ubiquitous and, in vertebrates, unaffected by the absence of the other two types of melanin in albinism. Spin in yellow flies is smaller by about a factor of two (Fig. 3).

6.2 Oxygen-Dependent Signal

An oft-repeated statement found in almost all accounts of biological ESR is that variations in the spin content of living organisms are too small to be measured in vivo. This is due to the high reactivity of biological free radicals,

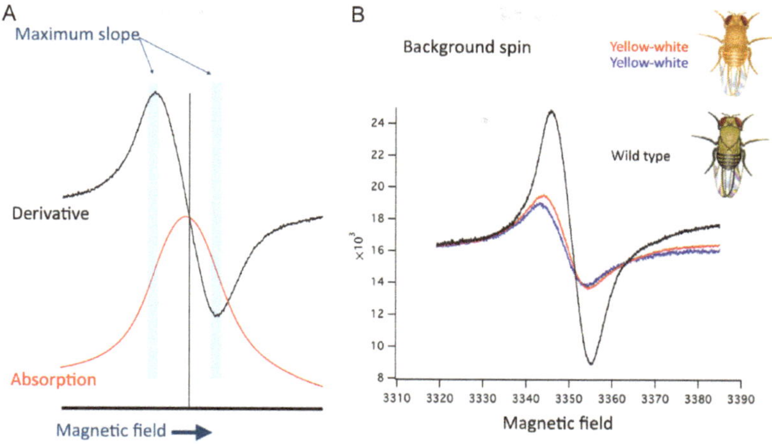

Fig. 3 (A) The spin signal measured using the derivative spectrometer (*above, black trace*) and its integral (*below, red trace*). The zero-crossing point of the derivative trace corresponds to the underlying absorption maximum, and the two regions of maximum slope (*blue shading*) in the absorption trace correspond to the maxima and minima of the derivative. (B) Background spin signal in strains of Drosophila containing (wild type, *black trace*) and devoid of (*yellow, white, red,* and *blue traces*) eumelanin. Over half of the spin signal is contributed by stable eumelanin free radicals.

to the presence of radical scavenging enzymes and cofactors, and to the use of less sensitive L-band (≈1 GHz) ESR spectrometers required for larger animals (typically mice) because of the direct relation between photon energy and tissue RF absorption (Swartz, Bolton, & Borg, 1972; Swartz & Clarkson, 1998). Accordingly, traps are almost always used to reveal the presence of free radicals in vivo. Traps are the chemical equivalent of an electronic integrator: they react with an unstable free radical, capture its unpaired electron, and store it in stable form.

Remarkably, this is not the case in Drosophila in the X-band, where clear spin variations in response to oxygen and anesthetics can be measured. They typically amount to about 5%–10% of total spin depending on the pigment background (whether wild type or yellow). Signal/noise ratio under normal experimental conditions is approximately 8–10 (Fig. 4). Given the relative size of the variable and constant signal, the question naturally arises as to whether the spin changes are due to small proportional changes in the melanin spin content or to a spin from a different origin.

This question can be addressed in the following ways.

6.3 Dependence of Variable Signal on Total Melanin Content

A comparison of responses to a given change in conditions can be made in wild-type and yellow flies. If, as is observed, the signal amplitude appears

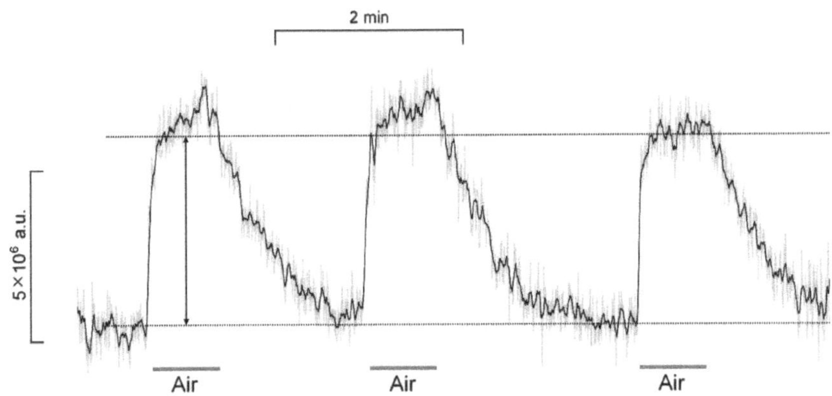

Fig. 4 Reproducibility and reversibility of the spin increase in response to air. The flies are held in nitrogen throughout. Three successive exposures to air cause responses of very similar magnitude. The upward drift during air exposure is a common feature of these recordings. For consistency the initial jump after air exposure is taken as the amplitude of the response.

constant, it is not due to eumelanin but could still be due to spin changes in pheomelanin and neuromelanin, which cannot be removed.

6.4 Dependence of Oxygen Signal on O_2 Concentration

Given the high density of spins in melanin, it is not beyond the possible that ambient gases and volatile pharmacological agents might alter the spin content. This, however, seems to be true only of photoexcited electronic states of melanin, which under illumination becomes an electron donor, capable of reducing dissolved oxygen to superoxide in solution.

In the dark, spin concentration in melanin varies little between 100% O_2 and O_2-free solutions. The dose–response curve of the spin signal to oxygen, by contrast, behaves quite differently (Fig. 5) and is consistent with a high-affinity (micromolar) binding site for oxygen. A large fraction, in the order of 75% of the oxygen-induced spin increase, also disappears upon cyanide poisoning. Taken together, these are clear indications that the oxygen spin signal is of mitochondrial origin, since the terminal oxidase in the respiratory chain exhibits both high oxygen affinity and sensitivity to cyanide.

6.5 Anesthetic-Dependent Signal

The extreme sensitivity of the Drosophila spin signal to dissolved oxygen was not clear to us when we published our first study, and had a confounding effect on some of the results reported. Once it became clear, however, that

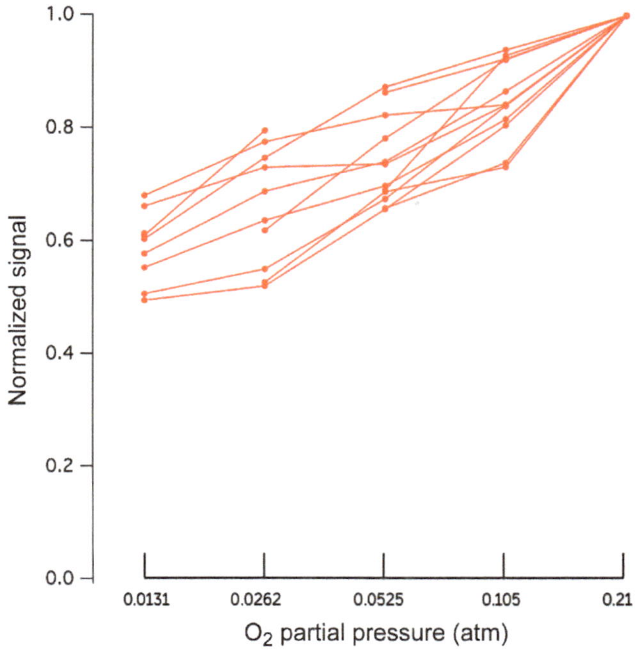

Fig. 5 Dependence of the oxygen signal on O_2 partial pressure, with the amplitude in 100% air (0.21 atmosphere partial pressure) normalized to 1. The abscissa is plotted on a logarithmic scale, and twofold dilutions give equal intervals. The relationship between log (O_2 partial pressure) and amplitude is linear in this range, showing no inflection point or saturation.

great care must be taken in excluding oxygen from anesthetic-containing atmospheres, the pattern of responses became much simpler. One of the puzzling features of our first study was the fact that chloroform alone (as compared to N_2O, SF_6, and Xe) was properly anesthetic to flies whereas all the anesthetic gases gave spin responses. In O_2-free samples of N_2O, SF_6, and Xe, no spin changes were seen. By contrast, oxygen-free chloroform gave a distinct spin signal (Fig. 6A). The spin-carrying species is as yet unknown, but it is known that reductive catabolism of $CHCl_3$ by NADPH can occur to give CCl_3 radicals (Testai & Vittozzi, 1986).

6.6 Mutants

The chief strength of Drosophila as a model organism is the existence of mutant strains selected by screens for a desired property. In the case of anesthesia resistance, the mutants are screened by dropping a population of mutagenized flies into a tall cylinder filled with a known concentration of

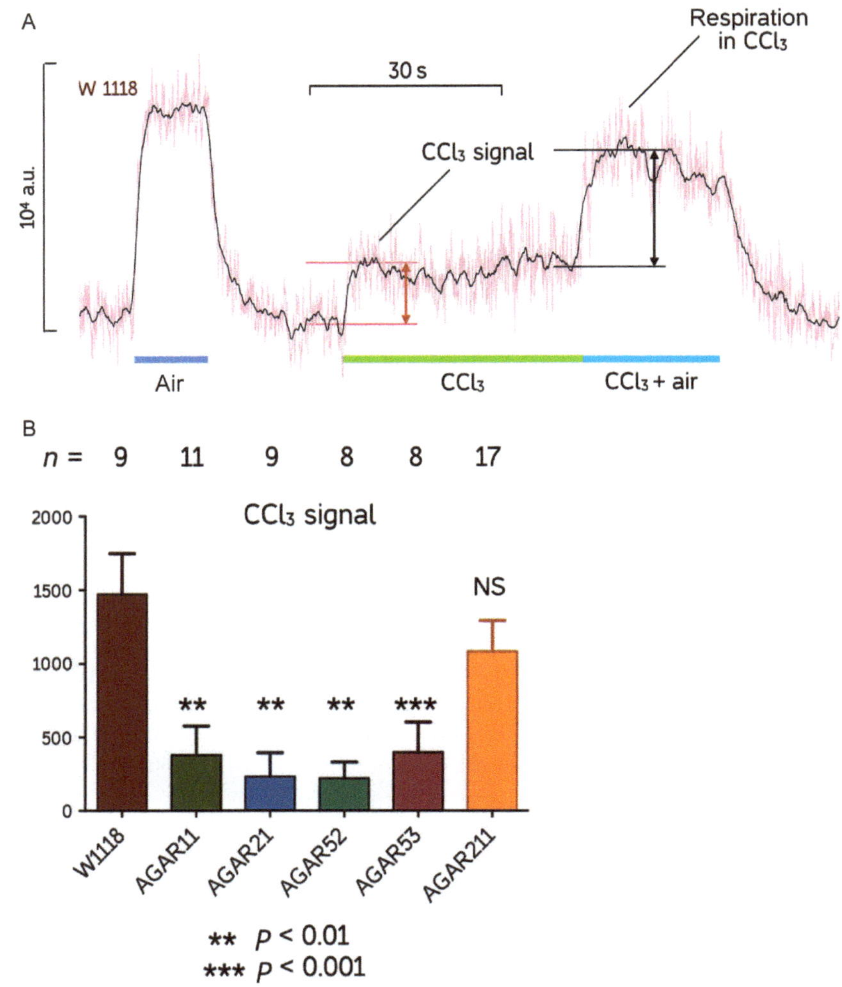

Fig. 6 A typical experiment displaying anesthesia-related spin changes. (A) The trace begins in nitrogen. A brief exposure to air causes a reversible increase in spin. The flies are then exposed to oxygen-free chloroform, and a signal is observed. Air is added back, and the spin increases without reaching the value of the original oxygen signal. (B) Mean and standard deviation of chloroform signal (*red arrow* in (A) in wild-type W1118 and five different anesthetic-resistant AGAR mutants. Experiment replication numbers are given above each bar, and *P* values from Mann–Whitney test comparing mutant to wild type. *From Turin, L., & Skoulakis, E. M. C. S. E. Electron spin and general anesthesia in Drosophila.* BioRXiv, *in preparation.*

anesthetic. The cylinder contains at intervals in the vertical dimension half-funnels of gauze which the flies cling to when awake. The flies that fall to the bottom are sensitive to the anesthetic and those retained by the "column" are insensitive. This fly "chromatography" has enabled the isolation of several mutants whose anesthetic resistance has been characterized (Guan, Scott, & Nash, 2000; Madhavan et al., 2000; Nash, Campbell, & Krishnan, 1991). Remarkably anesthetic resistance seems to correlate with a change in the chloroform-induced spin signal (Turin & Skoulakis, 2018; Turin et al., 2014). This manifests itself in two ways: (1) the appearance of a chloroform-induced increase in spin in the absence of oxygen (Fig. 6A) and (2) an interference between chloroform and the oxygen signal. Anesthetic-resistant mutants (except AGAR211) show a significantly smaller chloroform spin signal than wild-type controls (Fig. 6B).

7. CONCLUSIONS

Insofar as there is still a realistic prospect of discovering, if not exactly a unitary, at any rate a universal subcellular mechanism of GA, we believe that Drosophila is a good place to study it. The connection between mitochondria and neuronal function remains a strong contender for a mechanism of general anesthesia (Morgan, Hoppel, & Sedensky, 2002; Morgan, Sedensky, & Meneely, 1990) (see companion chapters "General genetic strategies" by Steele and Morgan and "Approaches to anesthetic mechanisms: The *C. elegans* model" by Steele and Sedensky in Vol. 602). ESR may provide insights into its genetics and could point to a solution of this outstanding mystery of neuroscience.

REFERENCES

Alkire, M. T., Hudetz, A. G., & Tononi, G. (2008). Consciousness and anesthesia. *Science*, *322*, 876–880.

Colleary, C., et al. (2015). Chemical, experimental, and morphological evidence for diagenetically altered melanin in exceptionally preserved fossils. *Proceedings of the National Academy of Sciences of the United States of America*, *112*, 12592–12597.

Eaton, G. R., Eaton, S. S., Barr, D. P., & Weber, R. T. (2010). *Quantitative EPR*. Springer Science & Business Media.

Eger, E. I., II (2014). A history of research in anesthesia. In *The wondrous story of anesthesia* (pp. 515–524). New York, NY: Springer.

Franks, N. P. (2014). The unfolding story of how general anesthetics act. In *The wondrous story of anesthesia* (pp. 597–608). New York, NY: Springer.

Grémiaux, A., Yokawa, K., Mancuso, S., & Baluška, F. (2014). Plant anesthesia supports similarities between animals and plants: Claude Bernard's forgotten studies. *Plant Signaling & Behavior*, *9*, e27886.

Guan, Z., Scott, R. L., & Nash, H. A. (2000). A new assay for the genetic study of general anesthesia in Drosophila melanogaster: Use in analysis of mutations in the X-chromosomal 12E region. *Journal of Neurogenetics*, *14*, 25–42.

Hallwell, B., & Gutteridge, J. (1985). *Free radicals in biology and medicine*. USA: Oxford University Press.

Madhavan, M. C., Kumar, R. A., & Krishnan, K. S. (2000). Genetics of anesthetic response: Autosomal mutations that render Drosophila resistant to halothane. *Pharmacology, Biochemistry, and Behavior, 67*, 749–757.

McGinness, J., Corry, P., & Proctor, P. (1974). Amorphous semiconductor switching in melanins. *Science, 183*, 853–855.

Meredith, P., & Sarna, T. (2006). The physical and chemical properties of eumelanin. *Pigment Cell Research, 19*, 572–594.

Morgan, P. G., Hoppel, C. L., & Sedensky, M. M. (2002). Mitochondrial defects and anesthetic sensitivity. *Anesthesiology: The Journal of the American Society of Anesthesiologists, 96*(5), 1268–1270.

Morgan, P. G., Sedensky, M., & Meneely, P. M. (1990). Multiple sites of action of volatile anesthetics in *Caenorhabditis elegans*. *Proceedings of the National Academy of Sciences of the United States of America, 87*, 2965–2969.

Nash, H. A., Campbell, D. B., & Krishnan, K. S. (1991). New mutants of Drosophila that are resistant to the anesthetic effects of halothane. *Annals of the New York Academy of Sciences, 625*, 540–544.

Perouansky, M. (2012). The quest for a unified model of anesthetic action: A century in Claude Bernard's shadow. *Anesthesiology, 117*, 465–474.

Poole, C. P. (1967). *Electron spin resonance: A comprehensive treatise on experimental techniques*. New York: John Wiley and Sons.

Pryor, W. (2012). *Free radicals in biology*. Elsevier.

Swartz, H. M., Bolton, J. R., & Borg, D. C. (1972). *Biological applications of electron spin resonance*. New York: Division of John Wiley, 1937.

Swartz, H. M., & Clarkson, R. B. (1998). The measurement of oxygen in vivo using EPR techniques. *Physics in Medicine and Biology, 43*, 1957–1975.

Testai, E., & Vittozzi, L. (1986). Biochemical alterations elicited in rat liver microsomes by oxidation and reduction products of chloroform metabolism. *Chemico-Biological Interactions, 59*, 157–171.

Trapp, C., Waters, B., Lebendiger, G., & Perkins, M. (1983). Electron spin resonances of a living system (Drosophila) on normal and carcinogenic diets. *Biochemical and Biophysical Research Communications, 112*, 602–605.

Turin, L. (1996). A spectroscopic mechanism for primary olfactory reception. *Chemical Senses, 21*, 773–791.

Turin, L., & Skoulakis, E. M. C. S. E. (2018). Electron spin and general anesthesia in Drosophila. *BioRXiv*. in preparation.

Turin, L., Skoulakis, E. M. C., & Horsfield, A. P. (2014). Electron spin changes during general anesthesia in Drosophila. *Proceedings of the National Academy of Sciences of the United States of America, 111*, E3524–33.

Yokawa, K., Kagenishi, T., Pavlovič, A., Gall, S., Weiland, M., Mancuso, S., & Baluška, F. (2017). Anaesthetics stop diverse plant organ movements, affect endocytic vesicle recycling and ROS homeostasis, and block action potentials in Venus flytraps. *Annals of Botany*, mcx155. https://doi.org/10.1093/aob/mcx155.

Zecca, L., Youdim, M. B. H., Riederer, P., Connor, J. R., & Crichton, R. R. (2004). Iron, brain ageing and neurodegenerative disorders. *Nature Reviews. Neuroscience, 5*, 863–873.

CHAPTER EIGHT

Giant Plasma Membrane Vesicles: An Experimental Tool for Probing the Effects of Drugs and Other Conditions on Membrane Domain Stability

Zoe Gerstle, Rohan Desai, Sarah L. Veatch[1]
University of Michigan, Ann Arbor, MI, United States
[1]Corresponding author: e-mail address: sveatch@umich.edu

Contents

Abstract

Giant plasma membrane vesicles (GPMVs) are isolated directly from living cells and provide an alternative to vesicles constructed of synthetic or purified lipids as an experimental model system for use in a wide range of assays. GPMVs capture much of the compositional protein and lipid complexity of intact cell plasma membranes, are filled with cytoplasm, and are free from contamination with membranes from internal organelles. GPMVs often exhibit a miscibility transition below the growth temperature of their parent cells. GPMVs labeled with a fluorescent protein or lipid analog appear uniform on the micron-scale when imaged above the miscibility transition temperature, and separate into coexisting liquid domains with differing membrane compositions and physical properties below this temperature. The presence of this miscibility transition in isolated GPMVs suggests that a similar phase-like heterogeneity occurs in intact plasma

Methods in Enzymology, Volume 603
ISSN 0076-6879
https://doi.org/10.1016/bs.mie.2018.02.007

membranes under growth conditions, albeit on smaller length scales. In this context, GPMVs provide a simple and controlled experimental system to explore how drugs and other environmental conditions alter the composition and stability of phase-like domains in intact cell membranes. This chapter describes methods to generate and isolate GPMVs from adherent mammalian cells and to interrogate their miscibility transition temperatures using fluorescence microscopy.

1. INTRODUCTION

It has long been known that general anesthetics, themselves hydrophobic molecules, partition into lipid membranes and influence their physical properties (Meyer, 1899; Mullins, 1954; Seeman, 1972). Much of the past work in this area has focused on simple, single-component membranes. These membranes undergo a main chain-melting transition between a liquid-phase and gel- or solid-phases, whose properties have been extensively studied by a variety of experimental methods. In all cases, incorporating hydrophobic general anesthetics into single-component membranes lowers the main chain transition temperature (T_M) and disorders lipid chains (makes their motions more isotropic) at fixed temperature (Heimburg & Jackson, 2007; Kamaya, Ueda, Moore, & Eyring, 1979; Lorincz, Mihaly, Nemeth, Wacha, & Bota, 2015; Miller, Firestone, Alifimoff, & Streicher, 1989; Shieh, Ueda, Lin, & Eyring, 1976). These molecules also can impact the lateral pressure profiles, bending rigidity, and water permeability of simple membranes (Gruner & Shyamsunder, 1991; Jorgensen, Ipsen, Mouritsen, Bennett, & Zuckermann, 1991; Jorgensen, Ipsen, Mouritsen, & Zuckermann, 1993; Mountcastle, Biltonen, & Halsey, 1978; Trudell, Payan, Chin, & Cohen, 1975). Over the years, it has been argued that these types of physical effects are unlikely to contribute to the molecular mechanisms of general anesthetics, as they tend to produce small perturbations at physiologically relevant concentrations (e.g., $\Delta T_M < 0.5°C$) (Franks & Lieb, 1982, 1986, 1990). In fact, many observations can be simply explained through the general lens of "freezing point depression," where any hydrophobic impurity will have a qualitatively similar effect on membrane properties that simply depend on T_M. However, many hydrophobic molecules are not general anesthetics, though they do impact the main chain transition temperature of simple membranes (Franks & Lieb, 1986, 1991; Krasowski et al., 2001). Because membrane physical properties that are simply related to a changing T_M can be produced by both anesthetics and nonanesthetics, it is unlikely that they contribute to the molecular mechanisms of general anesthetic function.

In addition to the main chain transition, multicomponent membranes that incorporate a sterol, such as cholesterol, can experience a miscibility transition between a single and two coexisting liquid phases (Veatch & Keller, 2005) (Fig. 1). One of these liquid phases, referred to as liquid-disordered or L_d, closely resembles the liquid crystalline phase present in a single-component membrane above T_M. The second liquid, called liquid-ordered or L_o, has chain order resembling the solid phase but with individual molecules retaining a high level of translational and rotational mobility (Ipsen, Karlstrom, Mouritsen, Wennerstrom, & Zuckermann, 1987). This miscibility transition occurs at a temperature called T_{mix} and was first observed experimentally in giant vesicles and black lipid membranes in 2001 (Dietrich et al., 2001; Samsonov, Mihalyov, & Cohen, 2001), and is also observed in extracted and isolated biological membranes (Baumgart et al., 2007; Dietrich et al., 2001) (Fig. 1). While all hydrophobic impurities are expected to lower the main chain transition temperature to roughly the same extent, the magnitude and sign of an impurity's effect on the miscibility transition temperature depends more strongly on the membrane composition and the detailed molecular structure of the impurity (Allender & Schick, 2017; Cornell et al., 2017). We recently found that several anesthetic *n*-alcohols lower T_{mix} in vesicles isolated from RBL-2H3 membranes, but two nonanesthetic *n*-alcohols do not (Gray, Karslake, Machta, &

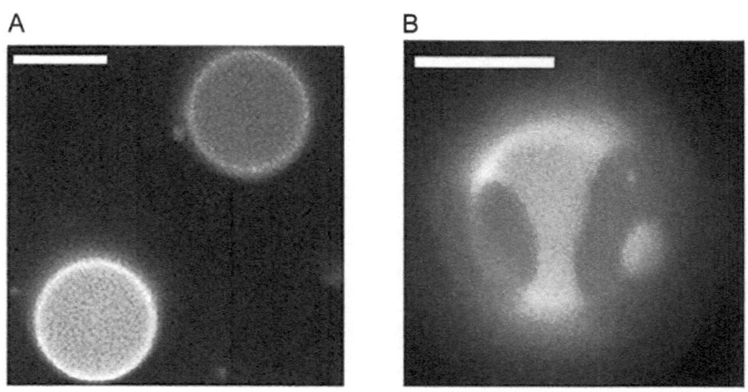

Fig. 1 Giant plasma membrane vesicles (GPMVs) can contain two coexisting liquid phases. (A) GPMVs imaged above their miscibility transition temperature (T_{mix}) are in a single liquid phase, as indicated by the uniform distribution of a fluorescent lipid analog across the vesicle surface. (B) GPMVs separate into coexisting liquid-ordered (L_o) and liquid-disordered (L_d) phases at a temperature below T_{mix}. The fluorescent dye added to the GPMVs incorporates into the L_d phase, facilitating easy distinction between the phases. *Scale bars* are 10 μm.

Veatch, 2013; Machta et al., 2016). This suggests that, at least in principle, membrane physical properties that depend on T_{mix} could contribute to the molecular mechanisms of *n*-alcohol general anesthesia.

Since the early observations of coexisting liquid phases in synthetic and isolated biological membranes (Baumgart et al., 2007; Dietrich et al., 2001; Veatch & Keller, 2005), there has been much speculation that this phase transition is in some way related to the submicron lipid raft domains hypothesized to exist in intact cell plasma membranes (Lingwood & Simons, 2010; Simons & Ikonen, 1997). In fact, the original proposal of the "raft" hypothesis (Simons & Ikonen, 1997) was in part based on even earlier observations of submicron heterogeneities (Sankaram & Thompson, 1991; Shimshick & McConnell, 1973; Vist & Davis, 1990) and detergent-resistant membrane fractions in purified membranes (Schroeder, London, & Brown, 1994), both of which are now thought to be related to the presence of a nearby miscibility transition (Heerklotz, 2002; Honerkamp-Smith, Veatch, & Keller, 2009). We have proposed that lipid raft domains are critical composition fluctuations that occur within a single-phase membrane that is biologically tuned to be close to a miscibility transition under physiological conditions (Burns, Wisser, Wu, Levental, & Veatch, 2017; Machta, Papanikolaou, Sethna, & Veatch, 2011; Machta, Veatch, & Sethna, 2012; Veatch et al., 2008; Zhao, Wu, & Veatch, 2013). If so, changes in T_{mix} are expected to impact the properties of these domains as well as the functions that they promote. Past work has shown that the ligand-gated ion channels thought to be responsible for general anesthesia are functionally modulated by other perturbations of lipid raft structure (Allen, Halverson-Tamboli, & Rasenick, 2007; Li, Serwanski, Miralles, Bahr, & De Blas, 2007; Sooksawate & Simmonds, 2001). Ongoing work is more directly probing the connections between the immiscibility transition in membranes- and lipid-mediated heterogeneity in intact cells in a wide range of cellular and signaling systems (Levental & Veatch, 2016; Stone, Shelby, Nunez, Wisser, & Veatch, 2017; Zhou et al., 2013).

The goal of this chapter is to describe methods used to generate and isolate giant plasma membrane vesicles (GPMVs) from intact cells. These vesicles closely mimic the protein and lipid compositions from the intact plasma membranes from which they were derived, but lack polymerized cytoskeletal elements (Fridriksson et al., 1999; Holowka & Baird, 1983). These vesicles experience a miscibility transition below growth temperatures, and microscopic liquid-ordered and liquid-disordered domains are easily visualized on the vesicle surface when membranes are doped with fluorescent lipid analogs and imaged using conventional epi-fluorescence or confocal microscopy. The described GPMV isolation method produces vesicles with a distribution of compositions and miscibility transition temperatures. The distribution of transition temperatures can shift to lower or higher values in the presence of drugs or upon the application of hydrostatic pressure (Gray et al., 2013; Machta et al., 2016; Raghunathan, Ahsan, Ray, Nyati, & Veatch, 2015). Methods are described to quantify the average miscibility transition temperatures over a population of vesicles.

2. GPMV FORMATION

Later, a procedure is described for isolating GPMVs that is built upon protocols described previously (Baumgart et al., 2007; Holowka & Baird, 1983; Scott, 1976; Sezgin et al., 2012). GPMV formation and isolation can be accomplished in roughly 2h, and it is recommended that vesicles be used within several hours of preparation. This is only one of several possible methods available to form GPMVs. Some other methods described involve long incubations in nutrient poor buffers (Lingwood, Ries, Schwille, & Simons, 2008), or incubation of cells in buffers of different salts and osmotic pressures (Sarabipour, Chan, Zhou, Di Paolo, & Hristova, 2015). Overall, we find that the method outlined later, sometimes along with suggested modifications, provides the fastest and most robust method to produce GPMVs. Typically, each adherent cell present produces one to several 5–20 μm diameter vesicles that are filled with cytoplasm, lack internal membranes, and whose membranes are rich in protein and lipid components associated with the plasma membrane. The methods described later are most effective on adherent cells. GPMVs can also be made from cells in suspension, but yields are frequently lower, most vesicles remain attached to cells, and it is challenging to isolate GPMVs from cells after their formation. We use rat basophilic leukemia (RBL-2H3) cells for most studies, but have had success with the described techniques in other cell types, including the mammalian cell lines CHO, HELA, HEK, CH27 B cells, as well as cell lines from fish (ZF4), amphibian (XtC-2), and fly (SF9).

2.1 Equipment

1. 37°C incubator with shaker
2. Cell culture flasks and dishes

2.2 Buffers and Reagents

1. Vesiculation buffer: 150 m*M* NaCl, 2 m*M* $CaCl_2$, and 20 m*M* HEPES in water, pH 7.4
2. Fluorescent lipid labeling buffer: vesiculation buffer containing 1% (v/v) methanol, 2 μg/mL of a fluorescently labeled lipid probe such as DiI-C_{12} which is available from ThermoFisher Scientific (prepare on day of GPMV formation)
   a. Dilute concentrated fluorophore stock in methanol to a final concentration of 200 μg/mL. For example, add 1 μL of a 10 mg/mL DiI-C_{12} solution to 49 μL methanol). If fluorophore is suspended in chloroform, evaporate solvent prior to resuspension in methanol.
   b. Add 10 μL of solution (a) to 1 mL of vesiculation buffer, resulting in a final concentration of 2 μg/mL of fluorophore in vesiculation buffer. Alternate concentrations of probe can be used as needed, but it is recommend to remain below 10 μg/mL and to maintain at least 1% methanol to facilitate transfer of the lipid probe into the membrane.
   c. Sometimes it is beneficial to include a lipid transporter protein (such as BSA) in the labeling buffer. This is useful for more hydrophobic lipid probes (e.g., ones with long and saturated hydrocarbon chains). Labeling with DiI-C_{12} does not require BSA for efficient incorporation.
3. Active vesiculation buffer: 1.9 m*M* DTT, 27.6 m*M* formaldehyde (HCHO), in vesiculation buffer (make on day of GPMV formation).
   a. Dissolve 15 mg of dithiothreitol (DTT) in 500 μL of vesiculation buffer. (This step is included to avoid measuring only 3 mg of DTT, since this quantity is difficult to measure accurately on most laboratory scales.)
   b. Add 100 μL of (a) to 10 mL of vesiculation buffer.
   c. Add 20.3 μL of 37% formaldehyde to (b).
4. Flask or dish of adherent cells.

2.3 Procedure

1. The cells used for GPMV isolation should be prepared so that they stably adhere to the surface. GPMV composition and properties such as transition temperature will be somewhat dependent on growth conditions (serum concentration, cell density, etc.) (Gray, Diaz-Vazquez, & Veatch, 2015); therefore, it is beneficial to maintain and/or characterize these variables throughout all trials of an experiment. Typically, we

prepare cells such that they are 50%–80% confluent and have adhered to the substrate at least overnight, and try to maintain details of cell maintenance (number of cells plated, time of incubation) throughout a given measurement.

2. Gently rinse cells twice with 1 mL of vesiculation buffer to remove dead cells and other cellular debris.
3. If labeling with a lipid fluorophore, add 1 mL of labeling buffer to the cell culture dish while gently shaking or rocking for 5–15 min, allowing the dye to incorporate into the cell membrane. This can be accomplished at room temperature on a benchtop rocker, at 37°C inside a shaking incubator (100 rev/min), or on a benchtop or incubator with occasional agitation. Labeled cells can be imaged prior to vesiculation to optimize fluorophore incorporation or to control for growth density. If needed, cells can be labeled more than once via this procedure.
4. To increase efficiency, the active buffer can be prepared while the cells are being labeled.
5. Rinse the cells several times with vesiculation buffer to remove excess fluorophores and once more with 1 mL of active buffer. Remove any remaining active buffer in the dish with a pipette, and add enough active buffer to cover cells completely. We find that 200–600 μL of buffer is sufficient for a 35-mm dish and 1 mL for a 25-cm^2 flask (see Note 2).
6. Incubate cells in buffer for 60–120 min at growth temperature with gentle agitation. This can be accomplished in a 37°C shaking incubator set at 100 rev/min, in a 37°C shaking water bath on a low setting, or in a stationary incubator, water bath, or hot plate with occasional manual agitation. During this time, cells vesiculate and GPMVs separate from cells.
7. Progress can be monitored by imaging the dish or flask. Under phase contrast, GPMVs appear dark and can be seen attached to cells or flowing over cells upon gentle agitation. GPMVs can be hard to image at this stage with fluorescence excitation as cells often appear much brighter than vesicles, since the fluorescent probe is often trafficked into internal membranes during the incubation (Fig. 2).
8. To harvest GPMVs, lightly rap the bottom of the dish to release additional cell-attached GPMVs. Decant the liquid containing active buffer and GPMVs into an appropriately sized tube.
9. If needed, GPMVs can be further separated from detached cells through a brief spin (we spin at $500 \times g$ for 5 min). GPMVs remain in solution

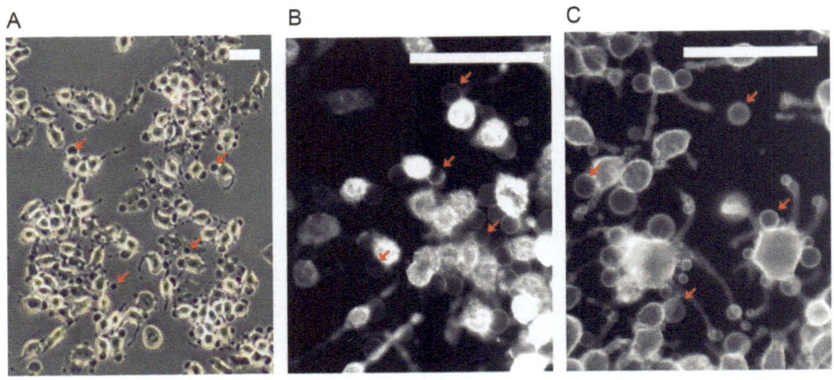

Fig. 2 Phase contrast (A) and fluorescent (B, C) images of cell attached GPMVs. (A) GPMVs appear *dark* in phase contrast images and can be seen attached to cells or flowing above the adherent cell layer (not shown). (B) Cell attached GPMVs are difficult to visualize when cells membranes are fluorescently tagged prior to incubation with active vesiculation buffer due to trafficking of fluorophores to internal membranes. (C) GPMVs and their parent cell membranes have roughly the same fluorescent intensity when cells are labeled after incubation with active vesiculation buffer. *Red arrows* point to GPMVs in all panels. *Scale bars* are 100 μm.

while detached cells will tend to pellet. To concentrate the sample of GPMVs (e.g., for lipidomics analysis), spin at 25–50 k × *g* for 20 min. It may be beneficial to wash the pellet and then repeat the spinning for another 20 min.

10. Chemicals used for GPMV generation can be dialyzed from vesicles using cassettes or dialysis tubing and multiple buffer exchanges. We used 10,000 MWCO tubing and 5 × 200 mL exchanges (see Note 8).

2.4 Notes

1. DiI-C_{12} is a fluorescent dye which labels the liquid-disordered phase of a lipid membrane and is excited maximally by 549 nm light. Other fluorescent dyes which selectively partition into either the L_o or the L_d phase may be used.
2. A minimal volume of active buffer should be used in order to increase the final density of prepared vesicles. A volume should be used that ensures no parts of the dish or flask surface will dry over the course of the incubation.
3. Sometimes it is beneficial to dilute the vesiculation buffer, either to match the osmotic strength of the growth media of the given cell type

(as is the case in Xtc2 cells), or to apply an osmotic gradient that favors vesiculation (as we found useful for ZF4 cells).

4. We and others find that most vesicles will form within 1 h of incubation with active buffer at 37°C, although in some cases longer incubations increase vesicle yields.
5. GPMVs can be fluorescently labeled postisolation with lipid dyes, lipid-binding proteins, or antibodies. The advantage of prelabeling with lipid fluorophores is that it enables washing of unbound probes, which is difficult to accomplish once vesicles are prepared. A disadvantage of prelabeling is that some probes will be trafficked to internal membranes during the incubation, leading to lower final concentrations of the probe in the isolated vesicles. Examples of pre- and postlabeled cell attached GPMVs are shown in Fig. 2B and C.
6. Our standard protocol uses 2 m*M* DTT and 25 m*M* HCHO in the vesiculation buffer to optimize the density of the vesicles. Table 1 shows a titration of DTT and HCHO, highlighting the necessity of HCHO in the vesiculation process.
7. DTT can be replaced with other reducing agents such as β-mercaptoethylamine (BME) or glutathione (GT) without significant modification to the described protocol. We recommend matching the reducing potential to achieve comparable yields (e.g., replace 2 m*M* DTT with 4 m*M* GT). Vesicle sizes are often smaller and yield lower when alternate reducing agents are used. It should be noted that DTT is membrane permeable and can chemically modify lipids or remove lipid modifications on proteins (Levental, Grzybek, & Simons, 2011; Sezgin et al., 2012).
8. GPMVs can also be prepared through a 5 min incubation in DTT (or another reducing agent) and HCHO containing buffer, followed by a wash and longer incubation in vesiculation buffer containing only HCHO. This method achieves similar yields of GPMVs.

Table 1 GPMV Yields From RBL-2H3 Cells as a Function of DTT and HCHO Concentration in the Active Vesiculation Buffer

	0 m*M* DTT	**1 m*M* DTT**	**2 m*M* DTT**
0 m*M* HCHO	None	None	None
12.5 m*M* HCHO	Low	Medium	Low
25 m*M* HCHO	Medium	High	High
50 m*M* HCHO	Medium	Very high	Very high

9. This protocol utilizes a minimal vesiculation buffer but other solutions can be used, as long as they are compatible with the cells being investigated. GPMVs have been successfully prepared with different buffering agents (e.g., phosphate buffer) or in solutions containing serum.

3. MEASURING GPMV MISCIBILITY TRANSITION TEMPERATURES

Miscibility transition temperatures of single GPMVs or the average transition temperature of a population of GPMVs can be accomplished by imaging vesicles over a range of temperatures using a fluorescence microscope equipped with a temperature-controlled stage. Vesicles should be examined on the same day that they are prepared to minimize the degradation of vesicle components.

3.1 Equipment

1. Temperature-controlled microscope stage. The stage should have a flat surface that is thermally isolated from the microscope base where the sample can be adhered using a thermal compound. It is helpful to measure temperature close to the sample, for example, with a thermistor fastened to the stage surface. GPMVs prepared as described in the previous section often have transition temperatures between 0°C and 30°C; therefore, the stage would ideally be able to achieve temperatures between −10°C and 40°C. This temperature range is well within accessible limits with a Peltier heating/cooling element. Our stage is assembled primarily from commercial components and is displayed in Fig. 3. Fully assembled stages and control units with similar functionality can be obtained commercially (e.g., the Heating Cooling Thermal Stages available from Brook Industries).
2. Fluorescence microscope with epifluorescence light source, filter cube matching the fluorophore used, an objective (20–60×), and a camera appropriate for low light fluorescence imaging (see Notes 1 and 2).

3.2 Procedure

1. GPMV samples are easily imaged when placed between two coverslips. This sample can be prepared by (1) lining the edges of a glass coverslip with a thin layer of Dow Corning high-vacuum grease or similar compound, (2) pipetting an appropriate volume of GPMV suspension (20–40 μL) onto this coverslip, and (3) covering with a second coverslip.

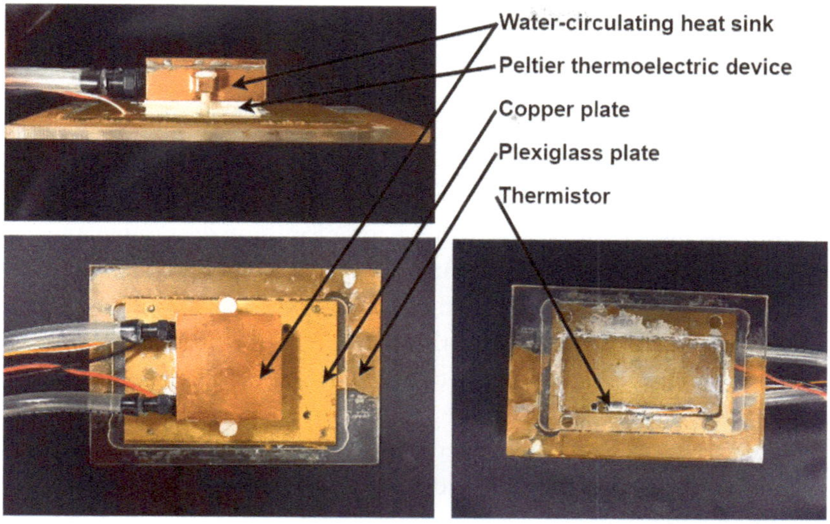

Fig. 3 Temperature can be controlled using a simple, home-built temperature stage. Our temperature stage consists of a water-circulating heat sink and Peltier thermoelectric device (Custom Thermoelectric, Bishopville, MD) and a PID-type controller unit (Oven Industries, Mechanicsburg, PA). Water is circulated through the heat sink using a garden fountain pump (available at hardware stores) submerged in a bucket. Temperature is measured with a thermistor probe calibrated to work with the controller mounted on the copper plate close to the sample. The sample, placed between two coverslips, is adhered to the copper plate near the thermistor and the assembly is placed, sample down, on a standard Olympus inverted microscope stage and is fastened with clips to prevent drift. We have also constructed stages where the Plexiglass thermal insulator was replaced by rubber or cork simply adhered to a metal plate.

The goal is to generate a thin layer of vesicles and fluid that is fully encapsulated within the chamber. The thin liquid layer greatly suppresses convective currents that lead to vesicle movement while imaging. When forming the chamber, avoid allowing liquid to leak out of the sealed area, as these channels frequently remain open and can lead to currents within the chamber that disrupt imaging. We use 22 × 22 mm number 1.5 coverslips for imaging.

2. When preparing a sample containing freshly prepared GPMVs mixed with a chemical perturbation such as an *n*-alcohol anesthetic, we recommend first mixing vesicles with chemicals in a separate container (e.g., a microcentrifuge tube) prior to assembling the sample. This ensures mixing and allows the system to equilibrate prior to imaging. Care should be taken when mixing hydrophobic chemicals with GPMVs (see Note 3). We recommend maintaining concentrations of chemicals

used to make GPMVs if still present in GPMV solutions while imaging, as some of these (namely DTT) can independently impact GPMV miscibility transition temperatures (see Note 4).

3. Secure the sealed coverslips to the temperature controller stage using a thermally conductive paste such as Arctic Silver polysynthetic thermal compound. The goal is to secure the coverslips to the stage both mechanically and thermally without compromising the sample integrity. Position the apparatus on the microscope in the proper orientation for viewing. Because GPMVs are filled with cytoplasm which has a higher density than the buffers used, they will settle to the bottom coverslip within roughly 5–10 min of becoming properly oriented.
4. Focus the objective lens on the vesicles settled at the bottom coverslip. Take care to not run the objective into the sample, as this could compromise its integrity. The most information-rich images often are focused on the bottom vesicle surfaces, where even vesicles of different sizes are coplanar. Images can also be taken of vesicle midplanes, which show the circumference, or the top surface of vesicles. For transition temperature measurements, the focus should be set such that the user can distinguish vesicles in which the probe is uniformly distributed from those containing domains. Sometimes it is useful to acquire images at multiple focal planes for more robust identification. Some examples of phase-separated GPMVs are shown in Fig. 4.
5. The acquisition time for the images should be set between 0.05 and 2 s, depending on the fluorescence intensity. Acquisition time or excitation light intensity may need to be adjusted as the temperature decreases. This is because the quantum yield of most fluorophores increases with decreasing temperature and because the probe will become concentrated within a subset of the vesicle surface when they separate into coexisting phases.
6. In order to measure the average transition temperature of a population of vesicles, start by acquiring images at a temperature where all vesicles are in the 1-phase state. Acquire images at a fixed temperature over several visual fields. It is recommended that at least 100 GPMVs be imaged per temperature. Lower the temperature in increments of 2–5°C and repeat the image acquisition procedure after allowing the sample to equilibrate for 2–5 min (or until the sample stops changing). If possible, continue lowering temperature until all or nearly all vesicles are in the 2-phase state. Use smaller increments when the temperature nears the transition temperature. Even though, the transition is fully reversible in most cases,

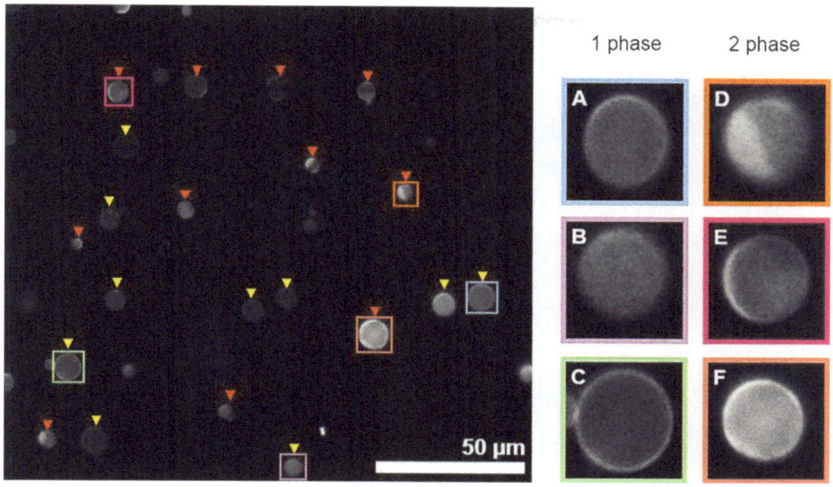

Fig. 4 Assigning phase state of individual vesicles within a population imaged at fixed temperature. (*Left*) An example image showing a field of GPMVs in which most vesicles can be unambiguously assigned to contain either 1 or liquid 2 phases. *Yellow triangles* point toward 1-phase vesicles while *red triangles* point toward 2-phase vesicles. (*Right*) Isolated view of selected vesicles shown at the *left*. GPMVs shown in (A)–(C) contain a single phase. (A) and (B) show GPMVs where the image is focused on a vesicle surface, and the brightness of the fluorophore is uniform across this surface. The GPMV in (C) is in focus along the midplane of the vesicle, and the brightness of fluorophore is uniform around the circumference. GPMVs shown in (D) and (E) contain coexisting liquid phases. The vesicle in (D) where the image is focused on the vesicle surface and a clear phase boundary is observed. The vesicle in (E) is focused close to the midplane, and the fluorophore intensity varies along the perimeter. (F) is difficult to assign. However, both the surface and the circumference vary in brightness so it was classified as containing two phases. In this example, some GPMVs in the field were not labeled because they are too small or too out of focus. In general, it is most important to remain consistent regarding assignments and the characteristics of vesicles that the investigator chooses not to assign. One reason to image at least 100 vesicles is to reduce the possible contributions of user bias in assigning vesicles.

we recommend imaging vesicles from high to low temperature for the reasons described in Note 5. We recommend blocking the excitation source when not acquiring images to avoid probe photobleaching and membrane photooxidation, which itself can impact transition temperatures in purified membranes (Veatch & Keller, 2003, 2005; Zhao et al., 2007), although in our experience this is less of a concern in isolated GPMVs.

7. Following image acquisition, average transition temperatures can be quantified by tabulating the fraction of vesicles containing two phases

at each temperature investigated, then fitting these points to a sigmoidal curve to extrapolate the temperature where 50% of vesicles contain two coexisting liquid phases (Fig. 5). We accomplish this analysis using software written in Matlab in which a user identifies if a given vesicle contains one or two phases, and this assignment is stored in memory as the user cycles through all vesicles imaged (see Note 6). Once all vesicles in a given measurement have been assigned, the fraction of phase-separated vesicles is computed at each temperature and fit to the following sigmoidal form:

$$\%\,\text{seperated} = 100 \times \left(1 - \frac{1}{1 + e^{-(T - T_{\text{misc}})/B}}\right)$$

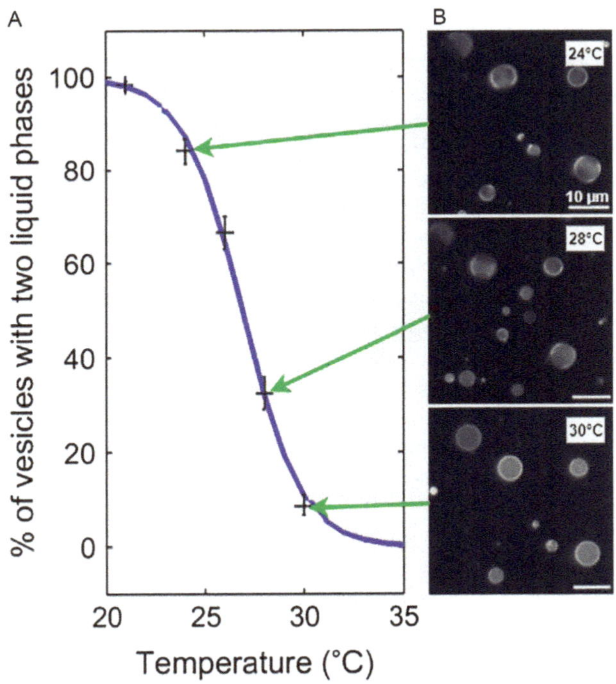

Fig. 5 Measurement of phase transition temperatures from GPMV images. (A) Plot summarizing the fraction of GPMVs assigned to contain two liquid phases as a function of temperature (*black crosses*). The *blue curve* is a fit to these points by the sigmoid function stated in the main text. For this example, the best fit value for the average transition temperature T_{mix} is 26.9°C. (B) A small subset of the GPMV images that were assigned and used to obtain the data points shown in part (A).

The best fit value of the parameter T_{mix} is the extrapolated temperature where 50% of vesicles contain two phases and the parameter B is a measure of the width of the transition.

8. Vesicles with a single liquid phase will have equal fluorescence across the plane when focused on the top or bottom vesicle surface, or equal intensity across the circumference when focused at the midplane. Vesicles in a two liquid state will present as having light and dark parts of the vesicles, as the dye selectively incorporates into the L_d phase. Typically, vesicles containing two liquid phases will have roughly equal surface fractions of the vesicle surface occupied by either phase. In some cases, liquid domains will remain dispersed or bud from or into the vesicle, deforming the vesicle surface. These vesicles are also categorized as containing two liquid phases. Sometimes, uniform vesicles appear heterogeneous because they are attached to smaller vesicles or probe aggregates. These should be categorized as uniform or skipped in the identification process if phase state is ambiguous. In some cases, GPMVs can present with coexisting gel and liquid phases, or even coexisting gel with multiple liquid phases. This is most common when cells or vesicles are treated in a way that reduces their cholesterol content. Gel domains tend to be noncircular and rotate as ridged bodies. In some cases, gel domains can form rigid networks that span the vesicle surface. Examples of GPMVs with different domain morphologies are shown in Fig. 6.
9. Errors in the measurement of the fraction of phase-separated vesicles at a given temperature can be determined by tabulating confidence intervals using binomial errors: $\mathrm{LB} = 1 - \beta_{\mathrm{inv}}(0.32/2, n - nk, nk + 1)$ and $\mathrm{UB} = 1 - \beta_{\mathrm{inv}}(1 - 0.32/2, n - nk + 1, nk)$, where LB and UB are the lower and upper bounds of the confidence interval, respectively, n is the total number of GPMVs investigated at that temperature, k is the average value for the number of GPMVs phase separated at that temperature, and β_{inv} is the β-inverse cumulative distribution function. The error in determining the average transition temperature is evaluated by computing the confidence intervals on the T_{mix} parameter in the nonlinear least squares fit. Typically, the same sample will produce equivalent T_{mix} values within error bounds when the same GPMV preparation is examined on the same day, but absolute measurements of T_{mix} can vary between different GPMV preparations. For this reason, we often report the impact of a biochemical perturbation in terms of the shift of T_{mix} from a control (ΔT_{mix}) instead of absolute values of T_{mix}.

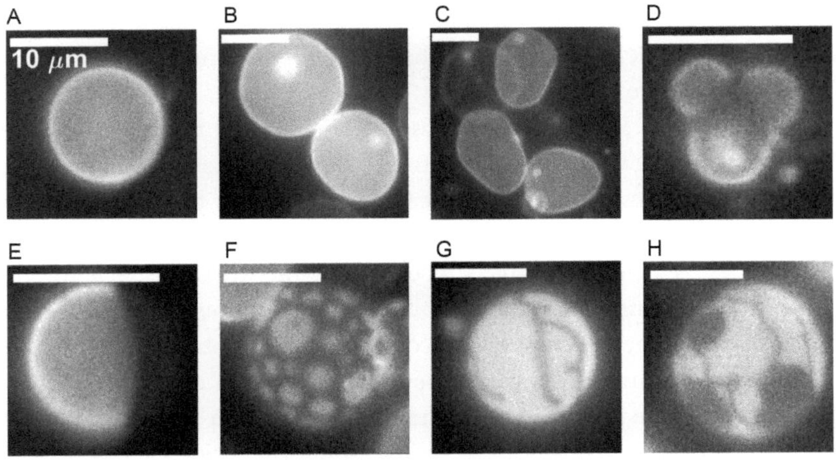

Fig. 6 Images of vesicles highlighting the range of possible shapes and domain morphologies. (A) Spherical vesicle with a single liquid phase. (B) Two single-phase vesicles that each contain internal smaller vesicles (*bright spots*). (C) Nonspherical vesicles imaged above T_{mix}. Vesicle edges undulate when vesicles are imaged in time (not shown). (D) Vesicle with bulging liquid-disordered domains. (E) Spherical vesicle exhibiting normal separation of two liquid phases where domains have fully coarsened through coalescence. (F) Liquid domains dispersed on the vesicle surface. In some cases domains do not coalesce, or do not coalesce on the timescale of the measurement. (G) Vesicle that contains a gel-phase domain. (H) Apparent coexistence of three phases within a single vesicle. Both gel- and liquid-disordered phases exclude the fluorescent probe. The vesicle in (D) was imaged in the presence of detergent which increased membrane permeability, enabling the dramatic shape observed. Vesicles in (G) and (H) are GPMVs isolated from cells incubated with methyl beta cyclodextrin to reduce their cholesterol content.

10. In order to measure the transition temperature of a single vesicle, first conduct a rough determination of the transition temperature by taking the large (2–5°C) temperature steps while visualizing the vesicle surface. This could be done in a separate sample to avoid additional temperature cycling. Then, set the temperature to be ~2°C above the transition and repeat the above procedure at smaller temperature increments, e.g., 0.5°C, 0.2°C, or 0.1°C until a temperature is reached, where two distinct phases are clearly visible. In many instances, flickering fluctuations are visible on the vesicle surface above the transition and undulating boundaries are visible below the transition (Veatch et al., 2008). These fluctuations can be quantified to precisely identify the transition using analytical methods described elsewhere (Honerkamp-Smith et al., 2008, 2009; Veatch et al., 2008).

11. Similar procedures can be followed for measuring transition temperatures for vesicles held at different hydrostatic pressures. This requires a specialized microscope stage that can stably maintain and manipulate both temperature and pressure, which has been described previously (McCarthy, Ces, Law, Seddon, & Brooks, 2015; Purushothaman, Cicuta, Ces, & Brooks, 2015).

3.3 Notes

1. Water or oil emersion objectives are not recommended when accurate temperature control is required, as they establish thermal contact between the sample and the objective or microscope base and can promote significant temperature gradients across the sample.
2. We use a 40× air objective on an IX81 inverted microscope (OLYMPUS) with a mercury lamp excitation source. A SCMOS camera (ANDOR) is used to collect images for analysis. DiI-C_{12} imaging is accomplished using a cy3 filter cube (Chroma).
3. When incubating GPMVs with hydrophobic compounds, it is possible to generate equilibrated or supersaturated solutions when using cosolvents such as DMSO or ethanol. To make equilibrated solutions, first add a small volume of a concentrated stock of the compound in solvent into the GPMV solution. This will often cause the compound to precipitate out of solution. Then, add enough cosolvent to redissolve the compound. It is recommended that the exact volumes of cosolvent required for resuspension of the compound be determined in a vesicle free solution, since small sample volumes prevent careful inspection for precipitates. To make supersaturated solutions, first dilute the stock solution in cosolvent so that when added to GPMVs, the desired final concentration of both the compound and the cosolvent is achieved. Then, add this diluted compound directly to GPMVs.
4. We find that DTT strongly impacts GPMV transition temperatures. GPMVs viewed in the absence of DTT often have transition temperatures 10–20°C below vesicles examined in the presence of 2 m*M* DTT.
5. GPMV membranes tend to reduce in size as temperature is lowered. This occurs because the average area per lipid in membranes is temperature dependent, and because the average area per lipid is further reduced upon the formation of the more condensed liquid-ordered phase. When temperature is lowered on freshly prepared vesicles for the first time, vesicle membranes become tense and then slowly relax by squeezing fluid

from the vesicle interior in a way that retains their integrity and spherical shape. Upon raising temperature, the vesicle membrane expands but typically the volume of fluid enclosed within the vesicle remains constant, giving the membrane excess area and allowing the vesicle to take on nonspherical geometries. We typically view vesicles from high to low temperature to avoid complications introduced by this geometric flexibility. Even so, we find that the miscibility transition temperature is typically reversible, although vesicles imaged after temperature cycling tend to be nonspherical above T_{mix} and contain bulging domains below T_{mix} (Fig. 6).

6. Due to the necessity of tabulating the fraction of vesicles containing two phases by hand, user bias is present at some degree, although in our experience this effect does not significantly impact transition temperature determination. To investigate the effects of user bias, we have had multiple researchers analyze the same set of data, and have randomized the images of GPMVs with the researcher blind to the temperature. We find that researchers that are more experienced in identifying vesicle states tend to generate sharper transitions because they are less likely to misidentify individual vesicles, but the midpoint of the transition is not significantly shifted when large numbers of vesicles are analyzed (Fig. 5).

4. SUMMARY AND CONCLUSION

GPMVs are simple to prepare and are easily visualized using wide field or confocal microscopy methods. Also, GPMV membranes can be concentrated and biochemically analyzed for their lipid and protein content (Burns et al., 2017; Fridriksson et al., 1999; Holowka & Baird, 1983; Levental et al., 2016), probed via spectroscopic methods to assay molecular scale structures (Ge et al., 2003), or deposited onto surfaces to make supported bilayers (Richards et al., 2016). Notably, most GPMV preparations contain two coexisting liquid phases when imaged at low temperature (Baumgart et al., 2007), which is remarkable given their biological origin and complex composition. The protocols detailed in the previous sections will lead to consistent yields of vesicles that have reproducible miscibility transition temperatures. These vesicles are model membranes that retain much of the compositional complexity of the intact cell plasma membranes from which they are derived, although have important differences that should be considered when interpreting results. For example, they are no longer coupled to cytoskeleton, and do not generally contain cytoskeleton-associated plasma membrane proteins. They also are depleted of $PI(4,5)P_2$ (Keller, Lorizate, &

Schwille, 2009), and have some signatures of loss of asymmetry, as indicated by PS lipids being exposed on the extracellular face (Baumgart et al., 2007). Even with these important differences, GPMVs are membrane models that much more closely reproduce intact cell plasma membranes than do giant unilamellar vesicles made of purified components. Understanding the similarities (Honerkamp-Smith et al., 2009) and differences (Cornell et al., 2017) between these two model membrane systems could shed light on the molecular mechanisms of membrane organization in cells.

ACKNOWLEDGMENTS

We thank Kathleen Wisser for assistance with cell culture. Research was funded by the NIH (R01GM110052) and the NSF (MCB 1552439).

REFERENCES

Allen, J. A., Halverson-Tamboli, R. A., & Rasenick, M. M. (2007). Lipid raft microdomains and neurotransmitter signalling. *Nature Reviews Neuroscience*, *8*(2), 128–140.

Allender, D. W., & Schick, M. (2017). The effect of solutes on the temperature of miscibility transitions in multi-component membranes. *Biophysical Journal*, *113*, 1814–1821.

Baumgart, T., Hammond, A. T., Sengupta, P., Hess, S. T., Holowka, D. A., Baird, B. A., et al. (2007). Large-scale fluid/fluid phase separation of proteins and lipids in giant plasma membrane vesicles. *Proceedings of the National Academy of Sciences of the United States of America*, *104*(9), 3165–3170.

Burns, M., Wisser, K., Wu, J., Levental, I., & Veatch, S. L. (2017). Miscibility transition temperature scales with growth temperature in a zebrafish cell line. *Biophysical Journal*, *113*, 1212–1222.

Cornell, C. E., McCarthy, N. L. C., Levental, K. R., Levental, I., Brooks, N. J., & Keller, S. L. (2017). n-Alcohol length governs shift in L_o-L_d mixing temperatures in synthetic and cell-derived membranes. *Biophysical Journal*, *113*, 1200–1211.

Dietrich, C., Bagatolli, L. A., Volovyk, Z. N., Thompson, N. L., Levi, M., Jacobson, K., et al. (2001). Lipid rafts reconstituted in model membranes. *Biophysical Journal*, *80*(3), 1417–1428.

Franks, N. P., & Lieb, W. R. (1982). Molecular mechanisms of general anaesthesia. *Nature*, *300*(5892), 487–493.

Franks, N. P., & Lieb, W. R. (1986). Partitioning of long-chain alcohols into lipid bilayers: Implications for mechanisms of general anesthesia. *Proceedings of the National Academy of Sciences of the United States of America*, *83*(14), 5116–5120.

Franks, N. P., & Lieb, W. R. (1990). Mechanisms of general anesthesia. *Environmental Health Perspectives*, *87*, 199–205.

Franks, N. P., & Lieb, W. R. (1991). Stereospecific effects of inhalational general anesthetic optical isomers on nerve ion channels. *Science*, *254*(5030), 427–430.

Fridriksson, E. K., Shipkova, P. A., Sheets, E. D., Holowka, D., Baird, B., & McLafferty, F. W. (1999). Quantitative analysis of phospholipids in functionally important membrane domains from RBL-2H3 mast cells using tandem high-resolution mass spectrometry. *Biochemistry*, *38*(25), 8056–8063.

Ge, M., Gidwani, A., Brown, H. A., Holowka, D., Baird, B., & Freed, J. H. (2003). Ordered and disordered phases coexist in plasma membrane vesicles of RBL-2H3 mast cells. An ESR study. *Biophysical Journal*, *85*(2), 1278–1288.

Gray, E. M., Diaz-Vazquez, G., & Veatch, S. L. (2015). Growth conditions and cell cycle phase modulate phase transition temperatures in RBL-2H3 derived plasma membrane vesicles. *PLoS One*, *10*(9), e0137741.

Gray, E., Karslake, J., Machta, B. B., & Veatch, S. L. (2013). Liquid general anesthetics lower critical temperatures in plasma membrane vesicles. *Biophysical Journal*, *105*(12), 2751–2759.

Gruner, S. M., & Shyamsunder, E. (1991). Is the mechanism of general anesthesia related to lipid membrane spontaneous curvature? *Annals of the New York Academy of Sciences*, *625*, 685–697.

Heerklotz, H. (2002). Triton promotes domain formation in lipid raft mixtures. *Biophysical Journal*, *83*(5), 2693–2701.

Heimburg, T., & Jackson, A. D. (2007). The thermodynamics of general anesthesia. *Biophysical Journal*, *92*(9), 3159–3165.

Holowka, D., & Baird, B. (1983). Structural studies on the membrane-bound immunoglobulin E-receptor complex. 1. Characterization of large plasma membrane vesicles from rat basophilic leukemia cells and insertion of amphipathic fluorescent probes. *Biochemistry*, *22*(14), 3466–3474.

Honerkamp-Smith, A. R., Cicuta, P., Collins, M. D., Veatch, S. L., den Nijs, M., Schick, M., et al. (2008). Line tensions, correlation lengths, and critical exponents in lipid membranes near critical points. *Biophysical Journal*, *95*(1), 236–246.

Honerkamp-Smith, A. R., Veatch, S. L., & Keller, S. L. (2009). An introduction to critical points for biophysicists; observations of compositional heterogeneity in lipid membranes. *Biochimica et Biophysica Acta*, *1788*(1), 53–63.

Ipsen, J. H., Karlstrom, G., Mouritsen, O. G., Wennerstrom, H., & Zuckermann, M. J. (1987). Phase equilibria in the phosphatidylcholine-cholesterol system. *Biochimica et Biophysica Acta*, *905*(1), 162–172.

Jorgensen, K., Ipsen, J. H., Mouritsen, O. G., Bennett, D., & Zuckermann, M. J. (1991). Anesthetics affect membrane heterogeneity. *Annals of the New York Academy of Sciences*, *625*, 747–750.

Jorgensen, K., Ipsen, J. H., Mouritsen, O. G., & Zuckermann, M. J. (1993). The effect of anaesthetics on the dynamic heterogeneity of lipid membranes. *Chemistry and Physics of Lipids*, *65*(3), 205–216.

Kamaya, H., Ueda, I., Moore, P. S., & Eyring, H. (1979). Antagonism between high pressure and anesthetics in the thermal phase-transition of dipalmitoyl phosphatidylcholine bilayer. *Biochimica et Biophysica Acta*, *550*(1), 131–137.

Keller, H., Lorizate, M., & Schwille, P. (2009). PI(4,5)P2 degradation promotes the formation of cytoskeleton-free model membrane systems. *Chemphyschem*, *10*(16), 2805–2812.

Krasowski, M. D., Jenkins, A., Flood, P., Kung, A. Y., Hopfinger, A. J., & Harrison, N. L. (2001). General anesthetic potencies of a series of propofol analogs correlate with potency for potentiation of gamma-aminobutyric acid (GABA) current at the GABA(A) receptor but not with lipid solubility. *The Journal of Pharmacology and Experimental Therapeutics*, *297*(1), 338–351.

Levental, I., Grzybek, M., & Simons, K. (2011). Raft domains of variable properties and compositions in plasma membrane vesicles. *Proceedings of the National Academy of Sciences of the United States of America*, *108*(28), 11411–11416.

Levental, K. R., Lorent, J. H., Lin, X., Skinkle, A. D., Surma, M. A., Stockenbojer, E. A., et al. (2016). Polyunsaturated lipids regulate membrane domain stability by tuning membrane order. *Biophysical Journal*, *110*(8), 1800–1810.

Levental, I., & Veatch, S. (2016). The continuing mystery of lipid rafts. *Journal of Molecular Biology*, *428*(24 Pt. A), 4749–4764.

Li, X., Serwanski, D. R., Miralles, C. P., Bahr, B. A., & De Blas, A. L. (2007). Two pools of Triton X-100-insoluble GABA(A) receptors are present in the brain, one associated to

lipid rafts and another one to the post-synaptic GABAergic complex. *Journal of Neurochemistry, 102*(4), 1329–1345.
Lingwood, D., Ries, J., Schwille, P., & Simons, K. (2008). Plasma membranes are poised for activation of raft phase coalescence at physiological temperature. *Proceedings of the National Academy of Sciences of the United States of America, 105*(29), 10005–10010.
Lingwood, D., & Simons, K. (2010). Lipid rafts as a membrane-organizing principle. *Science, 327*(5961), 46–50.
Lorincz, A., Mihaly, J., Nemeth, C., Wacha, A., & Bota, A. (2015). Effects of ursolic acid on the structural and morphological behaviours of dipalmitoyl lecithin vesicles. *Biochimica et Biophysica Acta, 1848*(5), 1092–1098.
Machta, B. B., Gray, E., Nouri, M., McCarthy, N. L. C., Gray, E. M., Miller, A. L., et al. (2016). Conditions that stabilize membrane domains also antagonize n-alcohol anesthesia. *Biophysical Journal, 111*(3), 537–545.
Machta, B. B., Papanikolaou, S., Sethna, J. P., & Veatch, S. L. (2011). Minimal model of plasma membrane heterogeneity requires coupling cortical actin to criticality. *Biophysical Journal, 100*(7), 1668–1677.
Machta, B. B., Veatch, S. L., & Sethna, J. P. (2012). Critical Casimir forces in cellular membranes. *Physical Review Letters, 109*(13), 138101.
McCarthy, N. L., Ces, O., Law, R. V., Seddon, J. M., & Brooks, N. J. (2015). Separation of liquid domains in model membranes induced with high hydrostatic pressure. *Chemical Communications (Cambridge, England), 51*(41), 8675–8678.
Meyer, H. (1899). Zur Theorie der Alkoholnarkose. *Archiv für Experimentelle Pathologie und Pharmakologie, 42*(2–4), 109–118.
Miller, K. W., Firestone, L. L., Alifimoff, J. K., & Streicher, P. (1989). Nonanesthetic alcohols dissolve in synaptic membranes without perturbing their lipids. *Proceedings of the National Academy of Sciences of the United States of America, 86*(3), 1084–1087.
Mountcastle, D. B., Biltonen, R. L., & Halsey, M. J. (1978). Effect of anesthetics and pressure on thermotropic behavior of multilamellar dipalmitoylphosphatidylcholine liposomes. *Proceedings of the National Academy of Sciences of the United States of America, 75*(10), 4906–4910.
Mullins, L. J. (1954). Some physical mechanisms in narcosis. *Chemical Reviews, 54*(2), 289–323.
Purushothaman, S., Cicuta, P., Ces, O., & Brooks, N. J. (2015). Influence of high pressure on the bending rigidity of model membranes. *The Journal of Physical Chemistry. B, 119*(30), 9805–9810.
Raghunathan, K., Ahsan, A., Ray, D., Nyati, M. K., & Veatch, S. L. (2015). Membrane transition temperature determines cisplatin response. *PLoS One, 10*(10), e0140925.
Richards, M. J., Hsia, C. Y., Singh, R. R., Haider, H., Kumpf, J., Kawate, T., et al. (2016). Membrane protein mobility and orientation preserved in supported bilayers created directly from cell plasma membrane blebs. *Langmuir, 32*(12), 2963–2974.
Samsonov, A. V., Mihalyov, I., & Cohen, F. S. (2001). Characterization of cholesterol-sphingomyelin domains and their dynamics in bilayer membranes. *Biophysical Journal, 81*(3), 1486–1500.
Sankaram, M. B., & Thompson, T. E. (1991). Cholesterol-induced fluid-phase immiscibility in membranes. *Proceedings of the National Academy of Sciences of the United States of America, 88*(19), 8686–8690.
Sarabipour, S., Chan, R. B., Zhou, B., Di Paolo, G., & Hristova, K. (2015). Analytical characterization of plasma membrane-derived vesicles produced via osmotic and chemical vesiculation. *Biochimica et Biophysica Acta, 1848*(7), 1591–1598.
Schroeder, R., London, E., & Brown, D. (1994). Interactions between saturated acyl chains confer detergent resistance on lipids and glycosylphosphatidylinositol (GPI)-anchored proteins: GPI-anchored proteins in liposomes and cells show similar behavior. *Proceedings of the National Academy of Sciences of the United States of America, 91*(25), 12130–12134.

Scott, R. E. (1976). Plasma membrane vesiculation: A new technique for isolation of plasma membranes. *Science, 194*(4266), 743–745.

Seeman, P. (1972). The membrane actions of anesthetics and tranquilizers. *Pharmacological Reviews, 24*(4), 583–655.

Sezgin, E., Kaiser, H. J., Baumgart, T., Schwille, P., Simons, K., & Levental, I. (2012). Elucidating membrane structure and protein behavior using giant plasma membrane vesicles. *Nature Protocols*, 7(6), 1042–1051.

Shieh, D. D., Ueda, I., Lin, H., & Eyring, H. (1976). Nuclear magnetic resonance studies of the interaction of general anesthetics with 1,2-dihexadecyl-sn-glycero-3-phosphorylcholine bilayer. *Proceedings of the National Academy of Sciences of the United States of America, 73*(11), 3999–4002.

Shimshick, E. J., & McConnell, H. M. (1973). Lateral phase separation in phospholipid membranes. *Biochemistry, 12*(12), 2351–2360.

Simons, K., & Ikonen, E. (1997). Functional rafts in cell membranes. *Nature, 387*(6633), 569–572.

Sooksawate, T., & Simmonds, M. A. (2001). Effects of membrane cholesterol on the sensitivity of the GABA(A) receptor to GABA in acutely dissociated rat hippocampal neurones. *Neuropharmacology, 40*(2), 178–184.

Stone, M. B., Shelby, S. A., Nunez, M. F., Wisser, K., & Veatch, S. L. (2017). Protein sorting by lipid phase-like domains supports emergent signaling function in B lymphocyte plasma membranes. *eLife, 6*, e19891.

Trudell, J. R., Payan, D. G., Chin, J. H., & Cohen, E. N. (1975). Antagonistic effect of an inhalation anesthetic and high-pressure on phase-diagram of mixed dipalmitoyl-dimyristoylphosphatidylcholine bilayers. *Proceedings of the National Academy of Sciences of the United States of America, 72*(1), 210–213.

Veatch, S. L., Cicuta, P., Sengupta, P., Honerkamp-Smith, A., Holowka, D., & Baird, B. (2008). Critical fluctuations in plasma membrane vesicles. *ACS Chemical Biology, 3*(5), 287–293.

Veatch, S. L., & Keller, S. L. (2003). Separation of liquid phases in giant vesicles of ternary mixtures of phospholipids and cholesterol. *Biophysical Journal, 85*(5), 3074–3083.

Veatch, S. L., & Keller, S. L. (2005). Seeing spots: Complex phase behavior in simple membranes. *Biochimica et Biophysica Acta, 1746*(3), 172–185.

Vist, M. R., & Davis, J. H. (1990). Phase equilibria of cholesterol/dipalmitoylphosphatidylcholine mixtures: 2H nuclear magnetic resonance and differential scanning calorimetry. *Biochemistry, 29*(2), 451–464.

Zhao, J., Wu, J., Shao, H., Kong, F., Jain, N., Hunt, G., et al. (2007). Phase studies of model biomembranes: Macroscopic coexistence of Lalpha + Lbeta, with light-induced coexistence of Lalpha + Lo phases. *Biochimica et Biophysica Acta, 1768*(11), 2777–2786.

Zhao, J., Wu, J., & Veatch, S. L. (2013). Adhesion stabilizes robust lipid heterogeneity in supercritical membranes at physiological temperature. *Biophysical Journal, 104*(4), 825–834.

Zhou, Y., Maxwell, K. N., Sezgin, E., Lu, M., Liang, H., Hancock, J. F., et al. (2013). Bile acids modulate signaling by functional perturbation of plasma membrane domains. *The Journal of Biological Chemistry, 288*(50), 35660–35670.

SECTION III

In vivo Technologies; Established and Evolving

CHAPTER NINE

Anesthetic Drug Discovery and Development: A Case Study of Novel Etomidate Analogs

Megan McGrath, Douglas E. Raines[1]
Massachusetts General Hospital, Boston, MA, United States
[1]Corresponding author: e-mail address: draines@partners.org

Contents

Abstract

All currently available general anesthetic agents possess potentially lethal side effects requiring their administration by highly trained clinicians. Among these agents is etomidate, a highly potent imidazole-based intravenous sedative-hypnotic that deleteriously suppresses the synthesis of adrenocortical steroids in a manner that is both potent and persistent. We developed two distinct strategies to design etomidate analogs that retain etomidate's potent hypnotic activity, but produce less adrenocortical suppression than etomidate. One strategy seeks to reduce binding to 11β-hydroxylase, a critical enzyme in the steroid biosynthetic pathway, which is potently inhibited by etomidate. The other strategy seeks to reduce the duration of adrenocortical suppression after etomidate administration by modifying the drug's structure to render it susceptible to rapid metabolism by esterases. In this chapter, we describe the methods used to evaluate the hypnotic and adrenocortical inhibitory potencies of two lead compounds designed using the aforementioned strategies. Our purpose is to provide a case study for the development of novel analogs of existing drugs with reduced side effects.

Methods in Enzymology, Volume 603
ISSN 0076-6879
https://doi.org/10.1016/bs.mie.2018.01.026

1. INTRODUCTION

Etomidate is an intravenous sedative-hypnotic agent that was originally invented by Janssen Pharmaceuticals in the 1960s (Fig. 1) (Godefroi, Janssen, Vandereycken, Vanheertum, & Niemegeers, 1965). It emerged from their antifungal agent development program, which involved the synthesis of novel imidazole-containing compounds designed to suppress the biosynthesis of the fungal steroid ergosterol by inhibiting the cytochrome P450 enzyme 14α-demethylase. When tested in rats, etomidate exhibited both potent anesthetic activity and a very high therapeutic index (Godefroi et al., 1965; Janssen, Niemegeers, & Marsboom, 1975). Consequently, it was developed as an anesthetic induction and maintenance agent for use in the operating room, and as a sedative for critically ill patients in the intensive care unit. However, a decade after its introduction into clinical practice in 1972, etomidate was discovered to also produce a profound and persistent suppression of adrenocortical steroid synthesis that was associated with a significant increase in the mortality of critically ill patients who had received prolonged infusions (Ledingham & Watt, 1983; Watt & Ledingham, 1984). Such suppression is caused by etomidate binding to and inhibiting the function of 11β-hydroxylase, a cytochrome P450 enzyme that is found in the adrenal cortex and required for the biosynthesis of cortisol, corticosterone, and aldosterone (Allolio, Stuttmann, Leonhard, Fischer, & Winkelmann, 1984; de Jong, Mallios, Jansen, Scheck, & Lamberts, 1984; Duthie, Fraser, & Nimmo, 1985; Wagner & White, 1984; Wagner, White, Kan, Rosenthal, & Feldman, 1984). So ironically, a drug whose roots were as an antifungal agent intended to act by suppressing the production of ergosterol—a steroid that is absent in animals, but a critical component of the fungal cell membrane—instead potently inhibits a homologous enzyme found in humans and other animals that is required for the biosynthesis of adrenocortical steroids. Because of this undesirable side

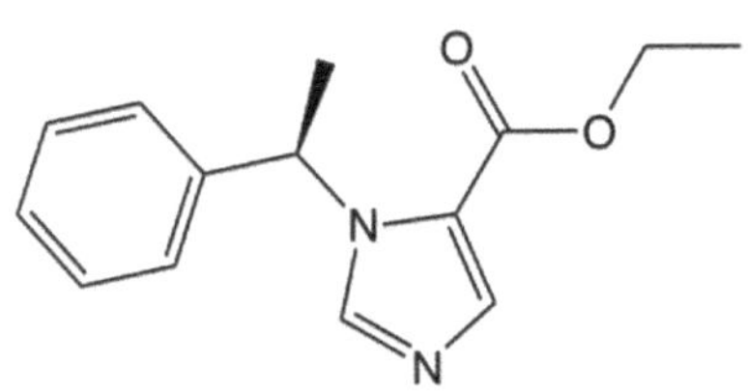

Fig. 1 Molecular structure of etomidate.

effect, etomidate is no longer used as a prolonged infusion to maintain sedation or anesthesia. However, it continues to be used as a single bolus to induce anesthesia because it maintains hemodynamic stability better than any other induction agent (Gooding & Corssen, 1977; Lamalle, 1976; Stockham et al., 1987). Nevertheless, significant concerns remain regarding even that brief use because although such a bolus produces anesthesia for several minutes, the associated suppression of steroid synthesis persists for hours or even days (den Brinker et al., 2008; Vinclair et al., 2007).

Etomidate produces anesthesia by binding to and enhancing the function of γ-aminobutyric acid type A ($GABA_A$) receptors (Forman, 2011). Thus, its anesthetic and steroidogenesis inhibitory activities are mediated by well-characterized but different protein targets with presumably distinct structure–activity relationships. This suggested to us that it might be possible to design etomidate analogs that retain etomidate-like anesthetic activity but are safer because they lack etomidate's ability to produce clinically significant adrenocortical suppression.

2. ETOMIDATE ANALOG DESIGN STRATEGIES

We developed two separate drug design strategies to produce etomidate analogs with reduced abilities to suppress adrenocortical function (Cotten et al., 2010, 2009; Raines, 2015). One strategy was pharmacodynamic in nature, whereas the other one was pharmacokinetic. The pharmacodynamic strategy was to alter etomidate's molecular structure to reduce binding affinity to 11β-hydroxylase, while retaining potent $GABA_A$ receptor positive modulatory activity. The pharmacokinetic strategy was to design a rapidly metabolized "soft" analog of etomidate so that any adrenocortical suppression that is produced by the drug would last for only minutes (rather than hours or days) following drug administration.

2.1 Pharmacodynamic Strategy

X-ray crystallographic structures of drug targets can suggest strategies for designing ligands (e.g., drugs) that bind—or in our case do not bind—with high affinity. Unfortunately, this was not available for 11β-hydroxylase at the time that we began to develop the pharmacodynamic strategy. However, there were published structures for several other P450 enzymes with imidazole-containing ligands bound. A common feature of these structures was the presence of a coordination bond between the aromatic nitrogen in the ligand's imidazole ring and the heme iron at the enzyme's active site

(Ouellet, Podust, & de Montellano, 2008; Scott et al., 2004; Seward, Roujeinikova, McLean, Munro, & Leys, 2006). By analogy, we (and others) hypothesized that a similar coordination bond between the aromatic nitrogen in etomidate's imidazole ring and 11β-hydroxylase's active site heme iron was responsible for etomidate's high binding affinity to this enzyme (Fig. 2) (Roumen et al., 2007). If correct, then etomidate's binding affinity to 11β-hydroxylase could be greatly reduced by replacing this nitrogen with another atom(s) that was unable to form such a bond. This led us to synthesize carboetomidate, an etomidate analog in which the aromatic nitrogen was replaced by a methine (CH) group (Fig. 3).

2.2 Pharmacokinetic Strategy

As noted earlier, a remarkable feature of etomidate-induced adrenocortical suppression is that it persists long after etomidate's anesthetic actions have worn off. This is attributable to etomidate's vastly (~100-fold) higher potency for inhibiting 11β-hydroxylase vs positively modulating $GABA_A$ receptors. A typical anesthetic induction dose of etomidate represents an extreme overdose with respect to the drug's ability to suppress

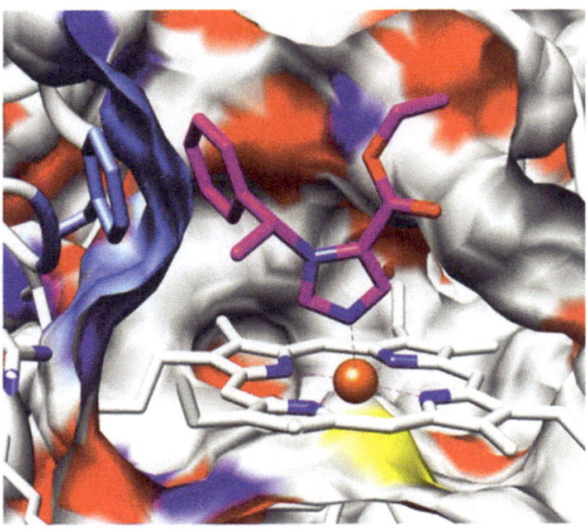

Fig. 2 Homology model of 11β-hydroxlase with etomidate computationally docked. A cross section through the surface of the active site cavity is shown. Etomidate is depicted in *stick representation* with *blue* nitrogens, *red* oxygens, and *purple* carbons. The heme iron is *orange* and coordinates with etomidate's aromatic imidazole nitrogen. The authors thank Drs. Keith W. Miller and Sivananthaperumal Shanmugasundararaj for producing this figure.

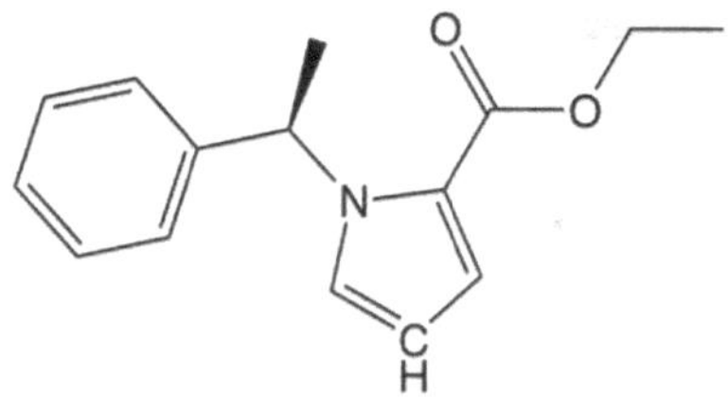

Fig. 3 Carboetomidate is an analog of etomidate in which the aromatic nitrogen in etomidate's imidazole ring has been replaced with a methine group. This disrupts the ability of etomidate to form a coordination bond with the heme iron at the active site of 11β-hydroxylase, lowering its affinity for the enzyme.

adrenocortical function. Thus following such a dose, it takes much longer to recover adrenocortical function than to emerge from anesthesia.

We hypothesized that we could significantly reduce the duration of adrenocortical suppression by modifying etomidate's structure to accelerate the rate at which the drug is metabolized. This "soft analog" approach, which is exemplified by remifentanil and esmolol, often entails the addition of a metabolically labile ester moiety to render the drug highly susceptible to hydrolysis by nonspecific esterases (Fig. 4A) (Bodor & Buchwald, 2000; Buchwald & Bodor, 2002). The resulting carboxylic acid metabolites typically have pharmacological activities that are orders of magnitude lower than their corresponding parent drugs. A key consideration when designing a soft analog is where on the molecule to append the ester moiety. As previous studies had demonstrated that large photoactivatable moieties could be appended distal to etomidate's existing (but metabolically stable) ester without disrupting either $GABA_A$ receptor modulatory or anesthetic activities, we elected to place the ester at this location (Fig. 4B) (Husain et al., 2006, 2010, 2003; Liao et al., 2005). Like remifentanil and esmolol, the ester was linked to the pharmacophore by a spacer composed of two methylene groups to reduce potential steric hindrance and facilitate esterase hydrolysis (Fig. 4C). We named this soft etomidate analog methoxycarbonyl-etomidate (MOC-etomidate).

3. TESTING NOVEL ETOMIDATE ANALOGS FOR HYPNOTIC ACTIVITY USING LOSS OF RIGHTING REFLEXES ASSAYS

When placed supine, animals will reorient themselves to return to their usual upright position. Because this righting reflex is lost when animals

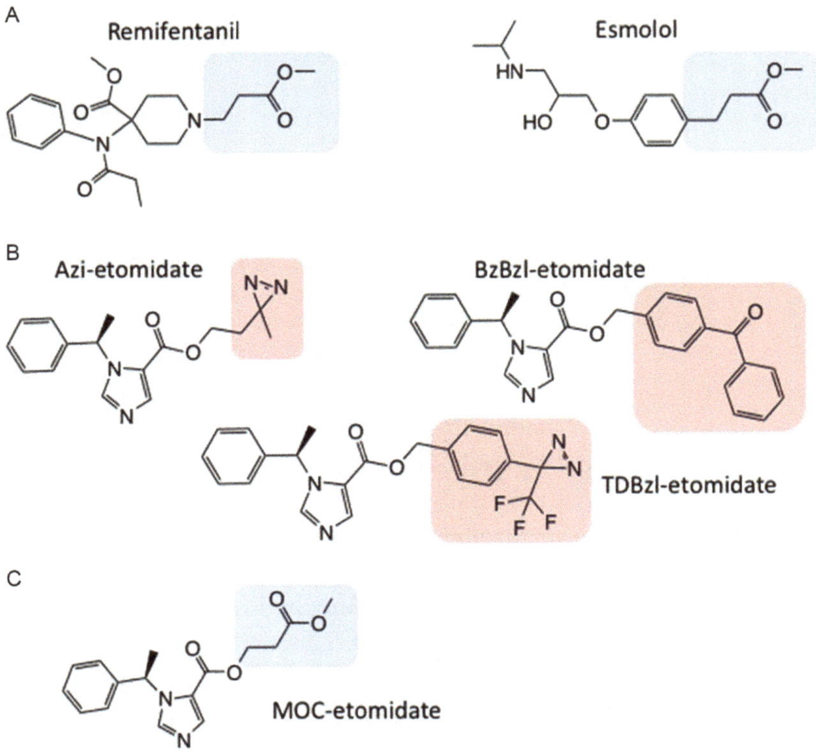

Fig. 4 (A) Molecular structures of remifentanil and esmolol. In each structure, the metabolically labile ester moiety along with the two-carbon linker are *highlighted.* (B) Molecular structures of photoactivatable etomidate analogs. These analogs retain etomidate's potent $GABA_A$ receptor positive modulatory and anesthetic activities. In each structure, the photoactivatable moiety is *highlighted.* (C) Molecular structure of methoxycarbonyl-etomidate (MOC-etomidate).

are anesthetized, it is commonly used as an animal correlate for hypnosis (unconsciousness). To confirm that carboetomidate and MOC-etomidate possessed hypnotic activities and to define their hypnotic potencies, we used tadpole and rat loss of righting reflexes (LoRR) assays.

3.1 Tadpole Loss of Righting Reflex Assay

Although a variety of animals can and have been used as subjects for LoRR assays, small aquatic animals (e.g., tadpoles) offer unique advantages. First, they are inexpensive and easy to care for. Consequently, they allow one to study large numbers of animals easily, quickly, and at low cost. Second, anesthetic actions can be assessed at well-defined steady-state anesthetic concentrations. This may be contrasted with assays in which anesthetics are injected into animals, where anesthetic concentrations in the brain vary constantly over time after administration due to anesthetic distribution, redistribution, and metabolism. Finally, protein binding minimally impacts measurements of anesthetic potencies because the free in vivo anesthetic concentrations approximate the anesthetic concentrations in the aqueous environment. It should be noted though that for very hydrophobic anesthetics, absorption into tadpoles may reduce the aqueous concentration to a value that is significantly below the nominal concentration calculated from the mass of added anesthetic and the aqueous volume. For such drugs, depletion can be minimized by keeping the number of tadpoles per unit aqueous volume low (e.g., 1 tadpole per 100 mL of water). For anesthetics possessing titratable chemical moieties, it is also preferable to buffer the pH at a constant near-physiological value (e.g., pH 7.4) to maintain a constant ratio of charged (i.e., hydrophilic) and uncharged (i.e., hydrophobic) molecular species regardless of the anesthetic concentration as only the uncharged species is likely to contribute significantly to hypnotic effects.

3.1.1 Equipment

- 100- to 1000-mL glass chamber(s)
- Bent glass rod or other device to turn tadpoles supine

3.1.2 Animals and Buffers

- Prelimb *Rana pipens* tadpoles (1–1.5 cm in length, ~100 in number to complete a full concentration–response curve)
- Tris–HCl buffer (2.5 m*M*; pH 7.4) in dechlorinated water

3.1.3 Procedure

1. Add 50–500 mL of the Tris–HCl buffer containing the desired concentration of test compound to the glass chamber.
2. Place five tadpoles in the buffer-containing chamber.
3. After a 10-min incubation time, test each tadpole for LoRR by turning it supine. A tadpole is defined as having LoRR if it does not turn back over to the prone position within 5 s. This test is repeated every 10 min until a stable behavioral response has been reached.
4. Remove tadpoles to fresh distilled water.
5. Tadpoles that had LoRR should be recovered separately to ensure reversibility and thus distinguish hypnosis from toxicity.

Notes

1. We have found that behavioral responses typically stabilize in ~30 min.
2. By defining the effects of a range of test compound concentrations, a concentration–response curve can be generated and a median effective concentration (EC_{50}) can be calculated.

3.2 Rat Loss of Righting Reflex Assay

After confirming that carboetomidate and MOC-etomidate possessed hypnotic activities in tadpoles (albeit at concentrations that were 5–10 times greater than that of etomidate), we evaluated their hypnotic activities in rats using an analogous LoRR. As with the tadpole LoRR assay, the outcome for each individual rat experiment is quantal: after receiving the anesthetic dose, the rat either has LoRR or its does not.

3.2.1 Equipment

- Rat restraint tube
- Volatile anesthetic delivery system (optional)
- 24-G intravenous catheters
- Syringes

3.2.2 Animals and Reagents

- Rats (e.g., Sprague–Dawley)
- Heparin (10 units/mL)

3.2.3 Procedure

1. Gently restrain the rat in the restraint tube with the tail exiting through the back of the tube.
2. Place a 24-G intravenous catheter in one of the lateral tail veins and secure. This may be accomplished in either an awake rat or one that has been anesthetized with a volatile anesthetic (e.g., 3% isoflurane). If a volatile anesthetic has been used to immobilize the rat during catheter placement, then allow at least 1 h for recovery before performing the LoRR test. After catheter placement, fill the catheter void volume with heparinized saline (10 units/mL) to prevent clotting.
3. Administer formulated test compound via syringe through the catheter followed by an 0.5 mL saline flush.
4. Turn the restraint tube containing the rat over to place the rat in a supine position. Open the restraining tube and slide the rat out, keeping the rat in the supine position.

5. A rat is defined as having LoRR if it does not turn back over to the prone position within 5 s.
6. The duration of LoRR is defined as the time from test compound injection until the time that the rat turns over onto all four paws.

Notes

1. Because anesthetics are typically hydrophobic, formulation can be challenging. We have commonly used dimethyl sulfoxide (DMSO) alone or cyclodextrin in water to solubilize test compounds for intravenous injection. Ideal intravenous injection volumes are 0.15–0.3 mL. As DMSO causes hemolysis, it is not uncommon for the urine to become pink tinged following its injection.
2. We attach a saline-flushed 5-in. Microbore Extension Set (Hospira, Lake Forest, IL) to the catheter to facilitate test compound injection and flushing.
3. The onset of LoRR after injecting a test compound with hypnotic activity is typically less than 10 s, although in some cases (e.g., carboetomidate) it may be longer.
4. By defining the effects of a range of test compound doses, a dose–response curve can be generated and a median effective dose (ED_{50}) can be calculated.

These LoRR assays confirmed that both carboetomidate and MOC-etomidate produce hypnosis in tadpoles and rats (Cotten et al., 2010, 2009). Of note, the durations of LoRR produced in rats by even very high doses of MOC-etomidate (e.g., 40 mg/kg, which is 8 × its ED_{50} for LoRR) were extremely short (~3 min), consistent with very rapid in vivo metabolism (Cotten et al., 2009). For comparison, a half-equivalent hypnotic dose of etomidate (4 mg/kg) produces nearly a half hour of LoRR in rats.

4. TESTING NOVEL ETOMIDATE ANALOGS FOR ADRENOCORTICAL INHIBITORY ACTIVITY

4.1 In Vivo Rat Adrenocortical Inhibition Assay

After having determined that carboetomidate and MOC-etomidate retain etomidate-like hypnotic action (and $GABA_A$ receptor positive modulatory activity, see chapter "Combining mutations and electrophysiology to map anesthetic sites on ligand-gated ion channels" by Forman in Vol. 602), we turned our attention to quantifying the extent to which these two etomidate analogs suppress adrenocortical function (Cotten et al., 2010,

2009). We used an adrenocorticotropic hormone (ACTH) stimulation test to evaluate their effects on corticosterone biosynthesis in rats. We pretreat each rat with dexamethasone prior to performing this assay to suppress endogenous ACTH production, thus reducing the baseline serum corticosterone concentration; if the baseline corticosterone concentration is already high due to (for example) the stress of handling, then interpreting the meaning of a similarly high corticosterone concentration after stimulating corticosterone biosynthesis with ACTH becomes difficult.

4.1.1 Equipment

- Rat restraint tube
- Volatile anesthetic delivery system (optional)
- 24-G intravenous catheters
- Centrifuge
- 96-Well plate reader

4.1.2 Animals and Reagents

- Rats (e.g., Sprague–Dawley)
- Dexamethasone
- $ACTH_{1-24}$
- Heparin (10 units/mL)
- Enzyme-linked immunoabsorbant assay (ELISA) kit for the desired adrenocortical steroid

4.1.3 Procedure

1. Place a 24-G intravenous catheter in one of the lateral tail veins and secure as described above under rat LoRR assay.
2. Administer dexamethasone (0.2 mg/kg intravenously) to the rat followed by an 0.5 mL saline flush.
3. Two hours after administering the initial dose of dexamethasone, administer a second dexamethasone dose and the test compound (e.g., carboetomidate or MOC-etomidate) followed by an 0.5 mL saline flush.
4. Administer $ACTH_{1-24}$ (25 μg/kg) intravenously either immediately after the test anesthetic (carboetomidate) or 15 min later (MOC-etomidate) followed by an 0.5 mL saline flush. For MOC-etomidate, the 15-min delay after administering $ACTH_{1-24}$ provides the brief time required for metabolism to convert the drug to its carboxylic acid metabolite.

5. Fifteen minutes after administering $ACTH_{1-24}$, draw a blood sample (~0.3 mL) from the intravenous catheter and allow it to clot at room temperature.
6. Centrifuge the clotted blood sample at 16,000 × *g* for 5–15 min to separate the serum from red blood cells and other cellular debris.
7. Determine the corticosterone concentrations in serum using an ELISA.

Notes

1. We use etomidate as a comparator for all etomidate analogs. An etomidate dose of 0.5–1 mg/kg will reduce the adrenocortical response by approximately half (Pejo et al., 2014).
2. After collection, serum may be frozen and stored at −20°C until the corticosterone concentration is measured. When we anticipate running multiple ELISAs on different days using a single blood sample, then we aliquot the blood into several vials that can be separately thawed to avoid multiple freeze–thaw cycles.
3. The concentrations of other adrenocortical steroids can also be measured from the same serum samples using the appropriate ELISA kit.
4. Many ELISAs can also be done with plasma. If using plasma, add the blood sample to a heparinized or EDTA-coated vacuum tube (e.g., BD Vacutainer). Gently invert the tube 5–10 times to ensure complete mixing with the anticoagulant. Immediately centrifuge at 16,000 × *g* for 5 min and then aliquot the plasma into Eppendorf tubes for freezing at −20°C until ready to use.

After administering a single bolus using a dose sufficient to produce LoRR, neither carboetomidate nor MOC-etomidate significantly reduced ACTH-stimulated serum corticosterone concentrations in rats.

4.2 In Vitro ^{3}H-Etomidate Competition Assay

We designed carboetomidate as an etomidate analog with low affinity for the etomidate binding site on 11β-hydroxylase, and results from both our rat adrenocortical inhibition assay and our computational docking studies suggested that we had achieved that goal (Cotten et al., 2010; Pejo et al., 2011; Shanmugasundararaj et al., 2013). As further confirmation, we developed a competitive binding assay between carboetomidate and ^{3}H-etomidate (Pejo, Zhou, Husain, & Raines, 2016). In this assay, ^{3}H-etomidate is used at a low nanomolar concentration to selectively bind to 11β-hydroxylase in adrenal tissue. The desired concentration of etomidate or carboetomidate (or any other test compound) is added, and after equilibration the mixture is filtered to separate free from enzyme-bound ^{3}H-etomidate. Bound ^{3}H-etomidate, which is retained on the filter, is then quantified by scintillation counting.

4.2.1 Equipment

- Centrifuge
- Liquid scintillation counter
- Electric homogenizer
- Glass and teflon homogenizer
- Vacuum filter
- Disposable culture tubes (borosilicate glass; $13 \times 100\,mm^2$)

4.2.2 Tissue, Buffers, and Reagents

- Fresh or freshly frozen adrenal glands
- Preparation buffer: 10 m*M* HEPES, 1 m*M* ethylenediaminetetracetic acid, 10 μg/mL leupeptin, 10 μg/mL chymostatin, 10 μg/mL pepstatin A, 2 μg/mL aprotinin, 1 m*M* polymethanesulfonyl fluoride, and 10 μL/mL ethanol
- Diisopropyl fluorophosphates (DFP)
- 10-mL syringe
- 23-G needle
- Bicinchoninic acid (BCA) protein assay kit
- ^{3}H-etomidate (specific activity ~10 Ci/m*M*)
- GF/B glass fiber filters
- Polyethylenimine
- Assay buffer: 11.9 m*M* phosphates, 137 m*M* NaCl, 200 m*M* KCl, 1 m*M* ethylenediaminetetracetic acid, and pH 7.4

4.2.3 Procedure

1. Preparation of adrenal homogenate
   a. Cut the adrenal tissue into small pieces. Add the pieces to ice-cold preparation buffer (10 mL buffer per 1 g of adrenal tissue, rounded up). Homogenize the mixture on ice first using an electric homogenizer (6 × for 5 s each) and then by hand using a glass and teflon homogenizer (20 passes).
   b. Add DFP (1 m*M* final concentration) and incubate for 60 min at room temperature to inactivate esterases.
   c. Centrifuge the mixture (4°C, 43,100 × *g*) for 30 min.

   d. Discard the supernatant and resuspend the pellet in fresh, cold preparation buffer (10 mL buffer per 1 g of adrenal tissue).
   e. Repeat the entire homogenization process with electric and manual homogenizers.
   f. Centrifuge the mixture (4°C, 43,100 × *g*) for 30 min.
   g. Resuspend the pellet in 5–8 mL of cold preparation buffer and briefly (5–10 s) mix with an electric homogenizer three times.
   h. Transfer the homogenate to a clean vial. Using a 10-mL syringe, draw and expel the homogenate through a 23-G needle three times to break up residual particulate matter.
   i. Aliquot 0.5–1 mL into 1.5-mL eppendorf tubes and store at −80°C until used. Save an 0.1 mL aliquot to quantify the protein concentration in the homogenate using the BCA kit.
2. Competition assay
   a. Prepare 25 mm GF/B glass fiber filters at least 2 h in advance by soaking them in water containing 0.5% polyethylenimine.
   b. Prepare a 2-mL solution of ^{3}H-etomidate (final concentration 2 n*M*), adrenal homogenate (final concentration 0.07 mg/mL), and the desired concentration of test compound in assay buffer.
   c. Gently mix the solution and allow it to equilibrate at room temperature for 15–30 min.
   d. Vortex adrenal solution for 5 s immediately before filtering. Pass 0.5 mL of the mixture through a presoaked 25 mm GF/B glass fiber filter under suction. Each solution can be run in triplicates by adding 0.5 mL to three different filters.
   e. Wash the filter immediately with 5 mL of buffer. Allow the buffer to pass through the filter and repeat wash.
   f. Dry the filters under a heat lamp for 2 h and transfer to a scintillation vial containing Liquiscint scintillation cocktail.
   g. Quantify tritium radioactivity by liquid scintillation counting.

Notes

1. DFP is highly toxic. Use extreme care and adequate protection.
2. Displacement of ^{3}H-etomidate by a high concentration of etomidate (e.g., 100 μ*M*) is >95%, indicative of minimal nonspecific ^{3}H-etomidate binding.
3. Do not subject adrenal homogenates to repeated freeze–thaw cycles as this will reduce ^{3}H-etomidate binding. We thaw homogenates in cold water and then keep them on ice until used (always within 6 h of thawing).

4. Membrane proteins are susceptible to aggregation and degradation. Therefore, we mix solutions frequently but gently, and minimize vortexing.
5. By defining the extent of ^{3}H-etomidate displacement produced by a range of test compound concentrations, a concentration–response curve can be generated and a half-inhibitory concentration (IC_{50}) can be calculated.

We determined that the half-inhibitory concentrations of etomidate and carboetomidate were 26 n*M* and 50 μ*M*, respectively. Thus, carboetomidate has an affinity for the etomidate binding site on 11β-hydroxylase that is 2000 × lower than etomidate.

5. SUMMARY AND CONCLUSION

In this chapter, we have presented two distinct strategies for producing etomidate analogs that retain etomidate's ability to positively modulate $GABA_A$ receptors and produce hypnosis, but with substantially less ability to suppress adrenocortical steroid synthesis. We used a tadpole LoRR assay to detect hypnotic activity and to define the steady-state concentrations required to produce hypnosis. Such information is critical for defining the relevant concentration ranges for performing in vitro studies (e.g., $GABA_A$ receptor physiology). We also used a rat LoRR to define the dosing requirement for producing hypnosis in a small mammal and to assess each analog's duration of action after single bolus administration. The abilities of etomidate analogs to bind to 11β-hydroxylase and suppress adrenocortical steroid synthesis were defined using an ^{3}H-etomidate displacement assay and an ACTH stimulation test, respectively. Together, these studies showed that carboetomidate positively modulates $GABA_A$ receptors and produces hypnosis, but does not suppress adrenocortical steroid synthesis because it binds to 11β-hydroxylase with an affinity that is three orders of magnitude lower than etomidate. MOC-etomidate similarly positively modulates $GABA_A$ receptors and produces hypnosis, but does not produce measurable suppression of adrenocortical steroid synthesis because it is so rapidly metabolized. This rapid metabolism also confers upon MOC-etomidate a very brief duration of hypnotic action in rats even after the administration of a very high bolus dose. Additional soft etomidate analogs have been made since the development of MOC-etomidate with properties that are optimized for clinical use. One of these analogs (CPMM, also known as ABP-700) has reached human trials (Struys et al., 2017).

ACKNOWLEDGMENTS

Funded by Grants GM087316 and GM58448 from the National Institutes of Health, Bethesda, MD, and the Department of Anesthesia, Critical Care, and Pain Medicine, Massachusetts General Hospital, Boston, MA.

REFERENCES

Allolio, B., Stuttmann, R., Leonhard, U., Fischer, H., & Winkelmann, W. (1984). Adrenocortical suppression by a single induction dose of etomidate. *Klinische Wochenschrift, 62*(21), 1014–1017.

Bodor, N., & Buchwald, P. (2000). Soft drug design: General principles and recent applications. *Medicinal Research Reviews, 20*(1), 58–101.

Buchwald, P., & Bodor, N. (2002). Physicochemical aspects of the enzymatic hydrolysis of carboxylic esters. *Pharmazie, 57*(2), 87–93.

Cotten, J. F., Forman, S. A., Laha, J. K., Cuny, G. D., Husain, S. S., Miller, K. W., et al. (2010). Carboetomidate: A pyrrole analog of etomidate designed not to suppress adrenocortical function. *Anesthesiology, 112*(3), 637–644.

Cotten, J. F., Husain, S. S., Forman, S. A., Miller, K. W., Kelly, E. W., Nguyen, H. H., et al. (2009). Methoxycarbonyl-etomidate: A novel rapidly metabolized and ultra-short-acting etomidate analogue that does not produce prolonged adrenocortical suppression. *Anesthesiology, 111*(2), 240–249.

de Jong, F. H., Mallios, C., Jansen, C., Scheck, P. A., & Lamberts, S. W. (1984). Etomidate suppresses adrenocortical function by inhibition of 11 beta-hydroxylation. *The Journal of Clinical Endocrinology and Metabolism, 59*(6), 1143–1147.

den Brinker, M., Hokken-Koelega, A. C., Hazelzet, J. A., de Jong, F. H., Hop, W. C., & Joosten, K. F. (2008). One single dose of etomidate negatively influences adrenocortical performance for at least 24h in children with meningococcal sepsis. *Intensive Care Medicine, 34*(1), 163–168.

Duthie, D. J., Fraser, R., & Nimmo, W. S. (1985). Effect of induction of anaesthesia with etomidate on corticosteroid synthesis in man. *British Journal of Anaesthesia, 57*(2), 156–159.

Forman, S. A. (2011). Clinical and molecular pharmacology of etomidate. *Anesthesiology, 114*(3), 695–707.

Godefroi, E. F., Janssen, P. A., Vandereycken, C. A., Vanheertum, A. H., & Niemegeers, C. J. (1965). Dl-1-(1-Arylalkyl)Imidazole-5-carboxylate Esters. A novel type of hypnotic agents. *Journal of Medicinal Chemistry, 8*, 220–223.

Gooding, J. M., & Corssen, G. (1977). Effect of etomidate on the cardiovascular system. *Anesthesia and Analgesia, 56*(5), 717–719.

Husain, S. S., Nirthanan, S., Ruesch, D., Solt, K., Cheng, Q., Li, G. D., et al. (2006). Synthesis of trifluoromethylaryl diazirine and benzophenone derivatives of etomidate that are potent general anesthetics and effective photolabels for probing sites on ligand-gated ion channels. *Journal of Medicinal Chemistry, 49*(16), 4818–4825.

Husain, S. S., Stewart, D., Desai, R., Hamouda, A. K., Li, S. G., Kelly, E., et al. (2010). *p*-Trifluoromethyldiazirinyl-etomidate: A potent photoreactive general anesthetic derivative of etomidate that is selective for ligand-gated cationic ion channels. *Journal of Medicinal Chemistry, 53*(17), 6432–6444.

Husain, S. S., Ziebell, M. R., Ruesch, D., Hong, F., Arevalo, E., Kosterlitz, J. A., et al. (2003). 2-(3-Methyl-3H-diaziren-3-yl)ethyl 1-(1-phenylethyl)-1H-imidazole-5-carboxylate: A derivative of the stereoselective general anesthetic etomidate for photolabeling ligand-gated ion channels. *Journal of Medicinal Chemistry, 46*(7), 1257–1265.

Janssen, P. A., Niemegeers, C. J., & Marsboom, R. P. (1975). Etomidate, a potent non-barbiturate hypnotic. Intravenous etomidate in mice, rats, guinea-pigs, rabbits and dogs. *Archives Internationales de Pharmacodynamie et de Thérapie*, *214*(1), 92–132.

Lamalle, D. (1976). Cardiovascular effects of various anesthetics in man. Four short-acting intravenous anesthetics: Althesin, etomidate, methohexital and propanidid. *Acta Anaesthesiologica Belgica*, 27(suppl), 208–224.

Ledingham, I. M., & Watt, I. (1983). Influence of sedation on mortality in critically ill multiple trauma patients. *Lancet*, *1*(8336), 1270.

Liao, M., Sonner, J. M., Husain, S. S., Miller, K. W., Jurd, R., Rudolph, U., et al. (2005). R (+) etomidate and the photoactivable R (+) azietomidate have comparable anesthetic activity in wild-type mice and comparably decreased activity in mice with a N265M point mutation in the gamma-aminobutyric acid receptor beta3 subunit. *Anesthesia and Analgesia*, *101*(1), 131–135.

Ouellet, H., Podust, L. M., & de Montellano, P. R. (2008). Mycobacterium tuberculosis CYP130: Crystal structure, biophysical characterization, and interactions with antifungal azole drugs. *The Journal of Biological Chemistry*, *283*(8), 5069–5080.

Pejo, E., Feng, Y., Chao, W., Cotten, J. F., Le Ge, R., & Raines, D. E. (2011). Differential effects of etomidate and its pyrrole analogue carboetomidate on the adrenocortical and cytokine responses to endotoxemia. *Critical Care Medicine*, *40*(1), 187–192.

Pejo, E., Santer, P., Jeffrey, S., Gallin, H., Husain, S. S., & Raines, D. E. (2014). Analogues of etomidate: Modifications around etomidate's chiral carbon and the impact on in vitro and in vivo pharmacology. *Anesthesiology*, *121*(2), 290–301.

Pejo, E., Zhou, X., Husain, S. S., & Raines, D. E. (2016). Sedative-hypnotic binding to 11beta-hydroxylase. *Anesthesiology*, *125*(5), 943–951.

Raines, D. E. (2015). The pharmacology of etomidate and etomidate derivatives. *International Anesthesiology Clinics*, *53*(2), 63–75.

Roumen, L., Sanders, M. P., Pieterse, K., Hilbers, P. A., Plate, R., Custers, E., et al. (2007). Construction of 3D models of the CYP11B family as a tool to predict ligand binding characteristics. *Journal of Computer-Aided Molecular Design*, *21*(8), 455–471.

Scott, E. E., White, M. A., He, Y. A., Johnson, E. F., Stout, C. D., & Halpert, J. R. (2004). Structure of mammalian cytochrome P450 2B4 complexed with 4-(4-chlorophenyl) imidazole at 1.9-A resolution: Insight into the range of P450 conformations and the coordination of redox partner binding. *The Journal of Biological Chemistry*, *279*(26), 27294–27301.

Seward, H. E., Roujeinikova, A., McLean, K. J., Munro, A. W., & Leys, D. (2006). Crystal structure of the Mycobacterium tuberculosis P450 CYP121-fluconazole complex reveals new azole drug-P450 binding mode. *The Journal of Biological Chemistry*, *281*(51), 39437–39443.

Shanmugasundararaj, S., Zhou, X., Neunzig, J., Bernhardt, R., Cotten, J. F., Ge, R., et al. (2013). Carboetomidate: An analog of etomidate that interacts weakly with 11beta-hydroxylase. *Anesthesia and Analgesia*, *116*(6), 1249–1256.

Stockham, R. J., Stanley, T. H., Pace, N. L., King, K., Groen, F., & Gillmor, S. T. (1987). Induction of anesthesia with fentanyl or fentanyl plus etomidate in high-risk patients. *Journal of Cardiothoracic Anesthesia*, *1*(1), 19–23.

Struys, M., Valk, B. I., Eleveld, D. J., Absalom, A. R., Meyer, P., Meier, S., et al. (2017). A phase 1, single-center, double-blind, placebo-controlled study in healthy subjects to assess the safety, tolerability, clinical effects, and pharmacokinetics-pharmacodynamics of intravenous cyclopropyl-methoxycarbonylmetomidate (ABP-700) after a single ascending bolus dose. *Anesthesiology*, *127*(1), 20–35.

Vinclair, M., Broux, C., Faure, P., Brun, J., Genty, C., Jacquot, C., et al. (2007). Duration of adrenal inhibition following a single dose of etomidate in critically ill patients. *Intensive Care Medicine*, *34*(4), 714–719.

Wagner, R. L., & White, P. F. (1984). Etomidate inhibits adrenocortical function in surgical patients. *Anesthesiology, 61*(6), 647–651.
Wagner, R. L., White, P. F., Kan, P. B., Rosenthal, M. H., & Feldman, D. (1984). Inhibition of adrenal steroidogenesis by the anesthetic etomidate. *The New England Journal of Medicine, 310*(22), 1415–1421.
Watt, I., & Ledingham, I. M. (1984). Mortality amongst multiple trauma patients admitted to an intensive therapy unit. *Anaesthesia, 39*(10), 973–981.

CHAPTER TEN

Optoanesthesia: Use of Anesthetic Photolabels In Vivo

Andrew R. McKinstry-Wu[1], Max B. Kelz
University of Pennsylvania, Philadelphia, PA, United States
[1]Corresponding author: e-mail address: andrew.mckinstry-wu@uphs.upenn.edu

Contents

Abstract

Investigation of how anesthetics produce hypnosis requires knowledge of their effects at the molecular, neuronal, circuit, and whole-brain network level. Anesthetic photolabels have long been used to explore how anesthetics bind and affect known protein targets, but they could potentially assist in investigation of anesthetic effects at higher organizational levels of the central nervous system. Here, we advocate the use and provide detailed methods for the application of anesthetic photolabels in slice electrophysiology and in intact animals as a means of investigating anesthetic effects on distinct circuits and brain centers.

1. INTRODUCTION

Investigating the mechanism of anesthetic action is a particularly thorny problem, not only because general anesthetics can effect changes at multiple organizational levels in the central nervous system (CNS), but also because discerning which changes are necessary or sufficient to produce

Methods in Enzymology, Volume 603
ISSN 0076-6879
https://doi.org/10.1016/bs.mie.2018.01.021

an anesthetic state are not yet clear. Any approach that hopes to tackle how anesthetics produce hypnosis necessarily must address effects at these various organizational levels: specific proteins affected, changes in neuronal function, circuit changes, and network changes. It must also tie those effects to the ultimate behavioral phenotype, hypnosis itself. Existing strategies to address complexities of anesthetic effects on higher-order structures such as circuits and networks are mostly comprised of local administrations by microinjection or infusions of anesthetics at specific structures in the CNS (Correa-Sales, Rabin, & Maze, 1992; Minert, Yatziv, & Devor, 2017; Nelson et al., 2002). This strategy accounts for anesthetic effects possibly arising from combined effects on multiple neuronal types and has previously proven relatively effective. The primary disadvantages of this approach are poor control of concentrations, and diffusion of anesthetic away from the area of investigation resulting in effects from drug at areas external to the region of interest (particularly relevant if the area under investigation is near to a ventricle).

Here, we outline methods for investigation of mechanisms of anesthetic-induced hypnosis using anesthetic photolabels in vivo in specific regions of intact animals or in slice, a top-down approach we have dubbed optoanesthesia. Anesthetic photolabels are hypnotics that convert to highly reactive species with exposure to a specific wavelength of light converting the typical extremely transient anesthetic protein–target interaction to a covalent linkage. While the inhaled anesthetic halothane has intrinsic photolabel properties, other photolabel anesthetics have been created by modifying the base structure of anesthetic compounds to include photoactive groups. These agents have been successfully used in vitro to identify anesthetic binding sites on known relevant targets of anesthetics (Chiara, Dangott, Eckenhoff, & Cohen, 2003; Chiara et al., 2013; Li et al., 2006; Yip et al., 2013). Photolabel anesthetics that additionally contain a chemically modifiable tag have been used to identify previously unrecognized protein targets of anesthetics (as well as labeling known targets) (Woll et al., 2016). Anesthetic photolabels have further been employed to explore functional changes in known protein targets of anesthetics using artificial expression in electrophysiologic models (Zhong, Rüsch, & Forman, 2008). Weiser et al. first described the use of anesthetic photolabels in an in vivo system, *Xenopus laevis* tadpoles, where illuminating the entire organism extended hypnosis in an illumination duration-dependent manner (Weiser, Kelz, & Eckenhoff, 2012). We have extended this in vivo approach to a mammalian (rodent) model where illumination of limited CNS regions causes irreversible covalent binding of anesthetic photolabel in those regions, while in all other regions drug diffuses away and is eliminated. We additionally propose an extension of this method in slice electrophysiology to limit drug effect to one portion of the slice. These approaches can be extended to use previously described anesthetic photolabel molecular target identification methods such that organismal effects may be translated to regional brain anesthetic effects and molecular targets within those relevant regions (Woll et al., 2016).

2. ANESTHETIC PHOTOLABEL AVAILABILITY

These novel compounds are not available commercially, but small amounts are available, generally as a collaboration, from a limited number of academic laboratories. Validated volatile anesthetic and propofol analogues can be obtained from the Eckenhoff Lab at the University of Pennsylvania; etomidate, barbiturate, and alcohol analogues from the Miller Lab at Harvard; and the neurosteroid analogues from the Evers Lab at Washington University. The amounts made available are typically tiny, and great care must be taken in their deployment.

3. ANESTHETIC EFFECTS IN A CIRCUMSCRIBED REGION

The use of photolabel anesthetics in slice and in vivo has the advantage of acting on anesthetic targets in a similar method to their parent compounds, while being able to limit the extent of that action to regions of photoadduction after washout of anesthetic drug from other parts of the brain or slice occurs (Woll, Dailey, Brannigan, & Eckenhoff, 2016). Our in vivo and slice photolabeling studies employ *m*-azipropofol (aziPm), an alkylphenol with a photoactive diazirine ring. Peak absorbance of the diazirine ring occurs at 367 nm, with sufficient near-ultraviolet (UVC) light converting the diazirine ring into nitrogen gas and a highly reactive carbene that undergoes rapid covalent binding with virtually any nearby moiety. The limit of the extent of photoadduction is thus primarily dependent on the spread, duration, and intensity of UVC illumination. Fortunately, brain tissue readily absorbs wavelengths in the UVC spectrum, such that spread by both penetration and scatter is limited, even with an intracerebral light source (Fig. 1). For slice electrophysiology, UVC light can either be directed through the microscope optics in order to illuminate only the portion of the slice within view when photoadducting the neuron under recording, or a separate external fiber may be mounted and used to illuminate a portion of the slice external to the field of view.

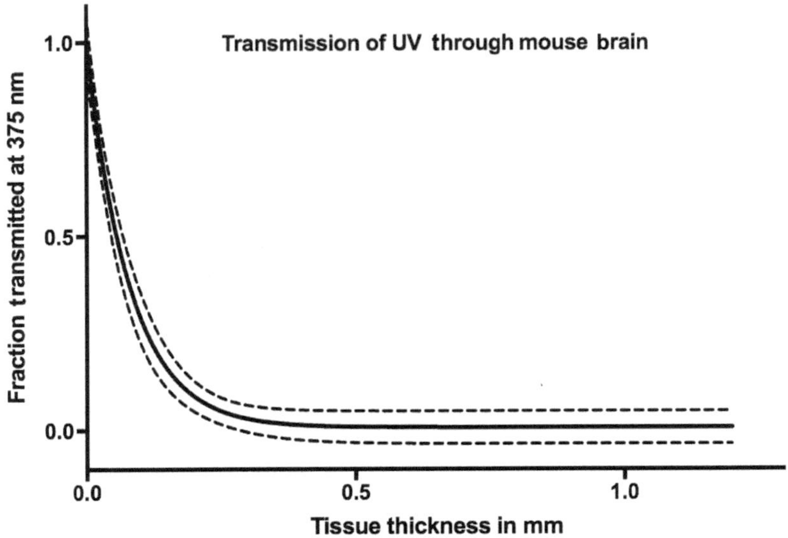

Fig. 1 Penetrance of ultraviolet light through brain tissue. 375 nm light absorbance as a function of tissue depth. A 50 mW laser was projected through varying thicknesses of murine brain tissue and the resulting decrease in power plotted.

4. SLICE ELECTROPHYSIOLOGY

Anesthetic photolabels facilitate investigation of circuit-level anesthetic effects through slice electrophysiology. Whereas drugs are generally introduced into the artificial cerebrospinal fluid and thus applied across the entire slice during recording, photoadducting drug in one portion of the slice and subsequently allowing washout can extend the drug effect for the illuminated portion of the slice while the remainder of the slice no longer experiences direct effects of the drug. UV illumination is achieved either through an integrated 120 mW 365 nm LED light source (Thorlabs M365L2-C1, Newton, NJ) that illuminates through the microscope optics, or through a multimode fiber attached to a second arm on the recording stage, which is in turn coupled to 200 mW 375 nm laser (Power Technology Inc., Alexander, AK). The investigation of two regions that project to each other requires careful selection of the plane in which to cut the slice so that one can preserve those connections (Tóth, Freund, & Miles, 1997). Slice electrophysiological recordings are generally performed as described previously (Beck, Pan, Akanwa, & Kirby, 2004; Moore et al., 2012), in brief:

- Sacrifice mice by cervical dislocation followed by decapitation, place the head in ice-cold preoxygenated sucrose solution containing of 248 m*M* sucrose, 2.5 m*M* KCl, 1.25 m*M* NaH_2PO_4, 2.0 m*M* $MgSO_4$, 2.5 m*M* $CaCl_2$, 10 m*M* dextrose, and 26 m*M* $NaHCO_3$.
- Rapidly remove the brain and slice 200-μm thick slices in the appropriate plane using a Leica microslicer (Leica, Allendale, NJ), while continuously oxygenating the brain in the ice-cold sucrose solution.
- Transfer sections to a holding chamber containing artificial cerebrospinal fluid (aCSF, consisting of 124 m*M* NaCl, 2.5 m*M* KCl, 1.25 m*M* NaH_2PO_4, 2.0 m*M* $MgSO_4$, 2.5 m*M* $CaCl_2$, 10 m*M* dextrose, and 26 m*M* $NaHCO_3$) that is continuously bubbled with 95% O_2/5% CO_2. Keep the slices in the chamber at 37°C for 1°h, followed by room temperature until recording.
- Whole-cell voltage and current clamp recordings are conducted as described previously (Beck et al., 2004; Moore et al., 2012). The slice is continuously perfused with oxygenated aCSF at 2 mL/min at 32°C. After visualizing a cell in the area of interest, make a gigaohm seal with a recording electrode and rupture the membrane to obtain a whole-cell recording. The electrodes should have a resistance of 5–8 MΩ when filled with an intracellular solution of 130 m*M* K-gluconate, 5 m*M* NaCl, 10 m*M* phosphocreatine disodium salt, 1 m*M* $MgCl_2$, 10 m*M* HEPES, 0.02 m*M* EGTA, 0.5 m*M* Na_2GTP, 2 m*M* MgATP, and 0.1% biocytin (pH 7.3, 280–290 mOsm).
- After patching, switch to current clamp recording cells should be characterized as previously described, followed by 10 min of baseline recording obtained.
- Photolabel anesthetics should be diluted from stock in fresh oxygenated aCSF immediately prior to use.
- After baseline recording, expose the cells of interest to UVC stimulation alone. For either source of illumination, the goal is a 30 s exposure at 1 Hz with an 80% duty cycle and an intensity of 60 mW/mm^2.
- Allow for another 3 min of recording after UVC exposure to ensure cell remains at baseline.
- Turn on injection pump and provide the slice with 3 min of anesthetic photolabel in aCSF, followed by a 20 min washout period.
- Once cell has returned to baseline, turn on the injection pump again to give 3 min of anesthetic photolabel in aCSF, this time illuminating for 30 s as above once slice has been exposed to drug for 1 min.
- Perform washout, recording for 2 h or cell is no longer viable.

5. PHOTOADDUCTION IN INTACT ANIMALS

5.1 Measures of Hypnosis and Sedation in a Mouse Model

The singular advantage of working in a top-down manner for investigating anesthetic action is the possibility of evaluating the most relevant anesthetic endpoint, hypnosis, which as a behavioral phenotype can *only* be evaluated at the level of the intact organism. While the classic behavioral measure in animal models that serves as a proxy for loss and return of consciousness is the righting reflex, it is limited. The binary output can fail to capture submaximal sedation effects that contribute to but are not sufficient to produce hypnosis, and may produce false positives if there is loss of motor tone without sedation or hypnosis. To address these limitations, we propose examining three behavioral measures of hypnosis and sedation. In addition to the righting reflex, we use video tracking to record distance traveled by, and speed of, the mouse, and a response to a very mild aversive stimulus, an adhesive sticker placed on the mouse's nose. We additionally use electroencephalographic (EEG) recording to look for known signatures of sedation, including increased power in the delta frequency range (0–4 Hz). Finally, we also continuously record cervical electromyelograms (EMG) in order to distinguish a general loss of motor tone from sedation or hypnosis.

5.2 Method for In Vivo Photoadduction With Behavioral and EEG Sedative-Hypnotic Endpoints

- Adult mice are implanted with stereotaxically targeted guide cannulae (PlasticsOne, Roanoke, VA) directed bilaterally to structures of interest and simultaneously implanted with dural EEG leads and with cervical electromyographic wires using techniques previously described, though with reduced lead number to accommodate the cannula (Wasilczuk, Proekt, Kelz, & McKinstry-Wu, 2016). We also implant an indwelling jugular catheter tunneled to the mouse's back, as described by Kmiotek, Baimel, and Gill (2012)
    - o If the targeted structures are cortical, rather than deep, bilateral cranial windows can be created by carefully drilling out the skull above the cortical region of interest and placing miniature circular coverslips with high-UVC transmittance (Warner Instruments, Hamden, CT) over the windows. The coverslips must be secured by careful placement of dental cement along their edges. Placement of these coverslips

will likely require altering your EEG lead locations to accommodate the windows.

- Mice are allowed to recover for 1 week. In the second week after surgery, they are habituated to the testing box (a plexiglass box 15 × 25 cm with an overhead video camera [Dinolite VGA, New Tapei City, Taiwan] capturing 5 fps of 640 × 480 pixel images) while tethered to an EEG recording cable. They additionally undergo daily training for nasal tape removal (Fig. 2).
- Note: Details of laser exposure and drug dosing are specific to aziPm, other photolabels may require different wavelengths, exposure power, or exposure time.
- Two weeks after surgery, the mouse is restrained, bilateral multimode fibers (Thorlabs, Newton, NJ) coupled to a 200 mW 375 nm laser (Power Technology Inc., Alexander, AK) are inserted into the guide cannulae and secured, the mouse is tethered to the recording cable via an amplifier chip also containing a 3-axis accelerometer (Intan Technologies, Los Angeles, CA) and the central venous catheter is attached to a syringe pump containing the anesthetic photolabel.
  - For cortical windows, the laser is manually targeted at the window rather than placing the fiber in the guide cannula.
  - All acquisition of biopotentials (EEG/EMG) used open-source Open-Ephys hardware (http://www.open-ephys.org works) via Intan amplifier chips.

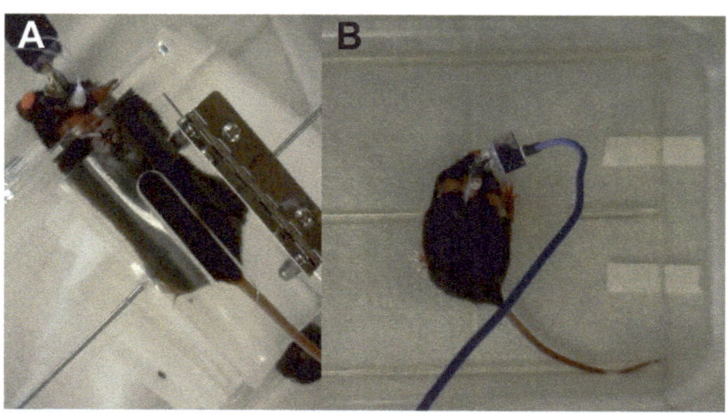

Fig. 2 Optoanesthesia in the intact animal. The mouse is restrained for optical fiber placement and attaching the syringe pump to the venous catheter (A) and subsequently allowed to move freely in the cage (B).

- The mouse is placed in the testing box, video recording is started, and baseline activity and EEG/EMG are recorded for 10min. At the end of the 10min, a sticker is placed on the mouse's nose for a tape test trial.
- The laser is turned on at 10mW at 1Hz with a 50% duty cycle for 1min. After laser administration, the tape test is repeated and another 10min of activity and EEG/EMG are recorded to ensure that UVC illumination alone does not affect the animal.
- 48mg/kg aziPm bolus (a hypnotic dose [ED95] of the anesthetic photolabel) is administered via the syringe pump, and once the mouse is unresponsive it is placed on its side and a sticker placed on its nose.
- In addition to continuous EMG/EEG, accelerometry, and video, the mouse's time to return of righting reflex and sticker removal are recorded.
- The mouse is allowed to recover under constant monitoring before being detached and returned to its home cage (at least 1h to allow full recovery).
- The mouse is allowed 2 days to recover before further testing.
- On the second testing day, the mouse is attached to the recording, drug, and laser setup. Another 10min baseline is taken, followed by anesthetic photolabel administration with simultaneous laser administration (as detailed earlier) to photoadduct the drug at the region of interest.
- The mouse is placed on its side and a sticker placed on its nose, and recovery endpoints are recorded as above. Note that we record for a minimum of 2h after the removal of the sticker, as we have seen continued behavioral sedation, including slowed movement, for significant amounts of time after sticker removal.
- Video for return of right reflex and sticker removal is manually scored by blinded investigators. All video and biopotentials can be analyzed using MATLAB® 2016b with the Signal Processing Toolbox and Image Processing Toolbox.

6. POTENTIAL LIMITATIONS AND PITFALLS

A strength of using anesthetic photolabels in vivo, their effect on actual anesthetic targets in a defined physical space, is also a potential limitation. The anesthetic photolabels, like their parent compounds, affect many targets in many neuron types. This lack of cell specificity could potentially make it difficult to identify the populations contributing to the anesthetic effect. It is for this reason that this technique is truly complementary with existing cell type-specific methods, such as optogenetics and the use of designer receptors activated by designer drugs, that can target specific neuronal populations, but do not necessarily capture the effects of anesthetics on their native targets. While we did not encounter significant effects of near-UV illumination alone in either slice recordings or in intact animals, it remains a possibility that light alone may have an effect in certain regions or in specific neuronal populations, potentially damaging or killing neurons in the exposed areas. This is the reason we have included all the light alone controls, and it is also why it is critical to ensure that a cell or animal has in fact returned to a true baseline before progressing with testing. Finally, experimentally determining the minimum necessary photoadduction time and intensity for a given anesthetic photolabel is critical given the theoretic possibility of photoconcentration. Photolabeling in excess could produce functional concentrations of drug significantly above the physiologically relevant concentration. This could be similar to the problem encountered with microinjections, where at the injection point, the concentration of drug is higher than could ever be administered peripherally. The times and powers given here to minimize that possibility are for aziPm, but will vary with other compounds.

ACKNOWLEDGMENTS

These techniques were developed with the financial support of the NIH Institute of General Medical Sciences, via K08 GM123317, and the Foundation for Anesthesia Education and Research. Special thanks to Roderic Eckenhoff and William Dailey for the development of the anesthetic photolabels described here, as well as their advice in the biochemical and medical chemistry spheres. Thanks to Andrzej Wasilczuk, Jason Moore, Michael Chalifoux, and David Piel for their expertise in mouse surgery, MATLAB® analysis, and electrophysiology. Thanks to the Department of Anesthesiology and Critical Care at the University of Pennsylvania for its support.

REFERENCES

Beck, S. G., Pan, Y.-Z. Z., Akanwa, A. C., & Kirby, L. G. (2004). Median and dorsal raphe neurons are not electrophysiologically identical. *Journal of Neurophysiology, 91*, 994–1005.

Chiara, D. C., Dangott, L. J., Eckenhoff, R. G., & Cohen, J. B. (2003). Identification of nicotinic acetylcholine receptor amino acids photolabeled by the volatile anesthetic halothane. *Biochemistry, 42*, 13457–13467.

Chiara, D. C., et al. (2013). Specificity of intersubunit general anesthetic-binding sites in the transmembrane domain of the human α1β3γ2 γ-aminobutyric acid type A (GABAA) receptor. *The Journal of Biological Chemistry, 288*, 19343–19357.

Correa-Sales, C., Rabin, B. C., & Maze, M. (1992). A hypnotic response to dexmedetomidine, an alpha 2 agonist, is mediated in the locus coeruleus in rats. *Anesthesiology, 76*, 948–952.

Kmiotek, E. K., Baimel, C., & Gill, K. J. (2012). Methods for intravenous self administration in a mouse model. *Journal of Visualized Experiments, 70,* e3739. https://doi.org/10.3791/3739.

Li, G.-D. D., et al. (2006). Identification of a GABAA receptor anesthetic binding site at subunit interfaces by photolabeling with an etomidate analog. *The Journal of Neuroscience, 26,* 11599–11605.

Minert, A., Yatziv, S.-L. L., & Devor, M. (2017). Location of the mesopontine neurons responsible for maintenance of anesthetic loss of consciousness. *The Journal of Neuroscience, 37,* 9320–9331.

Moore, J. T., et al. (2012). Direct activation of sleep-promoting VLPO neurons by volatile anesthetics contributes to anesthetic hypnosis. *Current Biology, 22,* 2008–2016.

Nelson, L. E., et al. (2002). The sedative component of anesthesia is mediated by GABAA receptors in an endogenous sleep pathway. *Nature Neuroscience, 5,* 979–984.

Tóth, K., Freund, T. F., & Miles, R. (1997). Disinhibition of rat hippocampal pyramidal cells by GABAergic afferents from the septum. *The Journal of Physiology, 500*(Pt. 2), 463–474.

Wasilczuk, A. Z., Proekt, A., Kelz, M. B., & McKinstry-Wu, A. R. (2016). High-density electroencephalographic acquisition in a rodent model using low-cost and open-source resources. *Journal of Visualized Experiments, 117,* https://doi.org/10.3791/54908.

Weiser, B., Kelz, M., & Eckenhoff, R. (2012). In vivo activation of azipropofol prolongs anesthesia and reveals synaptic targets. *The Journal of Biological Chemistry, 288,* 1279–1285.

Woll, K. A., Dailey, W. P., Brannigan, G., & Eckenhoff, R. G. (2016). Shedding light on anesthetic mechanisms: Application of photoaffinity ligands. *Anesthesia and Analgesia, 123,* 1253–1262.

Woll, K. A., et al. (2016). A novel bifunctional alkylphenol anesthetic allows characterization of γ-aminobutyric acid, type A (GABAA), receptor subunit binding selectivity in synaptosomes. *The Journal of Biological Chemistry, 291,* 20473–20486.

Yip, G. M., et al. (2013). A propofol binding site on mammalian GABAA receptors identified by photolabeling. *Nature Chemical Biology, 9,* 715–720.

Zhong, H., Rüsch, D., & Forman, S. (2008). Photo-activated azi-etomidate, a general anesthetic photolabel, irreversibly enhances gating and desensitization of γ-aminobutyric acid type A receptors. *Anesthesiology, 108,* 103.

CHAPTER ELEVEN

Optogenetics and Chemogenetics

Ksenia Vlasov[*,†], **Christa J. Van Dort**[*,†,‡], **Ken Solt**[*,†,‡,1]
[*]Massachusetts General Hospital, Boston, MA, United States
[†]Massachusetts Institute of Technology, Cambridge, MA, United States
[‡]Harvard Medical School, Boston, MA, United States
[1]Corresponding author: e-mail address: ksolt@mgh.harvard.edu

Contents

Abstract

Optogenetics and chemogenetics provide the ability to modulate neurons in a type- and region-specific manner. These powerful techniques are useful to test hypotheses regarding the neural circuit mechanisms of general anesthetic end points such as hypnosis and analgesia. With both techniques, a genetic strategy is used to target expression of light-sensitive ion channels (opsins) or designer receptors exclusively activated by designer drugs in specific neurons. Optogenetics provides precise temporal control of neuronal firing with light pulses, whereas chemogenetics provides the ability to modulate neuronal firing for several hours with the single administration of a designer drug. This chapter provides an overview of neuronal targeting and experimental strategies and highlights the important advantages and disadvantages of each technique.

Methods in Enzymology, Volume 603
ISSN 0076-6879
https://doi.org/10.1016/bs.mie.2018.01.022

1. INTRODUCTION

Since the 1980s, there has been great progress in identifying the molecular targets of general anesthetics, and it is now evident that multiple molecular mechanisms are involved in general anesthesia (Covarrubias, Barber, Carnevale, Treptow, & Eckenhoff, 2015; Eckenhoff, 2001; Eger et al., 1997; Grasshoff, Rudolph, & Antkowiak, 2005). Although the receptor targets likely to be responsible for the behavioral actions of general anesthetics have been identified, it is still unclear how the molecular actions of these drugs lead to specific behavioral end points such as unconsciousness, analgesia, and amnesia. Recently, there has been growing interest in understanding the mechanisms of anesthetic actions at the level of neural circuits and systems, and emerging evidence suggests that anesthetic drugs produce their profound behavioral effects by acting at discrete neural circuits (Brown, Purdon, & Van Dort, 2011; Devor & Zalkind, 2001) such as endogenous pathways that modulate sleep and arousal (Franks, 2008; Mashour & Pal, 2012; Moore et al., 2012). The recent development of optogenetics and chemogenetics has revolutionized systems neuroscience by allowing for targeted excitation and inhibition of specific neuronal subpopulations in discrete brain regions. For both techniques, a genetic strategy is employed to insert photoactivatable ion channels (opsins, for optogenetics) or designer receptors exclusively activated by designer drugs (DREADDs, for chemogenetics) in specific neurons.

For optogenetics, a light-sensitive ion channel is expressed in targeted cells, allowing for neuronal depolarization or hyperpolarization with pulses of light (Deisseroth, 2011). Although a permanent intracranial implant is required to deliver the light pulses, one key advantage of optogenetics is the ability to have precise temporal control of neuronal activity. For chemogenetics, the expression of DREADDs is targeted to specific neurons. DREADDs are modified G protein-coupled "designer" receptors that are engineered to have low affinity for their native ligand, but high affinity for a synthetic, otherwise inert "designer" ligand that can be administered locally or systemically to induce neuronal activation or inhibition. Unlike optogenetics, chemogenetics does not offer the temporal resolution to rapidly activate or silence neurons, and the administration of a drug (i.e., designer ligand) is necessary. However, chemogenetics provides the advantages of not requiring chronic intracranial implants, and allowing for sustained neural excitation or inhibition for several hours with a single drug administration.

Studies that utilize optogenetics or chemogenetics to study sleep and arousal are emerging rapidly. For example, optogenetic activation of cholinergic neurons in the pedunculopontine tegmentum or laterodorsal tegmentum has been shown to induce rapid eye movement sleep (Van Dort et al., 2015). Although the use of optogenetics and chemogenetics has grown briskly in the neuroscience literature, relatively few studies have employed these techniques to study anesthetic mechanisms. It has been reported that chemogenetic activation of noradrenergic neurons in the locus coeruleus accelerates emergence from isoflurane anesthesia in mice (Vazey & Aston-Jones, 2014), and more recently, it was shown that optogenetic activation of dopamine neurons in the ventral tegmental area (VTA) restores the righting reflex in anesthetized mice (Taylor et al., 2016). These studies suggest that arousal-promoting noradrenergic and dopaminergic neurons are important for emergence from general anesthesia. However, there are many other arousal-promoting pathways and neurotransmitters in the brain (Brown et al., 2011; Franks, 2008; Van Dort, Baghdoyan, & Lydic, 2008), and it is important to understand their respective roles in anesthetic induction and emergence.

Another key component of general anesthesia is analgesia (Brown, Lydic, & Schiff, 2010), and recently, it was demonstrated that nociception is suppressed by chemogenetic activation of glutamatergic neurons, or inhibition of GABAergic neurons, in the ventrolateral periaqueductal gray area (Samineni et al., 2017). With neural circuit manipulations, a variety of behavioral end points may be selected to test for effects relevant to anesthesia (e.g., hypnosis, antinociception, and immobility). However, this chapter will focus primarily on optogenetic and chemogenetic methods to target specific neural circuits, rather than behavioral assays. A flowchart showing the typical sequence of procedures and experiments is provided in Fig. 1.

2. USING GENETIC TOOLS TO TARGET SPECIFIC NEURONS

In order to achieve specific neural circuit manipulations with optogenetics or chemogenetics, one of the most important aspects to consider is the targeting strategy. There are many ways to target subpopulations of neurons, their cell bodies, or projections, and Kim et al. provide an excellent overview of current methods (Kim, Adhikari, & Deisseroth, 2017). For both optogenetics and chemogenetics, genetically modified rodents that express the enzyme Cre recombinase (Cre) under the transcriptional control

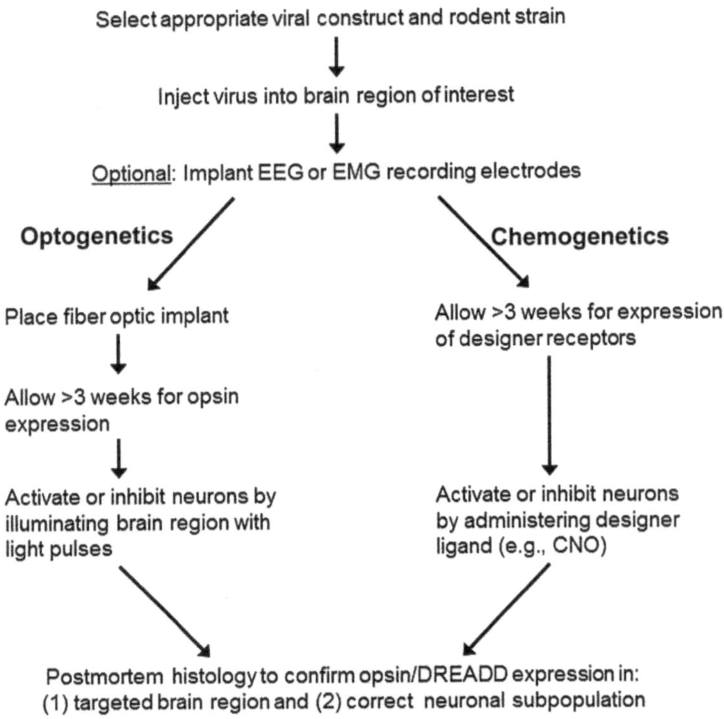

Fig. 1 Flowchart showing the typical sequence of events for performing optogenetics or chemogenetics experiments.

of a specific gene are typically used to target neuronal subpopulations. For example, vesicular GABA transporter (VGAT)-Cre mice express Cre only in GABAergic neurons that express VGAT. Many different mouse lines with stable and heritable expression of Cre are commercially available through Jackson Labs and other breeding facilities, allowing investigators to target and manipulate a variety of different neuronal subpopulations.

In order to gain anatomic specificity of opsin or DREADD expression, it is necessary to perform stereotaxic injections of viral vectors encoding these proteins in the brain regions of interest. Cre is an enzyme that catalyzes site-specific recombination between two LoxP sites, and modern Cre-driven viral vectors are constructed with "double-floxed" genes encoding the opsin or DREADD, causing gene expression only in transfected cells that contain Cre. A fluorescent tag is also encoded in the viral construct, allowing for postmortem histological confirmation of gene expression in the targeted cell type and brain region. Cre-inducible adeno-associated viruses (AAVs) are commercially available from Addgene and other sources. These viruses are replication deficient and not known to cause disease in humans. Different AAV strains (e.g., AAV5 or AAV8) have specific transfection patterns in brain tissue, so it is important to test the viral construct for proper expression in the region of interest. After virus injection, a minimum of 3 weeks is recommended prior to performing experiments, to allow sufficient time for opsin or DREADD expression in transfected cells.

3. OPTOGENETICS

As shown in Fig. 2, with optogenetics, stimulatory opsins such as Channelrhodopsin2 (ChR2) are commonly used to depolarize neurons (Boyden, Zhang, Bamberg, Nagel, & Deisseroth, 2005), whereas inhibitory opsins such as halorhodopsin or Jaws may be used to silence neurons (Chuong et al., 2014). In order to illuminate the opsin-containing neurons, it is necessary to place a fiber optic implant above the brain region of interest. The implantation is typically performed immediately after the virus injection during a single surgical operation, although it may also take place during a separate surgery several weeks after the virus injection, when transfection and protein expression have already taken place. Details of the stereotaxic surgery are described later. After neurosurgery, a minimum of 1 week is recommended to allow for sufficient recovery and healing prior to experimentation.

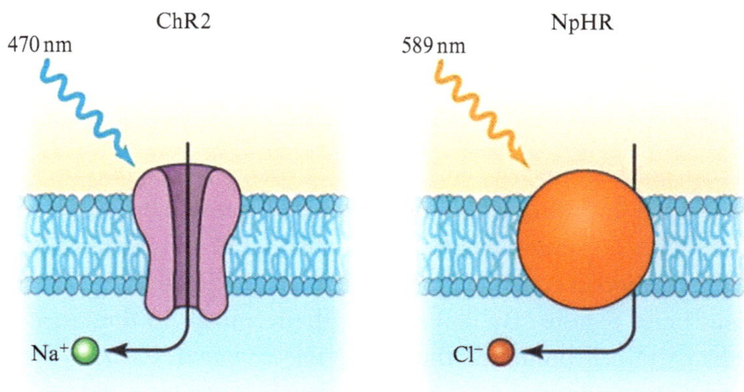

Fig. 2 Illumination of neurons that express channelrhodopsin2 (ChR2) with *blue light* (470 nm) allows Na^+ ions to enter the cell, causing depolarization. Illumination of neurons that express halorhodopsin (NpHR) with *yellow-orange light* (589 nm) causes Cl^- ions to be pumped into the cell, inducing hyperpolarization.

To perform optogenetics experiments, a light source such as a laser or LED and a method of controlling the timing of light pulses (e.g., a stimulus generator) are needed for precisely timed activation or inhibition of neurons. Each opsin is activated by specific wavelengths of light, so it is necessary to ensure that the correct wavelength is used to modulate the selected opsin. A patch cord is also necessary to connect the implanted fiber optic to the light source. Alternatively, an LED-mounted fiber optic may be used.

There are three main parameters to consider for the light pulses: duration (5–15 ms), frequency (1–50 Hz), and intensity (1–10 mW). For neuronal activation, it is often desirable to mimic "normal" neurophysiology, so for initial settings it may be helpful to consider the natural firing rates and patterns of the targeted neurons. ChR2 is the most commonly used opsin and can reliably cause neurons to fire at 20–40 Hz (Boyden et al., 2005). For faster firing rates, new ChR2 mutants such as ChETAH have been introduced, which allow neurons to fire at up to 200 Hz (Gunaydin et al., 2010). A typical set of parameters for ChR2 is to use a blue light 473 mm laser, with 5 ms light pulses at a frequency of 5 Hz, and a light intensity of 5 mW.

For neuronal inhibition, there are several inhibitory opsins available such as halorhodopsin (eNpHR3.0) which is a chloride pump that induces cellular hyperpolarization with yellow or green light (Gradinaru et al., 2010). A red-shifted cruxhalorhodopsin (Jaws) was recently introduced that has greater photocurrrents than halorhodopsin, and allows for the possibility of noninvasive photoinhibition as red light penetrates brain tissue more efficiently than blue light (Chuong et al., 2014).

Cell bodies may be illuminated to activate or inhibit all of the targeted neurons in a given brain region (e.g., VTA dopamine neurons), but terminal field illumination may also be used to test the effect of modulating a neural circuit that projects to a specific brain region. For example, by injecting a viral construct encoding ChR2 in the VTA of tyrosine hydroxylase (TH)-Cre mice, VTA dopamine neurons that contain TH will express ChR2 not only in the cell bodies contained in the VTA but also in the axons that project to various brain regions. By illuminating the terminals in the target region rather than the cell bodies in the VTA, selective activation of dopamine neurons projecting from the VTA to specific brain regions can be achieved. This approach was recently utilized to demonstrate that photostimulation of dopamine neurons projecting from the VTA to the nucleus accumbens promotes wakefulness (Eban-Rothschild, Rothschild, Giardino, Jones, & de Lecea, 2016).

4. CHEMOGENETICS

With chemogenetics, DREADDs are used to activate or inhibit targeted neurons, rather than opsins. As shown in Fig. 3, the stimulatory DREADD hM3Dq is a modified human M3 muscarinic receptor that has low affinity for the native ligand acetylcholine, but high affinity for the synthetic ligand clozapine-*N*-oxide (CNO) (Armbruster, Li, Pausch, Herlitze, & Roth, 2007). When this Gq-coupled receptor is activated with local or systemic administration of CNO, targeted neurons undergo burst firing (Alexander et al., 2009). For neuronal inhibition, the Gi-coupled hM4Di DREADD (a modified human M4 muscarinic receptor) is often used to silence targeted neurons with CNO (Urban & Roth, 2015). CNO causes a downstream signaling cascade leading to either burst firing (for hM3Dq)

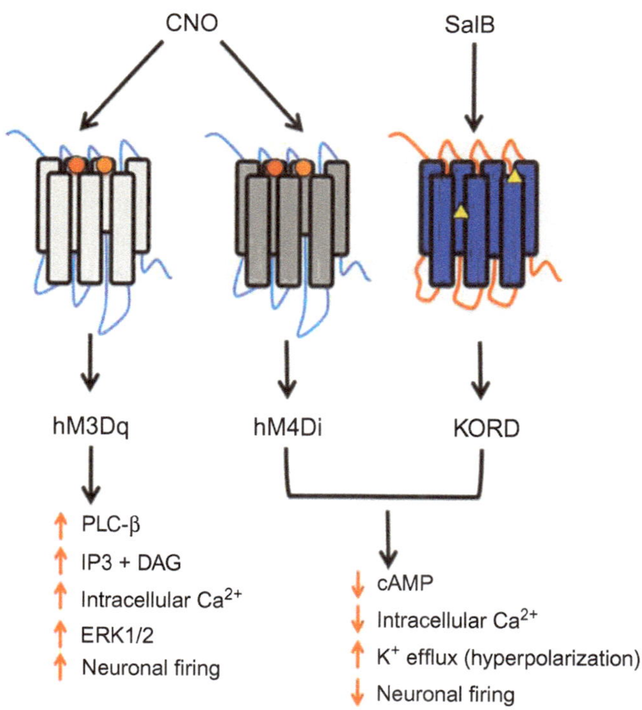

Fig. 3 Human muscarinic (hM)-based DREADDs are activated by clozapine-*N*-oxide (CNO), whereas kappa opioid receptor (KOR)-based DREADDs are activated by Salvinorin B (SalB). CNO causes burst firing of neurons that express hM3Dq, and inhibition of neurons that express hM4Di. SalB inhibits neurons that express KORD, allowing for bidirectional chemogenetics in neurons that express both hM3Dq and KORD.

or silencing (for hM4Di) of the targeted neurons, allowing for prolonged neuronal excitation or inhibition. In mice, the effects of intraperitoneally injected CNO are observed in 10–15 min, peak at 45–50 min, and slowly return to baseline over approximately 9 h (Alexander et al., 2009). CNO is soluble in aqueous solutions such as normal saline, and for mice investigators typically use 0.1–3 mg/kg CNO injected intraperitoneally (Roth, 2016), although intracranial microinjections have also been used (Vazey & Aston-Jones, 2014).

Recently, it was reported that hM3Dq and hM4Di DREADDs are activated by clozapine (a metabolite of CNO that crosses the blood–brain barrier) rather than CNO (Gomez et al., 2017). Because clozapine is an antipsychotic drug that may cause sedation at high doses, this finding stressed the importance of using low doses of CNO for chemogenetics experiments, and performing proper control experiments by administering CNO to animals that do not express hM3Dq and hM4Di DREADDs (Roth, 2016).

In 2015, a new kappa opioid receptor-based DREADD (KORD) was reported that allows for neuronal silencing with salvinorin B (SalB) (Vardy et al., 2015). This new DREADD has made it possible to conduct bidirectional chemogenetics by activating or inhibiting the same neurons with different ligands. However, SalB (available through Sigma-Aldrich, Cayman Chemicals, etc.) is relatively insoluble in aqueous solution, and therefore chemical solvents such as dimethyl sulfoxide (DMSO) may be required for parenteral administration. Depending on the experimental design, additional controls may need to be performed to test if the vehicle (e.g., DMSO) produces any significant effects. The kinetics of SalB are also quite different from CNO, with behavioral effects appearing shortly after injection and lasting about 1 h (Vardy et al., 2015). Although the KORD/SalB system provides the unique advantage of bidirectional chemogenetic control of the same neurons (when used in conjunction with hM3Dq/CNO), the hM4Di/CNO system is more useful when longer term neuronal inhibition is desired. Other DREADD constructs include β-arrestin-preferring DREADD and Gs DREADD, though these are less commonly used (Roth, 2016).

Similar to terminal field illumination with optogenetics, axonal projections may also be targeted selectively with chemogenetics. There are generally two approaches to achieve this aim. The first is to administer the designer ligand (e.g., CNO) locally (rather than systemically) via microinjection in the target site. The second approach is to use a genetic strategy to restrict DREADD expression to a specific circuit, which allows for selective activation or inhibition with systemic administration of CNO. The latter method provides the distinct advantage of not requiring an invasive intracranial implant, which is useful when a freely behaving animal is desired for behavioral analysis. To achieve selective DREADD expression in a specific circuit, a canine adenovirus (CAV2-Cre) may be injected in the projection site, causing retrograde transport to cell bodies, leading to the expression of Cre. When a Cre-driven AAV encoding the DREADD is injected in the location containing cell bodies, only cells that project to the target site of interest will express Cre, leading to selective DREADD expression in neurons that project to that site (Urban & Roth, 2015). Because the CAV2-Cre viral construct is used to induce expression of Cre, another advantage of this approach is that genetically modified animals are not needed.

5. STEREOTAXIC SURGERY

Below is a sample protocol for stereotaxic neurosurgery, including steps for virus injection as well as optional implantation of fiber optics and electroencephalogram (EEG) electrodes.

5.1 Major Equipment

(1) Stereotaxic frame for rodents with anesthesia mask
(2) Microinjector
(3) Micro bead sterilizer or autoclave
(4) Anesthetic induction chamber, vaporizer, and scavenging system
(5) Micropipette puller
(6) Electric microdrill for making craniotomies
(7) Light source (laser or LED, for optogenetics only)
(8) Stimulus generator (for optogenetics only)
(9) Recording system for EEG (optional)

5.2 Preoperative Preparation

(1) Sterilize surgical tools in a micro bead sterilizer or autoclave
(2) Gather materials:
  (a) Sterile gauze
  (b) Sterile cotton tip applicators
  (c) Eye lubricant
  (d) Extra fine tip pen or pencil

(e) Electric drill and drill bits
(f) Bone anchor screws and compatible screw driver
(g) Silicone cup to mix dental acrylic
(h) Dental acrylic (powder + liquid)
(i) Suture
(j) Fiber optic implant and EEG electrodes (optional)

(3) Fill four small (25 mL) beakers with
(a) 70% ethanol (place bone anchor screws and drill bit in the alcohol to sterilize)
(b) 3% hydrogen peroxide
(c) Povidone-iodine solution
(d) Saline

(4) If using isoflurane as the anesthetic, ensure that the isoflurane vaporizer and oxygen tank are full
(5) Turn on fiber optic illuminator and heating pad to preheat (this will be used to maintain the core body temperature at approximately 37°C)

5.3 Preparation of Virus

(1) Choose an area adjacent to the surgical area that will be dedicated to loading the micropipettes with virus, and cover it with absorbent lab bench paper.
(2) Place an aliquot of virus in a container of ice in this area, and allow the virus to thaw on ice as the surgery is being performed.
(3) Set up a waste container of 10% bleach in the dedicated virus handling area for disposing of pipette tips, etc., that come in contact with the virus.
(4) Pull a glass Wiretrol micropipette to a tip diameter of approximately 20 μm. Fill a Hamilton syringe (Model 701 LT), a microelectrode adaptor (WPI catalog #MPH6S10), and the glass micropipette with mineral oil and secure the apparatus in a microinjector.

5.4 Animal Preparation

(1) Weigh animal and place in an anesthetic induction chamber
(2) Administer 2.5%–3% isoflurane in oxygen
(3) Wait for the animal to lose the righting reflex, and then wait another 60 s before handling
(4) Remove the animal from the induction chamber and clip the scalp hair with an electric shaver, from between the eyes (anteriorly) to the neck (posteriorly)

(5) Place the animal in the stereotaxic frame with the nose on the bite bar and the ear bars in the ear canal, ensuring that the animal remains adequately anesthetized
  (a) Ear bar readings should be the same on both sides to ensure that the head is correctly positioned in the center
(6) Apply eye lubricant and insert a rectal temperature probe to ensure that the animal's core body temperature is maintained at 37°C
(7) Once the animal is positioned, sterilize the shaved scalp area where the incision will be made with 70% ethanol and povidone-iodine solution. Start in the center and work your way out in circles. Alternate cleaning with alcohol and povidone-iodine solution 3 × each
(8) Assess anesthetic depth regularly by observing the respiratory rate and testing for movement in response to paw pinch

5.5 Surgery

(1) Put on sterile gloves prior to making the incision
(2) Make a vertical incision in the center of the scalp
(3) Clean the skull surface with a cotton-tipped applicator soaked in saline or hydrogen peroxide
(4) Level the head by measuring the vertical height of bregma and lambda (aim for a difference in height between bregma and lambda of less than 0.05 mm)
(5) Calculate the stereotaxic targets and mark the locations with a fine tip pen for viral injection sites as well as bone anchor screws, fiber optics, and EEG electrodes (as needed)
(6) Drill craniotomies for viral injection sites and bone anchor screws, fiber optics and EEG electrodes (as needed)
  (a) In rats, the dura is thicker than mice, so it is important to make sure that the dura is cleared using fine-tipped tweezers or a small hooked 27G needle before placing the delicate glass pipette for viral injection or bendable electrodes.
  (b) If bleeding occurs use sterile cotton tip applications and saline to stop the bleeding
  (c) If implanting fiber optics or EEG electrodes, aim for six bone anchor screw holes distributed around the skull bones

5.6 Virus Injection

(1) Pipette 1.0 μL of virus for a 0.5 μL injection onto a small piece of parafilm (typical viral injection volumes are 0.2–0.5 μL). The optimal viral

titer depends on the AAV serotype, promoter, and size of the target brain region. Therefore, the optimal viral titer must be determined empirically for each set of experiments.

(2) Place the tip of the micropipette into the viral stock and manually withdraw the plunger

(3) If drawing viral stock into the micropipette is difficult, the micropipette tip can be enlarged slightly by cutting it

(4) Move plunger down until a small amount of virus is dispelled from the tip of the micropipette, and remove this small drop with a cotton tip applicator and discard in a virus waste container

(5) Apply a drop of mineral oil to the tip of the micropipette to prevent clogging as the micropipette is lowered into the brain

(6) Using *X*- and *Y*-axis stereotaxic coordinates from bregma, position the micropipette over the area to be injected. Very slowly lower the micropipette (at an approximate rate of 1 mm/min) to the proper *Z*-axis position

(7) Enter the desired injection parameters into the microinjector controller box (SYS Micro4) and initiate the injection (e.g., 0.5 μL over 10 min)

(8) When the injection is complete, leave the pipette for 5–15 min to prevent efflux of virus during removal

(9) Very slowly remove the micropipette from the brain (approximately 1 mm/min).

(10) If no further fiber optic or EEG implant is planned, suture the skin closed

(11) Administer a postsurgical analgesic such as the nonsteroidal antiinflammatory drug (NSAID) ketoprofen or equivalent at the end of surgery, and for at least 2 days postoperatively. If a more potent analgesic is necessary, buprenorphine may be used.

5.7 Optional Steps for Implantation of Fiber Optics or EEG Electrodes

(1) Place all bone anchor screws using a small screwdriver. Turn screws 1.5–2.5 turns, so they are stable but not protruding into the brain. Ensure there is enough space for acrylic between the screw head and skull surface: we use 0.7 mm diameter anchor screws from Antrim for mice and Stoelting bone anchor screws for rats.

(2) For EEG recordings, check ground screw impedance (should be less than 20 kΩ when measured between the ground screw and the other anchor screws) and place EEG electrodes just under the skull.

(3) For fiber optic implants, stereotaxically place the fiber optic over the brain area of interest.
(4) Cover bone anchor screws and fiber optic or EEG electrodes with dental acrylic, making sure to leave smooth edges in order to avoid skin irritation.
(5) If the skin is loose around the implant, add a suture just behind the implant to close the skin.
(6) Administer a postsurgical analgesic such as the NSAID ketoprofen or equivalent at the end of surgery, and for at least 2 days postoperatively. If a more potent analgesic is necessary, buprenorphine may be used.

5.8 Cleanup

(1) Rinse the glass micropipette with 10% bleach and discard it in a sharps container
(2) Discard waste container
(3) Discard lab bench paper into biohazard bin and wipe down all surfaces and instruments that may have come in contact with the virus with 10% bleach
(4) Unused virus may be refrozen, but keep in mind that repeated freeze/thaw cycles will cause virus degradation
(5) Allow at least 1 week for full recovery from surgery, at which time baseline behavioral and EEG data may be collected
(6) Allow approximately 2–3 weeks for viral transfection and gene expression to occur before attempting neural circuit manipulations

6. CONTROL EXPERIMENTS

For both optogenetics and chemogenetics experiments, controls are needed to ensure the effect is specific to the opsin or DREADD. Ideally, the control animals have the same genetic background and receive an injection of a control AAV encoding only a fluorescent tag without opsins or DREADDs. For chemogenetics experiments in particular, these controls are essential to assess potential nonspecific effects of the designer ligand. In addition, for chemogenetics a vehicle control for the ligand (e.g., normal saline) is also useful. For optogenetic experiments, a similar control strategy is also necessary (i.e., illumination of the same brain region in animals that received an injection of a control AAV that does not encode the opsin) to ensure that any observed behaviors are not due to nonspecific effects such as heating of the brain tissue, or the viral injection.

7. HISTOLOGY AND VERIFICATION OF RECEPTOR FUNCTION

After completing all behavioral experiments, it is necessary to show that the opsin or DREADD was expressed in the targeted cell type and the correct brain region using histological analysis. The exact histological method depends on the cell type targeted and scientific question. For most applications, immunohistochemistry (IHC) is used to label specific cell types. In cases where IHC does not work well (e.g., targeting GABAergic or glutamatergic neurons in the brainstem), other methods like fluorescent in situ hybridization (FISH) can be used to target RNA. In order to perform IHC, animals are euthanized and transcardially perfused with formalin to fix the brain. The fixed brain is then sliced into thin sections (40–60 μm) using a vibratome. For FISH, the animal is deeply anesthetized and the brain is removed and frozen fresh, then sectioned at 10–15 μm on a cryostat. Brain sections are imaged using a fluorescent or confocal microscope. Once the brain sections are stained with IHC or FISH and imaged, colocalization of the opsin or DREADD and the stained marker is confirmed. For optogenetic experiments and chemogenetic experiments involving electrophysiological recordings, histology is used to identify the anatomic location of electrodes and fiber optic implants. Finally, to confirm that illumination (for optogenetics) or designer ligand application (for chemogenetics) induces the expected changes in neuronal firing, fresh brain slices from experimental animals can be used to perform whole-cell patch clamp recordings.

8. SUMMARY AND CONCLUSION

Optogenetics and chemogenetics provide powerful new tools to probe the neural circuits and systems in the brain where anesthetics act to produce their profound behavioral effects. These techniques also provide the ability to test hypotheses regarding whether the neural circuits that induce sleep are also involved in anesthetic hypnosis. Using different behavioral assays, investigators may also employ neural circuit manipulations to test for other anesthetic end points such as analgesia and amnesia. Future studies using neural circuit manipulations will likely provide many new mechanistic insights into how the behavioral end points that define general anesthesia are produced.

REFERENCES

Alexander, G. M., Rogan, S. C., Abbas, A. I., Armbruster, B. N., Pei, Y., Allen, J. A., et al. (2009). Remote control of neuronal activity in transgenic mice expressing evolved G protein-coupled receptors. *Neuron*, *63*, 27–39.

Armbruster, B. N., Li, X., Pausch, M. H., Herlitze, S., & Roth, B. L. (2007). Evolving the lock to fit the key to create a family of G protein-coupled receptors potently activated by an inert ligand. *Proceedings of the National Academy of Sciences of the United States of America*, *104*, 5163–5168.

Boyden, E. S., Zhang, F., Bamberg, E., Nagel, G., & Deisseroth, K. (2005). Millisecond-timescale, genetically targeted optical control of neural activity. *Nature Neuroscience*, *8*, 1263–1268.

Brown, E. N., Lydic, R., & Schiff, N. D. (2010). General anesthesia, sleep, and coma. *The New England Journal of Medicine*, *363*, 2638–2650.

Brown, E. N., Purdon, P. L., & Van Dort, C. J. (2011). General anesthesia and altered states of arousal: A systems neuroscience analysis. *Annual Review of Neuroscience*, *34*, 601–628.

Chuong, A. S., Miri, M. L., Busskamp, V., Matthews, G. A., Acker, L. C., Sorensen, A. T., et al. (2014). Noninvasive optical inhibition with a red-shifted microbial rhodopsin. *Nature Neuroscience*, *17*, 1123–1129.

Covarrubias, M., Barber, A. F., Carnevale, V., Treptow, W., & Eckenhoff, R. G. (2015). Mechanistic insights into the modulation of voltage-gated ion channels by inhalational anesthetics. *Biophysical Journal*, *109*, 2003–2011.

Deisseroth, K. (2011). Optogenetics. *Nature Methods*, *8*, 26–29.

Devor, M., & Zalkind, V. (2001). Reversible analgesia, atonia, and loss of consciousness on bilateral intracerebral microinjection of pentobarbital. *Pain*, *94*, 101–112.

Eban-Rothschild, A., Rothschild, G., Giardino, W. J., Jones, J. R., & de Lecea, L. (2016). VTA dopaminergic neurons regulate ethologically relevant sleep-wake behaviors. *Nature Neuroscience*, *19*, 1356–1366.

Eckenhoff, R. G. (2001). Promiscuous ligands and attractive cavities: How do the inhaled anesthetics work? *Molecular Interventions*, *1*, 258–268.

Eger, E. I., 2nd, Koblin, D. D., Harris, R. A., Kendig, J. J., Pohorille, A., Halsey, M. J., et al. (1997). Hypothesis: Inhaled anesthetics produce immobility and amnesia by different mechanisms at different sites. *Anesthesia and Analgesia*, *84*, 915–918.

Franks, N. P. (2008). General anaesthesia: From molecular targets to neuronal pathways of sleep and arousal. *Nature Reviews. Neuroscience*, *9*, 370–386.

Gomez, J. L., Bonaventura, J., Lesniak, W., Mathews, W. B., Sysa-Shah, P., Rodriguez, L. A., et al. (2017). Chemogenetics revealed: DREADD occupancy and activation via converted clozapine. *Science*, *357*, 503–507.

Gradinaru, V., Zhang, F., Ramakrishnan, C., Mattis, J., Prakash, R., Diester, I., et al. (2010). Molecular and cellular approaches for diversifying and extending optogenetics. *Cell*, *141*, 154–165.

Grasshoff, C., Rudolph, U., & Antkowiak, B. (2005). Molecular and systemic mechanisms of general anaesthesia: The 'multi-site and multiple mechanisms' concept. *Current Opinion in Anaesthesiology*, *18*, 386–391.

Gunaydin, L. A., Yizhar, O., Berndt, A., Sohal, V. S., Deisseroth, K., & Hegemann, P. (2010). Ultrafast optogenetic control. *Nature Neuroscience*, *13*, 387–392.

Kim, C. K., Adhikari, A., & Deisseroth, K. (2017). Integration of optogenetics with complementary methodologies in systems neuroscience. *Nature Reviews. Neuroscience*, *18*, 222–235.

Mashour, G. A., & Pal, D. (2012). Interfaces of sleep and anesthesia. *Anesthesiology Clinics*, *30*, 385–398.

Moore, J. T., Chen, J., Han, B., Meng, Q. C., Veasey, S. C., Beck, S. G., et al. (2012). Direct activation of sleep-promoting VLPO neurons by volatile anesthetics contributes to anesthetic hypnosis. *Current Biology*, *22*, 2008–2016.

Roth, B. L. (2016). DREADDs for neuroscientists. *Neuron, 89*, 683–694.
Samineni, V. K., Grajales-Reyes, J. G., Copits, B. A., O'Brien, D. E., Trigg, S. L., Gomez, A. M., et al. (2017). Divergent modulation of nociception by glutamatergic and GABAergic neuronal subpopulations in the periaqueductal gray. *eNeuro, 4*, 0129–16.
Taylor, N. E., Van Dort, C. J., Kenny, J. D., Pei, J., Guidera, J. A., Vlasov, K. Y., et al. (2016). Optogenetic activation of dopamine neurons in the ventral tegmental area induces reanimation from general anesthesia. *Proceedings of the National Academy of Sciences of the United States of America, 113*(45), 12826–12831.
Urban, D. J., & Roth, B. L. (2015). DREADDs (designer receptors exclusively activated by designer drugs): Chemogenetic tools with therapeutic utility. *Annual Review of Pharmacology and Toxicology, 55*, 399–417.
Van Dort, C. J., Baghdoyan, H. A., & Lydic, R. (2008). Neurochemical modulators of sleep and anesthetic states. *International Anesthesiology Clinics, 46*, 75–104.
Van Dort, C. J., Zachs, D. P., Kenny, J. D., Zheng, S., Goldblum, R. R., Gelwan, N. A., et al. (2015). Optogenetic activation of cholinergic neurons in the PPT or LDT induces REM sleep. *Proceedings of the National Academy of Sciences of the United States of America, 112*, 584–589.
Vardy, E., Robinson, J. E., Li, C., Olsen, R. H., DiBerto, J. F., Giguere, P. M., et al. (2015). A new DREADD facilitates the multiplexed chemogenetic interrogation of behavior. *Neuron, 86*, 936–946.
Vazey, E. M., & Aston-Jones, G. (2014). Designer receptor manipulations reveal a role of the locus coeruleus noradrenergic system in isoflurane general anesthesia. *Proceedings of the National Academy of Sciences of the United States of America, 111*, 3859–3864.

CHAPTER TWELVE

Genetic Reporters of Neuronal Activity: c-Fos and G-CaMP6

Andrew E. Hudson[1]
UCLA, Los Angeles, CA, United States
[1]Corresponding author: e-mail address: ahudson@mednet.ucla.edu

Contents

Abstract

The majority of 20th century investigations into anesthetic effects on the nervous system have used electrophysiology. Yet some fundamental limitations to electrophysiologic recordings, including the invasiveness of the technique, the need to place (potentially several) electrodes in every site of interest, and the difficulty of selectively recording from individual cell types, have driven the development of alternative methods for detecting neuronal activation. Two such alternative methods with cellular

Methods in Enzymology, Volume 603
ISSN 0076-6879
https://doi.org/10.1016/bs.mie.2018.01.023

scale resolution have matured in the last few decades and will be reviewed here: the transcription of immediate early genes, foremost *c-fos*, and the influx of calcium into neurons as reported by genetically encoded calcium indicators, foremost GCaMP6. Reporters of *c-fos* allow detection of transcriptional activation even in deep or distant nuclei, without requiring the accurate targeting of multiple electrodes at long distances. The temporal resolution of *c-fos* is limited due to its dependence upon the detection of transcriptional activation through immunohistochemical assays, though the development of RT-PCR probes has shifted the temporal resolution of the assay when tissues of interest can be isolated. GCaMP6 has several isoforms that trade-off temporal resolution for signal to noise, but the fastest are capable of resolving individual action potential events, provided the microscope used scans quickly enough. GCaMP6 expression can be selectively targeted to neuronal populations of interest, and potentially thousands of neurons can be captured within a single frame, allowing the neuron-by-neuron reporting of circuit dynamics on a scale that is difficult to capture with electrophysiology, as long as the populations are optically accessible.

1. INTRODUCTION

Experimental techniques must be carefully selected to balance their strengths and weaknesses in order to address the questions of interest. As our understanding of the effects of anesthetics on isolated neurons has matured, the field has begun to ask how those changes that occur in isolated neurons combine together in the brain to produce the features of general anesthesia, be they unarousable unresponsiveness, amnesia, or analgesia. The modulation of the dynamics of neuronal assemblies may be the sine qua non of anesthetic drugs (Collins, Mann, Hill, Dommett, & Greenfield, 2007). As but one example, consciousness appears to involve distributed processes in multiple neuronal circuits. Thus, studies of how anesthetics disrupt consciousness must necessarily assess the effects of anesthetics at a circuit or network level. Hence, probing the effects of anesthetics on circuit function depends upon our having methods for sampling multiple neurons in the circuit simultaneously, with certainty about the recorded neurons identity.

The fundamental unit of information transmission in the brain is the action potential, a stereotyped electrical event. It should come as no surprise that the dominant method for studying the function of the nervous system has been electrical recording of neuronal activity. And yet, depending upon the spatial layout of the neuronal circuit of interest, some investigations at a network level are difficult at best with standard electrophysiologic tools. To address these shortcomings, neuroscientists have developed multiple alternative methods for assaying neuronal activation, with their own associated advantages and limitations. Two of the most well-developed assays of neuronal activation utilize either the transcription of the so-called immediate early genes (IEGs, most frequently *c-fos*) as a reporter of transcriptional activation or on the change in fluorescence of genetically encoded calcium indicators (GECIs, of which the dominant reporter at the time of this writing is GCaMP6).

The near-universal ability of volatile anesthetics to place organisms across multiple kingdoms into a behaviorally quiescent state offers a wide range of models for the effects of anesthetics, beginning with Claud Bernard's vitalist notions that life itself can be defined as susceptibility to anesthetics (Perouansky, 2012). More molecularly selective agents, such as propofol, etomidate, and ketamine, require more specific nervous system design principles; an organism that does not express a γ-aminobutyric acid A receptor (GABA-A) should not be expected to be anesthetized by a GABA-ergic hypnotic agent. While single-cell organisms may offer limited insights about how anesthetics disrupt human consciousness, different model organisms with variably complicated nervous systems can be exploited to explore the effects of anesthetic drugs. Intuitively, however, the more complicated the model system, the more useful complementary assays of neuronal activity may prove.

The focus of this chapter is the motivation for and application of these nonelectrophysiologic methods to studies of anesthetic mechanisms at the circuit level, and interested readers should explore one of several excellent reviews of both IEG reporters and GECIs that exist. While other alternative assays of neuronal activation also exist (for example, functional imaging with fMRI), both IEGs and GECIs offer the ability to detect activation with single neuron resolution, motivating their grouping as distinct from population-level assays. Before delving into the methods by which these techniques can be used, it is worth exploring when and why these methods should be considered in the first place, whether they are used to replace or augment more established electrophysiologic techniques. I will begin with two of the major difficulties faced by electrophysiology that would suggest incorporating either IEGs or GECIs to assist with a study of circuit mechanisms of anesthesia, not to fault electrophysiology techniques, but to point out the relative strengths of alternative assays.

1.1 Measures of Neuronal Activation

Concurrent with the development of these alternative technologies has been a broadening of the notion of neuronal activation (Fig. 1A). Historically, investigators have focused on the action potential as the primary unit of neuronal activity, which had the advantage of being particularly well suited to the rapidly maturing technology that allowed electrophysiologic investigation. A glass microelectrode offers a clear view of the voltage within a single neuron. But intracellular recordings are difficult to obtain in vivo and

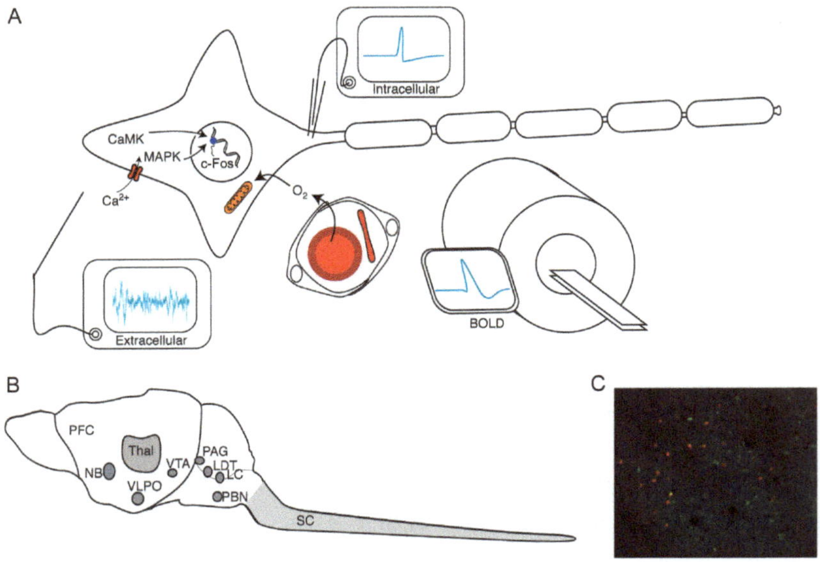

Fig. 1 (A) Different measures of neuronal activation include intracellular and extracellular electrical recordings, intracellular calcium concentration measured through fluorescent reporters or through downstream transcriptional activation of IEGs like *c-fos* through CAM kinase (CAMK) and MAP Kinase (MAPK) pathways, and measures of metabolism such as uptake of oxygen in the BOLD response. (B) Selected brain regions of interest in anesthesiology and pain research. Deep targets indicated by *gray shading*. The wide spatial distribution through the central nervous system makes IEG reporter approaches appealing for simultaneous sampling. Similarly, deep targets are difficult to access with electrophysiology and microscopy methods. *LC*, locus ceruleus; *LDT*, lateral dorsal tegmentum; *NB*, nucleus basalis of Meynert; *PAG*, periaqueductal gray; *PBN*, parabrachial nucleus; *PFC*, prefrontal cortex; *SC*, spinal cord; *Thal*, thalamus; *VLPO*, ventral lateral preoptic nucleus; *VTA*, ventral tegmental area. (C) Layer 2/3 of frontal cortex in a mouse expressing the red fluorescent protein Td-tomato in parvalbumin-expressing interneurons and the green fluorescent protein GCAMP6 in pyramidal neurons. Electrophysiologic methods cannot easily target specific cell types during recording, whereas genetically encoded calcium indicators can target expression to specific cell lines.

difficult to maintain for long periods of time. Extracellular electrical recordings are able to identify the stereotypic form of an action potential with good signal to noise by high-pass filtering the recorded electrical signal above approximately 300 Hz. In vivo, and even ex vivo in slice preparations, extracellular recordings are actually detecting a spatial summation of the activity of all nearby neurons. By positioning an extracellular electrode within approximately 50 μm of an individual neuron (Buzsáki, 2004), it is possible to obtain a reasonably isolated assessment of the response of a single neuron in vivo in real-time. To sample more members of a circuit, multiple electrodes can be recorded simultaneously, and the shape of the recorded action potential at each different electrode can be used to putatively assign that action potential to a distinct neuron generator (Abeles & Goldstein, 1977), though each added electrode risks traumatizing the neurons of interest. The most frequent electrode configuration employed for extracellular recording is a tetrode, to allow sorting of action potentials into different putative units on the basis of spike morphology at each of the different electrodes (Harris, Henze, Csicsvari, Hirase, & Buzsáki, 2000; Takahashi, Anzai, & Sakurai, 2003), without causing a great deal of damage to the neurons of interest. Yet spike variability over time, especially with different behavioral conditions, can seriously compromise the ability of spike sorting to properly attribute spike activity when compared with intracellular recordings (Stratton et al., 2012).

To combat this attribution problem while asking questions about population responses, more investigators have begun to consider the content of the low frequency below 300 Hz, the so-called local field potential (LFP), as a worthy topic of study in itself, reflecting the local average of the dendritic potentials and hence providing a measure of the computational input to a recorded neuronal population. The energetic demands of maintaining the resting potential in the face of constant dendritic depolarizations explain the high correlation between the LFP and the functional imaging measures of blood flow and blood oxygen level-dependent (BOLD) response detected with magnetic resonance imaging (Logothetis, 2003). The LFP thus forms a bridge between individual neuronal activity and population level assessment of circuit activity.

The development of calcium indicators as reporters of neuronal activity hinges upon the influx of calcium into the neuron that occurs, either via the NMDA receptor or the L-type voltage-sensitive calcium channel (Chaudhuri, Zangenehpour, Rahbar-Dehgan, & Ye, 2000). As a result, calcium influx can occur during both action potential propagation and also dendritic depolarization that does not necessarily lead to an action potential firing. Hence, measures of intracellular calcium in the dendritic tree approximate the synaptic inputs to an area, while in the cell body it represents the integrated inputs to the cell before final application of the spike thresholding nonlinearity that converts a continuous input into the point-process event of the action potential. Finally, if there is sufficient or prolonged enough calcium influx to activate the MAP kinase pathway, neurons will change their protein expression profile, undergoing transcriptional activation in response to particular input patterns, beginning with transcription of the IEGs, in yet another sense of neuronal "activation" (Chung, 2015).

2. DIFFICULTIES WITH ELECTROPHYSIOLOGY

2.1 Accessibility and Scale

Investigation of anesthetic mechanisms at a systems level requires the simultaneous sampling of multiple, identified circuit members. Neuronal circuits of interest exist at multiple scales, spanning spatial scales from micrometers to meters, and studies of anesthetic drug effects on circuit members must begin by identifying what cell populations participate in a particular circuit of interest. This is not always trivial, particularly given the mutual reciprocity of connectivity throughout most of the mammalian central nervous system. Anesthetic drugs have documented effects on consciousness, arousal, and pain perception, and alter processing at all anatomic levels of the central nervous system, including spinal cord, brainstem, and cortex. While Moruzzi and Magoun's initial work focused on the role of the reticular formation in maintenance of arousal and theorized that decreased reticular formation tone could account for the somnolence that occurs with anesthesia (Moruzzi & Magoun, 1949), other distributed circuits are clearly affected by general anesthetics and may contribute to the functional state of general anesthesia. The thalamus serves to relay most of the cortex's sensory afferents and motor efferents, while it receives reciprocal feedback from different cortical areas (Guillery & Sherman, 2002; Sherman & Guillery, 2002). Within cortex itself, the laminar organization of cortex suggests that particular cell populations may play generic roles in certain types of computation by a given cortical area (Douglas, Martin, & Whitteridge, 1989). Finally, immobility with volatile anesthetics appears to result from a suppression of spinal cord central pattern generators (Antognini & Schwartz, 1993), and modulation of painful sensory stimuli by general anesthetics begins in the spinal cord (Hao, Takahata, Mamiya, & Iwasaki, 2002).

When one considers the shear scale of the problem, with the large number of neurons being affected by general anesthetics, one of the major limitations of electrophysiologic methods that depend upon proximity to small electrodes for localization becomes clear (Fig. 1B). Placing electrodes in close proximity to neurons in deep nuclei of the brainstem is technically challenging, and simultaneously placing many electrodes in order to sample multiple sites in affected networks compounds the problem. For superficial targets, this has been somewhat mitigated by the development of high density electrode arrays, often made of silicon, but traversing large tracts of tissue with multiple perforating shafts is rife with risks of vascular and neuronal injury. This is where assays based on expression of the IEGs, most frequently *c-fos*, shine, because they in theory provide a record that allows post hoc examination of the entire nervous system, or at least as much of the nervous system as the investigators are willing to section and stain. Of course, in exchange for the ability to probe essentially the entire brain simultaneously for expression of IEGs, the temporal resolution is severely limited compared to electrophysiologic methods.

2.2 Neuronal Identity and Sampling Bias

In what may be the most reductionistic framing, consciousness is the ability to variably map the same stimulus onto multiple different responses, depending upon the organism's past experiences and present state. The capability of a conscious organism to shift its behavioral response to particular circumstances results from flexibility in the coupling between neurons, where the sensitivity to an individual neuron to a particular input can be increased or decreased based upon those circumstances to suit the behavioral task at hand (Nicolelis & Fanselow, 2002). This shifting in sensitivity is often measured as a change in the gain of the particular input and the process that changes the gain is known as gain control in the neuroscience literature. Gain control can increase or attenuate the resulting correlation between two different neuronal populations, increasing or decreasing the so-called functional connectivity between the regions. (This is in distinction to the active formation or pruning of synapses, or anatomic connectivity, between the regions and can be distinguished by the ability of the system to rapidly shift between two different functional connectivity states, without the time delay needed for protein expression necessary for the formation or elimination of synapses.) Shifts in cortical gain result in changes in summated synchronized synaptic activity and reflect the changes in effective connectivity seen with the characteristic shifts in EEG with different depths of anesthesia (Friston, Bastos, Pinotsis, & Litvak, 2015; Lee et al., 2013).

Yet understanding how a flexible network shifts its responsiveness depends upon understanding how different members of the circuit interact with each other. Different cell types in cortex are well established to have different characteristic firing properties, though these properties can themselves shift (Steriade, 2004), making it difficult to identify particular cell types from electrophysiologic recordings. Moreover, there is no reliable way to identify whether a particular cell in an extracellular electrophysiologic recording is inhibitory or excitatory.

The ability to identify and sample large numbers of particular, genetically distinct cell populations in vivo is the forte of the new genetically encoded indicators of neuronal activity (Lin & Schnitzer, 2016). Moreover, the maturation of microscopy techniques allows the simultaneous sampling of up to thousands of neurons and the same neurons can be recorded over days to weeks to look at time evolution of circuitry. The explosion of techniques for sampling neuronal activity with genetically encoded indicators has resulted from the development of a dizzying array of fluorescent protein sensors (Sanford & Palmer, 2017) combined with advanced microscopy techniques. While genetically encoded voltage indicators have been developed (reviewed in Yang & St-Pierre, 2016), the most reliable signal to noise over long experimental time frames has been obtained with GECIs, of which GCaMP6 (Chen et al., 2013) has some of the best performance characteristics at the time of this writing.

2.3 Additional Limits of Single-Unit Electrophysiology

Beyond the limitations outlined earlier, with respect to difficulty with identifying a recorded cell's type and morphology, and the invasiveness of the sampling at each spatial location of interest there are other, more subtle disadvantages of relying exclusively on extracellular recordings of single-unit electrophysiology for probing circuit function. The quality of single-unit isolation falls rapidly as the signal-to-noise ratio (SNR) falls (Stratton et al., 2012). To combat this problem, most experimentalists adjust electrode placement in vivo to optimize observed signal to noise. This introduces two sampling bias problems. First, large cells with large dipoles, predominantly pyramidal cells in cortex with their large apical dendrites and long axons, will be preferentially isolated if maximal SNR is sought by an experimentalist. Second, cells with high firing rates are more likely to be identified, because it is easy to move past slowly firing neurons without recording enough spikes to realize the proximity of the electrode tip to the neuron. As a result of this second bias, the reported firing rates from extracellular recording studies tend to be substantially higher than the average firing rate seen with intracellular recording studies and calcium imaging methods (Shoham, O'Connor, & Segev, 2006).

3. *c-fos* AS A REPORTER OF NEURONAL ACTIVITY

Regulatory IEGs are involved in stimulus-transcription coupling (Kovács, 2008). The most frequently used IEG reporter, *c-fos* mRNA is transcribed and c-Fos protein is transiently expressed in neurons after synaptic stimulation (Sagar, Sharp, & Curran, 1988), providing an indirect readout of neuronal activity. As alluded to above, c-fos is directly involved in transcription induction in the AP-1 pathway of target cells.

Activation of the AP-1 pathway begins with calcium influx into the neuron through a combination of the *N*-methyl-D-aspartate glutamine receptor (NDMAR) and voltage-dependent calcium channels (e.g., L-type calcium channels). Calcium influx can activate multiple kinases, including CaMKII, CaMKIV, PKA, and MAPK. Fos expression is mainly mediated by MAPK activation, which is relatively slow and requires relatively high calcium increases when compared to the CaMKIV pathway (Chaudhuri et al., 2000; Chung, 2015), with MAPK activity thus reporting strong and continued neural activity (Deisseroth, Mermelstein, Xia, & Tsien, 2003; Murphy, Smith, Chen, Fingar, & Blenis, 2002). Thus, expression of *c-fos* is not triggered by neuronal depolarization per se, as it is expressed at low levels in intact brain, but instead reports phenotypic reprogramming of the cell by either changes in afferent inputs or external stimuli (Luckman, Dyball, & Leng, 1994). The gene *c-fos*, and its protein product c-Fos, is transiently expressed after activation of the CAMK pathway results in phosphorylation of the transcription factors CREB and Elk-1, which bind the *c-fos* promoter (Fig. 2) (Chung, 2015). The expressed Fos protein in the nucleus forms a dimer with Jun to make up the AP-1 complex that regulates the expression of "late" genes.

Expression of *c-fos* begins within 15 min of stimulation, with peak expression of *c-fos* mRNA approximately 30 min poststimulus and peak expression of c-Fos protein approximately 90–120 min poststimulus

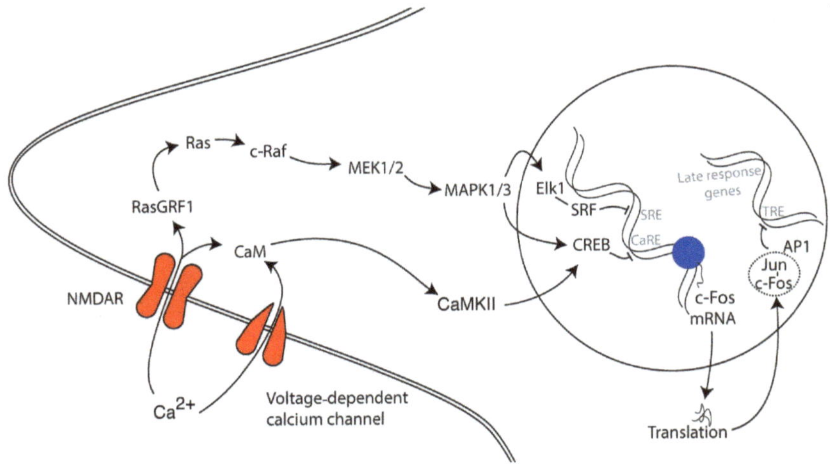

Fig. 2 Signal transduction pathways involving *c-fos*. Calcium fluxes into the neuron during action potentials via NMDARs and voltage-dependent calcium channels. The influx of Ca^{2+} binds calmodulin, activating CaMKII and then CREB. NMDAR activity also activates RasGRF1, activating the MAPK pathway via Ras and c-Raf activation, with resultant activation of CREB and Elk1. Elk1 binds SRF and acts as a transcription factor at SRE, while activated CREB acts as a transcription factor at CaRE, with resultant expression of *c-fos* mRNA. Once the c-Fos is transcribed, it binds to Jun to form the transcription factor AP-1, which results in transcription of late response genes via the TRE (Chaudhuri et al., 2000; Wang, Fibuch, & Mao, 2007). *AP-1*, activating protein 1; *CaM*, calmodulin; *CaMKII*, calcium/calmodulin-dependent protein kinase II; *CaV1.2*, L-type calcium channel; *CRE*, cAMP response elements; *CREB*, cyclic AMP response element-binding protein; *MEK1/2*, mitogen-activated protein kinase kinase 1 and 2; *MAPK1/3*, mitogen-activated protein kinase 1 and 3; *NMDAR*, *N*-methyl-D-aspartate glutamine receptor; *RasGRF1*: Ras-guanine nucleotide releasing factor; *SRE*, serum response element; *SRF*, serum response factor; *TRE*, 12-*O*-tetradecanoylphorbol-13-acetate response element.

(Kovács, 1998), but c-Fos expression can be delayed by as much as hours in some areas of brain with certain stimuli (Cullinan, Herman, Battaglia, Akil, & Watson, 1995). This double peak in expression time course can be exploited through careful experimental design if one stimulus can be placed at the c-Fos expression peak while the other is placed at the *c-fos* transcription peak.

3.1 *c-fos* in Studies of Anesthetics

Quite a few studies of anesthetics have utilized c-fos as a reporter, and only a cursory review is included here to give some indication of the range of applications for the technique to date. Anesthetic exposure to ether leads to increases *c-fos* expression in frontoparietal cortex, but decreases *c-fos* elsewhere compared to the stressor of open field maze exploration (Emmert & Herman, 1999). More frequently, studies have looked for anesthetics to modulate an unanesthetized response to a stimulus. As an example, the response to noxious stimuli has been robustly characterized throughout the CNS (Bullitt, 1990). The lack modulation of spinal cord responses to such stimuli by several sedative hypnotic agents (Shimode et al., 2003) suggests that the analgesic component of general anesthesia must occur elsewhere in the circuit. Anesthetic circuit mechanisms have also been studied as shifts in the balance of *c-fos* expression in sleep-related circuits with volatile anesthetics (Kelz et al., 2008), GABA-ergic agents (Nelson et al., 2002), ketamine (Lu et al., 2008), and dexmedetomidine (Garrity et al., 2015), and *c-fos* expression in parabracheal nucleus correlates with recovery of righting after isoflurane anesthesia (Muindi et al., 2016). This brief review suggests that investigators have been principally employing c-fos for its advantages in sampling inaccessible structures, such as brainstem nuclei, or for simultaneous sampling of regions that are too spatially separated for visualization in a single optical field of a microscope.

4. APPROACHES TO USE c-Fos AS A REPORTER OF NEURONAL ACTIVATION

4.1 Experimental Design

By necessity, mRNA transcription and protein expression take time to occur, and experimental designs must incorporate the appropriate post-stimulus delay in order to utilize these assays. It is also worth remembering that, in addition to the delay, c-fos responses integrate the degree of transcriptional activation over a window of time. Increases of expression of *c-fos* mRNA are maximal at roughly 30 min poststimulus in most cells. The subsequent translation into Fos protein occurs somewhat later, typically peaking at approximately 2 h poststimulus. Some clever experimental designs combine both mRNA and protein assays to test two stimuli in a single experiment, for example, combining immunocytochemistry for nuclear Fos expression 2 h poststimulus 1 with in situ hybridization for cytoplasmic c-fos mRNA levels 30 m poststimulus 2 (Chaudhuri, Nissanov, Larocque, & Rioux, 1997; Kovács, Csejtei, & Laszlovszky, 2001). Good experimental designs will incorporate multiple controls to integrate all of the possible exposures occurring during the hour or so prior to sacrifice. Fig. 3A illustrates an example design using an anesthetic emergence protocol

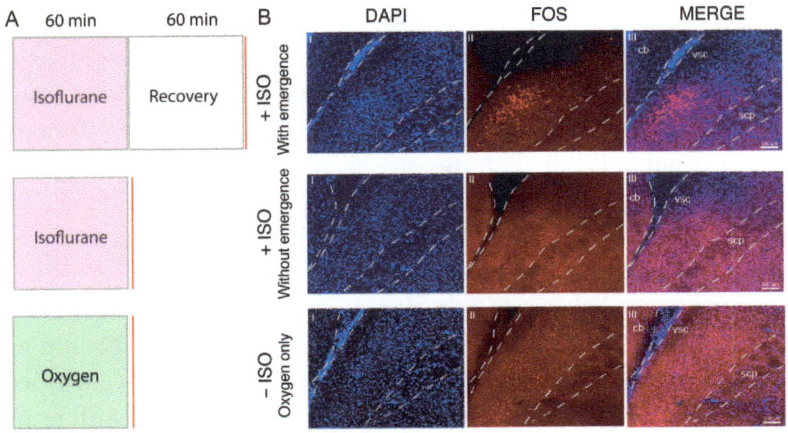

Fig. 3 (A) A sample c-fos experimental protocol. Animals were divided into three groups. Group one (+ISO with emergence) was exposed to isoflurane for 60 min, then allowed to recover for 60 min before being sacrificed for c-fos histology (*red line*). Group 2 (+ISO without emergence) was exposed to isoflurane for 60 min and then sacrificed. Group 3 (−ISO oxygen only) was exposed to oxygen alone for 60 min prior to sacrifice. (B) Sections through the parabrachial nucleus were stained for DAPI and c-fos, showing that exposure to isoflurane or oxygen alone were insufficient for increasing c-fos expression. Neurons in parabrachial nucleus increased c-fos expression when the animals were anesthetized and then emerged from anesthesia (seen as the cluster of cells in +ISO with emergence II and III). *cb*, cerebellum; *ISO*, isoflurane; *scp*, superior cerebellar peduncle; *vsc*, ventral spinocerebellar tract. *Data from Muindi, F., Kenny, J. D., Taylor, N. E., Solt, K., Wilson, M. A., Brown, E. N., et al. (2016). Electrical stimulation of the parabrachial nucleus induces reanimation from isoflurane general anesthesia.* Behavioural Brain Research, 306, *20–25. doi:10.1016/j.bbr.2016.03.021.*

from Muindi and colleagues, where the transition from anesthesia with isoflurane to free movement in oxygen is compared to both anesthesia with isoflurane and free movement in oxygen alone.

4.2 Detection and Visualization of c-Fos

Visualization of *c-fos* mRNA or protein levels in situ requires immunohistochemistry or in situ hybridization methods and expertise with a particular approach may reasonably drive the assay selection (Berghorn, Bonnett, & Hoffman, 1994; Krukoff, 1998). Tissue processing typically begins with fixation in formalin followed by sectioning, though protocols exist to process tissue directly from frozen section (Sundquist & Nisenbaum, 2005) to save substantial tissue processing time. If in situ visualization is not necessary for the particular experiment, it is possible to purchase RT-PCR probes for

FOS commercially (e.g., TaqMan from ThermoFisher, Waltham, MA; PrimePCR SYBR Green Assay; BioRad, Hercules, CA). Such technology has been used to study effects of propofol on isolated hippocampal slices in vivo as a model of the amnestic effects of the drug (Kidambi et al., 2010).

5. LIMITATIONS OF c-fos AS A REPORTER OF NEURONAL ACTIVATION

As a result of the imprecise onset of the transcriptional activation, classic c-Fos reporters, particularly based on protein expression, have low temporal resolution. This issue is compounded by the histological approaches that are necessary to visualize the c-fos expression, preventing in vivo utilization of *c-fos* to detect neuronal activation in real-time, though some investigators have attempted to address the nonreal-time component of *c-fos* by using a construct of GFP under a *c-fos* promoter to allow real-time imaging of transcriptional activation in vivo (Barth, Gerkin, & Dean, 2004).

Interpretation of *c-fos* results must also be tempered with some subtlety. Lack of a Fos response in a population does not preclude the involvement of that population in a functional circuit (Labiner et al., 1993). Indeed, there is no information on connectivity implied by *c-fos* activation; any such assessments of circuit connectivity have to be assessed via other means. IEGs are not expressed in chronically active neurons, and release of chronic inhibition variably results in an increase in IEG expression (Hoffman, Smith, & Verbalis, 1993). Moreover, neuronal activation can occur without induction of IEGs, possibly because the *c-fos* promoter has an AP-1 site, allowing for self-inhibition (Schönthal et al., 1989), or because the microRNA miR-7b interacts with the 3′ region of Fos mRNA inhibiting translation (Lee, Palkovits, & Young, 2006). Alternative technology for attempting to increase both the sensitivity and specificity of activity reporting relative to c-fos has been proposed as well (Sørensen et al., 2016).

6. REPORTING CALCIUM INFLUX WITH GECIs

Optical tools, including GECIs, give us spatial and genetic control over sampling not present with electrical methods (Scanziani & Häusser, 2009). Multiple fluorescent indicators have been developed for calcium imaging, including FRET- and single fluorophore-based indicators and several good comprehensive reviews of available technology exist (Grienberger & Konnerth, 2012; Liem et al., 2004).

6.1 Assessing Optical Indicators: Sensitivity, Kinetics, and Signal to Noise

The choice of a particular GECI will depend upon matching the signal-to-noise characteristics of the indicator with a temporal resolution sufficient for the response of interest. Concerns for temporal resolution can involve both the ability to resolve event timing and the duration over which the response evolves. In low-signal regimes, such as detecting calcium flux associated with a single action potential, this can amount to detection of only a few photons emitted from the fluorophore, making response detection an exercise in discerning shot noise from the background thermal noise. In practice, the optimization of GECIs for signal to noise has tended to favor GECIs that emit a large number of photons for a given calcium flux. Of late, isoforms of GCaMP6 (Chen et al., 2013) have been preferred due to their high SNR. GCaMP6 was developed by refining the single fluorophore-based G-CaMP (now GCaMP), which fused the C-terminus CaM-binding peptide to N-terminus of cpGFP145.

6.2 Imaging Strategies

The particulars in the choice of microscopic technique are beyond the scope of this chapter and will depend upon the time course of the phenomenon under study, the spatial scale that needs to be sampled, and the type of behavior that must be incorporated into the experiment, but a brief review of some of the issues that must be addressed will follow. To make calcium imaging practical in real world applications requires optimization to maximize not only the change in fluorescence with calcium influx relative to the baseline fluorescence but also maximizing the total number of photons generated and detected by the instrumentation (Scanziani & Häusser, 2009). Microscopy techniques based on two-photon methods, with their resultant optical sectioning, reduce the amount of background light out of which the response must be detected, increasing the contrast of the image and improving visualization (Svoboda & Yasuda, 2006). This offers a real world advantage over confocal microscopy for in vivo imaging, as maximizing contrast for a given illumination will limit photodamage to tissue and bleaching of the fluorophore in the GECI. That said, some photobleaching is inevitable and may begin to be significant by 1 h into an imaging session, leading to a varying baseline fluorescence level and unstable indicator performance over time. This can be partially addressed by normalizing responses to their background, reporting fluorescence signals as a fractional variation relative to their mean: $\Delta F/F_0$.

Temporal resolution, and the nature of motion artifacts captured during imaging, will depend upon the acquisition approach as well. Two-photon resonant-scanning systems with a traditional raster scan path directly trade-off spatial resolution or field of view for temporal resolution. Random or otherwise directed scan path technology is developing in an attempt to structure the optical sampling path to allow sampling of large numbers of neurons with faster frame rates, but these scan paths risk the introduction of complex aliasing problems with sample movement.

Image registration, also referred to as image stabilization, suppresses the inevitable movement introduced during in vivo sampling (Fig. 4). Such registration is the first preprocessing step necessary before neuronal isolation so that correlation structures in the neuronal population can be studied without artifactual contamination. In Fig. 4, there is partial movement between frame 17 (A) and frame 18 (B), highlighted in the difference between the two frames (D). This movement blurs the average of the frames in this movie (Fig. 4C). At high frame rates, where optical sampling is fast relative to sample movements, it is possible to model motion artifacts as rigid deformations of the mean image, and simply correct for the movement by shifting the entire image. In this case, there is a slight improvement in sharpness with a rigid body correction (Fig. 4F). When movements begin to approach the time scale of the scan path, the deformations become nonrigid and registration can take considerably more care. As an example, Fig. 4D is redrawn schematically in Fig. 4E; the neurons above the horizontal line moved before the laser scan path reached them, while the neurons below the horizontal line only moved after the scan path sampled them. This results in a nonrigid deformation of the image within the frame. For suppression of nonrigid deformations the method of Greenberg and Kerr works quite well, with a notable improvement in sharpness in Fig. 5G, but is computationally demanding (Greenberg & Kerr, 2009). As would be imagined, registration with random scan paths can be particularly difficult as the movement of the sample relative to the scan path can be hard to suppress.

6.3 Targeted Expression Strategies for GECIs

GCaMP6 is available in a number of forms. For experiments that do not require specific targeting of GECIs by the investigator, transgenic mouse

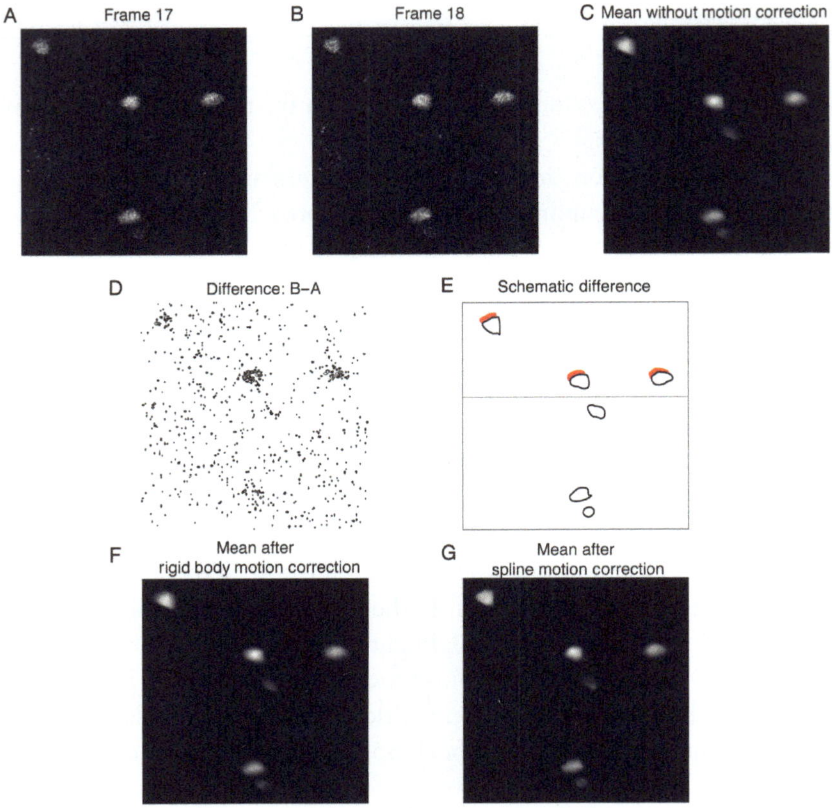

Fig. 4 Image registration for suppression of motion artifacts. Two frames of an acquired video of a fluorophore are shown with motion occurring between the acquisition of frame 17 (A) and frame 18 (B). (C) The motion results in blurring of the mean image. (D) The difference image, resulting from (B) to (A). (E) A schematic drawing of (D), showing that the movement occurred while the laser was scanning through the frame. This is indicated by the *thin horizontal line*, as those neurons above the line moved before the scan path reached them, while those below the line did not. (F) Assuming that all motion between frames occurs as a rigid translation does not substantially sharpen the mean image when compared to (C). (G) Using a spline approach to model movement as a nonrigid deformation of the image results in a notable sharpening of the mean image compared to (C), suggesting that superior registration has been obtained.

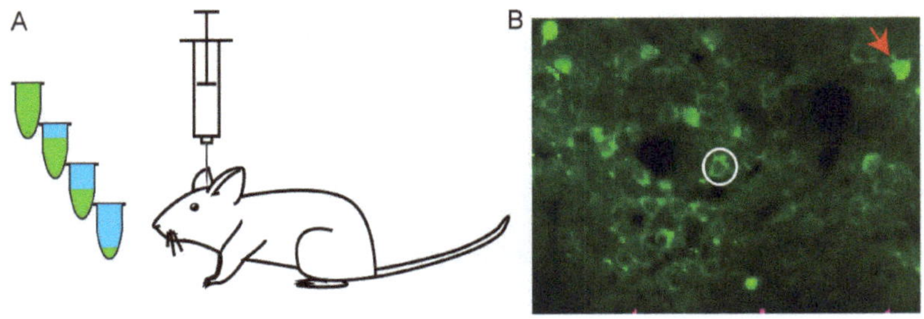

Fig. 5 (A) Selective expression of GCaMP6 is a cell line and tissue of interest will require titration studies, where serial dilutions are injected into different mice that are then imaged at the latest time point that will be imaged in the study. (B) Adequate expression is indicated by fluorescence signal that shows nuclear sparing (*circle*), whereas complete filling of the neuron (*arrow*) indicates overexpression.

lines are available from the Jackson Laboratories (Bar Harbor, ME) and transgenic flies are available from the Bloomington Drosophila Stock Center (Bloomington, IN). Plasmids are available through addgene.org (Cambridge, MA). Multiple isoforms of GCAMP6 are available in multiple constructs from the University of Pennsylvania Viral Vector Core (Philadelphia, PA), which is convenient for investigator targeting.

The simplest selective targeting strategy involves focal injection of the vector with an appropriate vector into the area of interest. More cell-type selective targeting can be achieved with specific promoters for cell types, as in targeting neurons with synapsin promoters. Additionally, floxed targeting with the lox-cre system is possible, allowing the use of transgenic mouse lines that have already been developed to target the populations of interest.

Successful investigator targeting of GCAMP6 expression will require titration studies to ensure adequate expression in the cells of interest for the duration of the sampling period. Initial titration studies with the promoter of interest will require dilution studies (Fig. 5) to find appropriate expression levels for a given vector and promoter at the most distant time point necessary for the study. Appropriate expression can be confirmed by fluorescence in the target populations that excludes the nucleus (e.g., the circled neuron in Fig. 5B). Nuclear staining with GCaMP (arrowhead in Fig. 5B) is suggestive of overexpression and could lead to cellular toxicity or dysfunction.

7. EXPERIMENTAL DESIGN WITH GCaMP6

A typical GCaMP experiment begins with targeted expression of the GCaMP in the area of interest. With several promoters in the Pennsylvania Viral Core multiple, small, slow microinjections in cortex spanning from 50 μm below the area of interest to 50 μm above the layer of interest produces reasonable expression. Most AAV-based GCAMP6 vectors will produce adequate expression to begin imaging after 10 days to 2 weeks.

Although it is possible to make injections through a burr hole and then open an imaging window later, the author has found it effective to perform a craniotomy for the injection leaving the dura otherwise intact. Covering the craniotomy with a glass cover slip held in place with methylmethacralate to provide a cranial window later. In this way, any small amount of bleeding associated with turning the bone flap will likely be resorbed by the time expression will allow imaging, increasing the experimental yield per animal used. A head bar to anchor the animal to the microscope headstage can be implanted in the same surgery or in a separate procedure, as is convenient.

Animals can begin to be habituated to the microscope a few days after implantation if any awake imaging is to be performed. To minimize the movement artifact that will result from imaging an awake animal, we have found both spherical treadmills and turntables can work well to dissipate the power generated by the animal's attempts at locomotion. Habituation to the unusual sensation of being stationary and moving the surface beneath it's feet may only take 1 h or so, depending upon the animal, but several sessions are probably prudent prior to imaging to maximize yield from imaging sessions. Habituation to the open environment has been acheived when the animal is no longer actively trying to run while attached to the microscope. An otherwise dark, quiet imaging room is also helpful in this regard.

Volatile anesthetic can be effectively delivered via mask to a head-fixed mouse, but some careful consideration should be given to rebreathing and leaking around the mask; monitoring exhaled gas concentrations is good practice. Delivery of intravenous medication will have to be made through catheter stable enough to survive as much movement as can be expected in the experimental design.

Finally, some attention must be made to active warming of the mouse while attached to the headstage, to avoid confounding results from hypothermia. If the animal is to be awake for a portion of this procedure a radiant system may be the most expedient, but experimental contamination with noise from any infrared emitters should be minimized with care. Alternatively, if the animal will not be awake for any portion of the procedure, a typical physiologic warming pad works well.

7.1 GECI Data Analysis

Extracting meaningful data from recorded videos requires several stages of processing. First, image registration is necessary. Following motion artifact suppression with registration, neuronal isolation can occur based upon the covariation of signal in nearby pixels. Subsequently, the signal from each putative neuron can be extracted as a raw fluorescence trace from the relevant pixels of the motion-corrected recording. The traces can then be normalized relative to the background fluorescence and analyzed as a fluorescence signal by itself as an estimate of the calcium in the imaged portion of the neuron. If the experiment is concerned with identifying action potentials, the impulse response can be modeled as an exponential decay and

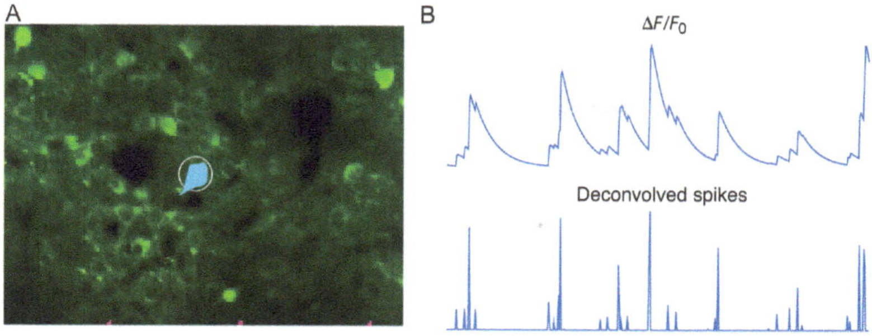

Fig. 6 (A) After registration, a neuron with appropriate expression levels of GCaMP6 (the circled neuron from Fig. 5A) is identified, and the correlation among pixels during the movie is used to isolate the pixels of interest in order to extract the fluorescence signal attributed to a single neuron, *highlighted in the blue area*. (B) The fluorescence signal is first normalized to is mean level (*top*), and the generating spike rate can be estimated through a deconvolution operation (*bottom*) to within a proportionality constant, assuming stable kinetics of the GCaMP6 signal.

the fluorescence trace can be deconvolved to produce an estimate of the action potential train that would be capable of generating the resultant fluorescence trace (Fig. 6). A computational suite for MATLAB exists that provides the neuronal isolation, normalization, and deconvolution steps in a reasonably efficient manner and is highly recommended (Pnevmatikakis et al., 2015). The quality of the resultant spike train estimate tends to degrade at high firing rates (Theis et al., 2016), which thankfully is less of an issue for anesthesia studies, given the usual low firing rates in the anesthetized state.

8. LIMITATIONS OF GCaMP6

Although targeted expression strategies usually work quite well, not all approaches have success. The author's own experience, corroborated by others (D.L. Ringach, personal communication), is that the fast isoform of GCaMP6, GCaMP6-f, does not produce adequate signal when expressed in GAD67 expressing cell lines. Whether this is due to the fast calcium cycling dynamics in this cell population, resulting in very small changes in the intracellular calcium changes or poor expression remains to be completely defined. That said, the higher signal to noise slow isoform of GCaMP6, GCaMP6-s, does produce adequate signal to noise to be used in this cell population, so sometimes several different strategies will need to be explored to settle upon one that allows adequate visualization of the properties of interest. Finally, it is necessary to ensure that the expression of GCaMP6 remains adequate to image throughout the time period of interest without accumulating to levels that prove toxic to neurons.

Many limitations of GCaMP6 have to do with needing optical access to the region of interest, restricting use to areas that are optically accessible, within a few hundred microns of the surface of the brain. Moreover, the deeper the area of interest, the greater the required illumination, increasing the risk of photodamage to tissue, increasing the rate of photobleaching of the fluorophore, and decreasing the signal to noise due to surface reflections of the illumination. Fiber photometry offers an opportunity to utilize GCaMP6 to image deeper structures, but the spatial resolution of the data suffers and most reported systems should be thought of as reporting an average calcium concentration for a local population rather than the subneuron resolution capable with two-photon resonant scanning microscopes (Deisseroth, 2016; Liang, Ma, Watson, & Zhang, 2017). This technology may still have potential when combined with novel imaging approaches, however, for simultaneous imaging of spatially separated areas (Kim et al., 2016) and is the focus of ongoing technique development.

Additionally, the potential dependence of an analysis on the deconvolution of neuronal activity estimates depends upon the accuracy of deconvolution algorithms, which may break down at higher firing rates (Theis et al., 2016). An additional concern is that calcium dynamics could shift their relationship to action potential firing in different behavioral regimes, which seems particularly likely when comparing anesthetized vs awake preparations, and this remains an area of active research.

9. CONCLUSIONS

Alternative assays for detecting neuronal activation will continue to be developed as investigators push against the limits of current technology. Two approaches that provide complementary insights into circuit function, and hence may be ideal for probing how anesthetics modulate of circuit function include the IEG *c-fos* and the GECI GCaMP6. Fos-based approaches detect transcriptional activation of the cell secondary to calcium influx and depend upon detection of gene products in neurons of interest, through immunohistochemical or RT-PCR reporters. Either isolated preparations of neurons of interest or neuronal populations in deep nuclei that are otherwise inaccessible but easily identified on sectioning make ideal targets for *c-fos* techniques. The temporal resolution of *c-fos* is, however, limited. In contrast, GCaMP6 generates a fluorescent signal proportional to intracellular calcium concentration. Imaging this signal requires optical access, usually using two-photon techniques in vivo, making it ideal for assaying activity patterns in large numbers of cortical neurons with defined identities. The temporal resolution of GCaMP6 can detect individual action potentials, but with most conventional microscopes the physical area scanned will decrease with increasing temporal resolution. Extending the reach of GCaMP to deeper areas and understanding the link between calcium dynamics and action potentials remain areas of active research.

ACKNOWLEDGMENT

This work was supported by a grant from the National Institute of General Medical Sciences (1K08GM121961).

REFERENCES

Abeles, M., & Goldstein, M. H. (1977). Multispike train analysis. *Proceedings of the IEEE*, *65*, 762–773. https://doi.org/10.1109/PROC.1977.10559.

Antognini, J., & Schwartz, K. (1993). Exaggerated anesthetic requirements in the preferentially anesthetized brain. *Anesthesiology*, *79*, 1244–1249.

Barth, A. L., Gerkin, R. C., & Dean, K. L. (2004). Alteration of neuronal firing properties after in vivo experience in a FosGFP transgenic mouse. *The Journal of Neuroscience*, *24*, 6466–6475.

Berghorn, K. A., Bonnett, J. H., & Hoffman, G. E. (1994). cFos immunoreactivity is enhanced with biotin amplification. *The Journal of Histochemistry and Cytochemistry*, *42*, 1635–1642. https://doi.org/10.1177/42.12.7983364.

Bullitt, E. (1990). Expression of c-fos-like protein as a marker for neuronal activity following noxious stimulation in the rat. *The Journal of Comparative Neurology*, *296*, 517–530. https://doi.org/10.1002/cne.902960402.

Buzsáki, G. (2004). Large-scale recording of neuronal ensembles. *Nature Neuroscience*, *7*, 446–451. https://doi.org/10.1038/nn1233.

Chaudhuri, A., Nissanov, J., Larocque, S., & Rioux, L. (1997). Dual activity maps in primate visual cortex produced by different temporal patterns of zif268 mRNA and protein expression. *Proceedings of the National Academy of Sciences of the United States of America*, *94*, 2671–2675.

Chaudhuri, A., Zangenehpour, S., Rahbar-Dehgan, F., & Ye, F. (2000). Molecular maps of neural activity and quiescence. *Acta Neurobiologiae Experimentalis (Wars)*, *60*, 403–410.

Chen, T.-W., Wardill, T. J., Sun, Y., Pulver, S. R., Renninger, S. L., Baohan, A., et al. (2013). Ultrasensitive fluorescent proteins for imaging neuronal activity. *Nature*, *499*, 295–300. https://doi.org/10.1038/nature12354.

Chung, L. (2015). A brief introduction to the transduction of neural activity into Fos signal. *Development & Reproduction*, *19*, 61–67. https://doi.org/10.12717/DR.2015.19.2.061.

Collins, T. F. T., Mann, E. O., Hill, M. R. H., Dommett, E. J., & Greenfield, S. A. (2007). Dynamics of neuronal assemblies are modulated by anaesthetics but not analgesics. *European Journal of Anaesthesiology*, *24*, 609–614.

Cullinan, W. E., Herman, J. P., Battaglia, D. F., Akil, H., & Watson, S. J. (1995). Pattern and time course of immediate early gene expression in rat brain following acute stress. *Neuroscience*, *64*, 477–505. https://doi.org/10.1016/0306-4522(94)00355-9.

Deisseroth, K. (2016). Form meets function in the brain: Observing the activity and structure of specific neural connections. In H. Kennedy, D. C. Van Essen, & Y. Christen (Eds.), *Micro-, meso- and macro-connectomics of the brain* (pp. 19–29). Cham: Springer International Publishing. https://doi.org/10.1007/978-3-319-27777-6_3.

Deisseroth, K., Mermelstein, P. G., Xia, H., & Tsien, R. W. (2003). Signaling from synapse to nucleus: The logic behind the mechanisms. *Current Opinion in Neurobiology*, *13*, 354–365. https://doi.org/10.1016/S0959-4388(03)00076-X.

Douglas, R. J., Martin, K. A. C., & Whitteridge, D. (1989). A canonical microcircuit for neocortex. *Neural Computation*, *1*, 480–488. https://doi.org/10.1162/neco.1989.1.4.480.

Emmert, M. H., & Herman, J. P. (1999). Differential forebrain c-fos mRNA induction by ether inhalation and novelty: Evidence for distinctive stress pathways. *Brain Research*, *845*, 60–67. https://doi.org/10.1016/S0006-8993(99)01931-9.

Friston, K. J., Bastos, A. M., Pinotsis, D., & Litvak, V. (2015). LFP and oscillations—What do they tell us? *Current Opinion in Neurobiology*, *31*, 1–6. https://doi.org/10.1016/j.conb.2014.05.004.

Garrity, A. G., Botta, S., Lazar, S. B., Swor, E., Vanini, G., Baghdoyan, H. A., et al. (2015). Dexmedetomidine-induced sedation does not mimic the neurobehavioral phenotypes of sleep in Sprague Dawley rat. *Sleep*, *38*, 73–84. https://doi.org/10.5665/sleep.4328.

Greenberg, D. S., & Kerr, J. N. D. (2009). Automated correction of fast motion artifacts for two-photon imaging of awake animals. *Journal of Neuroscience Methods*, *176*, 1–15. https://doi.org/10.1016/j.jneumeth.2008.08.020.

Grienberger, C., & Konnerth, A. (2012). Imaging calcium in neurons. *Neuron*, *73*, 862–885. https://doi.org/10.1016/j.neuron.2012.02.011.

Guillery, R. W., & Sherman, S. M. (2002). The thalamus as a monitor of motor outputs. *Philosophical Transactions of the Royal Society of London Series B, Biological Sciences*, *357*, 1809–1821. https://doi.org/10.1098/rstb.2002.1171.

Hao, S., Takahata, O., Mamiya, K., & Iwasaki, H. (2002). Sevoflurane suppresses noxious stimulus-evoked expression of Fos-like immunoreactivity in the rat spinal cord via activation of endogenous opioid systems. *Life Sciences*, *71*, 571–580. https://doi.org/10.1016/S0024-3205(02)01704-6.

Harris, K. D., Henze, D. A., Csicsvari, J., Hirase, H., & Buzsáki, G. (2000). Accuracy of tetrode spike separation as determined by simultaneous intracellular and extracellular measurements. *Journal of Neurophysiology*, *84*, 401–414.

Hoffman, G. E., Smith, M. S., & Verbalis, J. G. (1993). c-Fos and related immediate early gene products as markers of activity in neuroendocrine systems. *Frontiers in Neuroendocrinology*, *14*, 173–213. https://doi.org/10.1006/FRNE.1993.1006.

Kelz, M. B., Sun, Y., Chen, J., Cheng Meng, Q., Moore, J. T., Veasey, S. C., et al. (2008). An essential role for orexins in emergence from general anesthesia. *Proceedings of the National Academy of Sciences of the United States of America*, *105*, 1309–1314. https://doi.org/10.1073/pnas.0707146105.

Kidambi, S., Yarmush, J., Berdichevsky, Y., Kamath, S., Fong, W., & Schianodicola, J. (2010). Propofol induces MAPK/ERK cascade dependant expression of cFos and Egr-1 in rat hippocampal slices. *BMC Research Notes*, *3*, 3–8. https://doi.org/10.1186/1756-0500-3-201.

Kim, C. K., Yang, S. J., Pichamoorthy, N., Young, N. P., Kauvar, I., Jennings, J. H., et al. (2016). Simultaneous fast measurement of circuit dynamics at multiple sites across the mammalian brain. *Nature Methods*, *13*, 325–328. https://doi.org/10.1038/nmeth.3770.

Kovács, K. J. (1998). c-Fos as a transcription factor: A stressful (re)view from a functional map. *Neurochemistry International*, *33*, 287–297. https://doi.org/10.1016/S0197-0186(98)00023-0.

Kovács, K. J. (2008). Measurement of immediate-early gene activation- c-fos and beyond. *Journal of Neuroendocrinology*, *20*, 665–672. https://doi.org/10.1111/j.1365-2826.2008.01734.x.

Kovács, K. J., Csejtei, M., & Laszlovszky, I. (2001). Double activity imaging reveals distinct cellular targets of haloperidol, clozapine and dopamine D3 receptor selective RGH-1756. *Neuropharmacology*, *40*, 383–393. https://doi.org/10.1016/S0028-3908(00)00163-5.

Krukoff, T. L. (1998). c-Fos expression as a marker of functional activity in the brain. In *Cell neurobiology Techniques. Neuromethods*: Vol. 33. Totowa: Humana Press.

Labiner, D. M., Butler, L. S., Cao, Z., Hosford, D. A., Shin, C., & McNamara, J. O. (1993). Induction of c-fos mRNA by kindled seizures: Complex relationship with neuronal burst firing. *The Journal of Neuroscience*, *13*, 744–751.

Lee, U., Ku, S., Noh, G., Baek, S., Choi, B., & Mashour, G. A. (2013). Disruption of frontal–parietal communication by ketamine, propofol, and sevoflurane. *Anesthesiology*, *118*, 1264–1275.

Lee, H.-J., Palkovits, M., & Young, W. S. (2006). miR-7b, A microRNA up-regulated in the hypothalamus after chronic hyperosmolar stimulation, inhibits Fos translation. *Proceedings of the National Academy of Sciences of the United States of America*, *103*, 15669–15674. https://doi.org/10.1073/pnas.0605781103.

Liang, Z., Ma, Y., Watson, G. D. R., & Zhang, N. (2017). Simultaneous GCaMP6-based fiber photometry and fMRI in rats. *Journal of Neuroscience Methods*, *289*, 31–38. https://doi.org/10.1016/j.jneumeth.2017.07.002.

Liem, E. B., Lin, C.-M., Suleman, M.-I., Doufas, A. G., Gregg, R. G., et al. (2004). Anesthetic requirement is increased in redheads. *Anesthesiology*, *101*, 279–283.

Lin, M. Z., & Schnitzer, M. J. (2016). Genetically encoded indicators of neuronal activity. *Nature Neuroscience*, *19*, 1142–1153.

Logothetis, N. K. (2003). MR imaging in the non-human primate: Studies of function and of dynamic connectivity. *Current Opinion in Neurobiology*, *13*, 630–642. https://doi.org/10.1016/j.conb.2003.09.017.

Lu, J., Nelson, L. E., Franks, N., Maze, M., Chamberlin, N. L., & Saper, C. B. (2008). Role of endogenous sleep-wake and analgesic systems in anesthesia. *The Journal of Comparative Neurology*, *508*, 648–662. https://doi.org/10.1002/cne.21685.

Luckman, S. M., Dyball, R. E., & Leng, G. (1994). Induction of c-fos expression in hypothalamic magnocellular neurons requires synaptic activation and not simply increased spike activity. *The Journal of Neuroscience*, *14*, 4825–4830.

Moruzzi, G., & Magoun, H. W. (1949). Brain stem reticular formation and activation of the EEG. *Electroencephalography and Clinical Neurophysiology*, *1*, 455–473. https://doi.org/10.1016/0013-4694(49)90219-9.

Muindi, F., Kenny, J. D., Taylor, N. E., Solt, K., Wilson, M. A., Brown, E. N., et al. (2016). Electrical stimulation of the parabrachial nucleus induces reanimation from isoflurane general anesthesia. *Behavioural Brain Research*, *306*, 20–25. https://doi.org/10.1016/j.bbr.2016.03.021.

Murphy, L. O., Smith, S., Chen, R. H., Fingar, D. C., & Blenis, J. (2002). Molecular, interpretation of ERK signal duration by immediate early gene products. *Nature Cell Biology*, *4*, 556–564. https://doi.org/10.1038/ncb822.

Nelson, L. E., Guo, T. Z., Lu, J., Saper, C. B., Franks, N. P., & Maze, M. (2002). The sedative component of anesthesia is mediated by GABA(A) receptors in an endogenous sleep pathway. *Nature Neuroscience*, *5*, 979–984. https://doi.org/10.1038/nn913.

Nicolelis, M. A. L., & Fanselow, E. E. (2002). Dynamic shifting in thalamocortical processing during different behavioural states. *Philosophical Transactions of the Royal Society of London. Series B, Biological Sciences*, *357*, 1753–1758. https://doi.org/10.1098/rstb.2002.1175.

Perouansky, M. (2012). The quest for a unified model of anesthetic action: A century in claude Bernard's shadow. *Anesthesiology*, *117*, 465–474. https://doi.org/10.1097/ALN.0b013e318264492e.

Pnevmatikakis, E. A., Soudry, D., Gao, Y., Machado, T. A., Merel, J., Pfau, D., et al. (2015). Simultaneous denoising, deconvolution, and demixing of calcium imaging data. *Neuron*, *89*, 285–299. https://doi.org/10.1016/j.neuron.2015.11.037.

Sagar, S. M., Sharp, F. R., & Curran, T. (1988). Expression of c-fos protein in brain: Metabolic mapping at the cellular level. *Science*, *240*, 1328–1331. 80.

Sanford, L., & Palmer, A. (2017). Recent advances in development of genetically encoded fluorescent sensors. In *Methods in Enzymology* (1st ed.). Elsevier Inc. https://doi.org/10.1016/bs.mie.2017.01.019

Scanziani, M., & Häusser, M. (2009). Electrophysiology in the age of light. *Nature*, *461*, 930–939. https://doi.org/10.1038/nature08540.

Schönthal, A., Büscher, M., Angel, P., Rahmsdorf, H., Ponta, H., Hattori, K., et al. (1989). The Fos and Jun/AP-1 proteins are involved in the downregulation of Fos transcription. *Oncogene*, *4*, 629–636.

Sherman, S. M., & Guillery, R. W. (2002). The role of the thalamus in the flow of information to the cortex. *Philosophical Transactions of the Royal Society of London. Series B, Biological Sciences*, *357*, 1695–1708. https://doi.org/10.1098/rstb.2002.1161.

Shimode, N., Fukuoka, T., Tanimoto, M., Tashiro, C., Tokunaga, A., & Noguchi, K. (2003). The effects of dexmedetomidine and halothane on Fos expression in the spinal dorsal horn using a rat postoperative pain model. *Neuroscience Letters*, *343*, 45–48. https://doi.org/10.1016/S0304-3940(03)00309-4.

Shoham, S., O'Connor, D. H., & Segev, R. (2006). How silent is the brain: Is there a "dark matter" problem in neuroscience? *Journal of Comparative Physiology. A, Neuroethology, Sensory, Neural, and Behavioral Physiology*, *192*, 777–784. https://doi.org/10.1007/s00359-006-0117-6.

Sørensen, A. T., Cooper, Y. A., Baratta, M. V., Weng, F.-J., Zhang, Y., Ramamoorthi, K., et al. (2016). A robust activity marking system for exploring active neuronal ensembles. *eLife*, *5*, e13918. https://doi.org/10.7554/eLife.13918.

Steriade, M. (2004). Neocortical cell classes are flexible entities. *Nature Reviews. Neuroscience*, *5*, 121–134. https://doi.org/10.1038/nrn1325.

Stratton, P., Cheung, A., Wiles, J., Kiyatkin, E., Sah, P., & Windels, F. (2012). Action potential waveform variability limits multi-unit separation in freely behaving rats. *PLoS One*, 7, e38482.

Sundquist, S. J., & Nisenbaum, L. K. (2005). Fast Fos: Rapid protocols for single- and double-labeling c-Fos immunohistochemistry in fresh frozen brain sections. *Journal of Neuroscience Methods*, *141*, 9–20. https://doi.org/10.1016/J.JNEUMETH.2004.05.007.

Svoboda, K., & Yasuda, R. (2006). Principles of two-photon excitation microscopy and its applications to neuroscience. *Neuron*, *50*, 823–839. https://doi.org/10.1016/j.neuron.2006.05.019.

Takahashi, S., Anzai, Y., & Sakurai, Y. (2003). Automatic sorting for multi-neuronal activity recorded with tetrodes in the presence of overlapping spikes. *Journal of Neurophysiology*, *89*, 2245 LP–2258.

Theis, L., Berens, P., Froudarakis, E., Reimer, J., Román Rosón, M., Baden, T., et al. (2016). Benchmarking spike rate inference in population calcium imaging. *Neuron*, *90*, 471–482. https://doi.org/10.1016/j.neuron.2016.04.014.

Wang, J. Q., Fibuch, E. E., & Mao, L. (2007). Regulation of mitogen-activated protein kinases by glutamate receptors. *Journal of Neurochemistry*, *100*, 1–11. https://doi.org/10.1111/j.1471-4159.2006.04208.x.

Yang, H. H., & St-Pierre, F. (2016). Genetically encoded voltage indicators: Opportunities and challenges. *The Journal of Neuroscience*, *36*, 9977–9989. https://doi.org/10.1523/JNEUROSCI.1095-16.2016.

CHAPTER THIRTEEN

Genomics and Proteomic Techniques

Jonathan Z. Pan[1]
University of California at San Francisco, San Francisco, CA, United States
[1]Corresponding author: e-mail address: jonathan.pan@ucsf.edu

Contents

Abstract

Although general anesthesia induced by inhaled anesthetics produces definitive phenotypes (e.g., loss of mobility, amnesia, analgesia), the underlying targets of these drugs are still not clear. Genomics and proteomic techniques are discussed for measurement of global transcriptional and translational changes after inhaled anesthetic exposures. The current discussion focuses primarily on the genomic and proteomic technical methodology. We also include a discussion of network and pathway analyses for data interpretation after identification of the targets.

1. INTRODUCTION

Inhaled anesthetics produce significant physiological effects, including immobility, amnesia, and analgesia. They bind specifically to a large number of brain proteins (enzymes, channels, and receptors), yet the critical drug targets remain uncertain (Eckenhoff, 2001; Herold, Andersen, & Hemmings, 2017; Howard, Trudell, & Harris, 2014; Weiser et al., 2014). To better understand their mechanisms, it might be helpful to profile the cellular and organismal response to these critically important drugs. Regulation of

Methods in Enzymology, Volume 603
ISSN 0076-6879
https://doi.org/10.1016/bs.mie.2018.02.008

such responses occurs at several levels, e.g., transcriptional, translational, and posttranslational (Culley et al., 2006; Eckenhoff, Chan, & Eckenhoff, 2002; Marota, Crosby, & Uhl, 1992). Therefore, both genomic (e.g., DNA microarray) and proteomic (DIGE, differential in-gel electrophoresis) approaches can be employed to elucidate the global response to primary (anesthesia) and secondary (side) effects of inhaled anesthetics (Pan et al., 2006, 2007, 2008).

In the genomic study, at least two experimental designs can be used (Pan et al., 2006). The first employs intact animals exposed to different inhaled anesthetics (e.g., halothane, isoflurane, sevoflurane) with sub EC_{50} (MAC) concentrations in order to minimize adverse physiological effects (hypotension, hypoxia, etc.), but repeated exposures in order to amplify responses. The second uses cultured primary neurons exposed to higher concentrations in order to assess direct effects in the complete absence of adverse physiology. An unbiased study can then be carried out using commercially available DNA microarrays (Affymetrix and Agilent gene chips). Because of the high likelihood of statistical type 1 error, genes with differential expression levels revealed by the microarray analyses, should then be validated using quantitative reverse transcriptase-polymerase chain reaction (qRT-PCR).

In the proteomic (translational/posttranslational) study, DIGE is one approach, that can be used with a similar experimental design to assess both cellular (direct and primary effect) and organismal (both direct/primary and indirect/secondary effect) proteomic responses to general anesthetics (Pan et al., 2008). Protein spots with significant changes can be excised, sequenced, and identified by mass spectrometry. Other still evolving approaches not discussed here are entirely based on mass spectrometry, where different samples are labeled with different isotopes, pooled, digested, and then peptide ratios determined.

Both of these large-scale gene or protein expression experiments produce thousands of independent results that require careful statistical analysis, largely to avoid type 1 error; typically analyses of variance with appropriate post hoc multiple comparison tests (Hochberg & Benjamini, 1990). Type 2 error is also possible if the most stringent test is used (i.e., Bonferroni multiple comparison test). Subsequent validation experiments (qPCR or western immunoblot) are desirable. Once changes are confirmed, both free and commercial platforms are available for use in network and pathway analyses, which allow data interpretation based on the full body of related literature (Pan et al., 2007).

2. PROCEDURES AND RESULTS

2.1 Genomics Study (DNA Microarray)

2.1.1 Volatile Anesthetics

2.1.1.1 Animal Exposure

Sub EC_{50} (~0.8 MAC) should be used for animal treatment to avoid physiological side effects that themselves can alter cellular responses. At this level of anesthetic, animals are typically unconscious. In our own experiments, animals were treated for 90 min twice daily with 3 h recovery time for 5 or 10 exposures, respectively, in an attempt to amplify responses, although it should be recognized that changes may also reflect cellular attempts at adaptation. Anesthetic gas phase concentrations in the chamber should be monitored (by mass, infrared, or Raman spectroscopy), as should oxygen, carbon dioxide, and temperature (see chapter "The mouse as a model organism for assessing anesthetic sensitivity" by Wasilczuk et al. in Vol. 602). These experiments can be expensive, so one must balance cost to replicates. We used group sizes of no less than three and did not pool the material in order to retain a measure of biological variability. After all the exposures and recovery, animals were euthanized, perfused with ice-cold phosphate-buffered saline, and whole brains were quickly removed and snap frozen in liquid nitrogen and stored at −80°C.

2.1.1.2 Cell Exposure

Fetal brains are obtained from pregnant rats late in gestation (typically day 17). Cortical neurons are mechanically dissociated and resuspended in serum-free B27/neurobasal medium. Cortical neuron cultures are then maintained in the culture medium in a humidified atmosphere (5% CO_2, 95% air) at 37°C. At in vitro day 16 (DIV 16), plates of cells can be exposed to higher concentrations of volatile anesthetic (e.g., 3 MAC) in a gas-tight chamber for different lengths of time, while separate plates are treated with CO_2/air as controls. The geometry of the chamber and cell culture dish, together with volumes, must be taken into consideration when determining actual exposure duration. The media may take hours to equilibrate, and it is best to actually measure the anesthetic in the media using either GC or HPLC. After recovery for 6 h, cells are harvested by scraping in RNAlater, pelleted, and stored at −80°C.

2.1.2 RNA Preparation and Hybridization

Total cellular RNA samples are isolated from whole brain tissue and cultured neurons. Different microarray platforms are available. In our previous experiments, we used the Affymetrix chip (RG-U34A, 8799 genes) for the whole brain studies. In this case, cDNA was reverse transcribed, followed by synthesis of cRNA, which was then hybridized to the DNA microarray chip. The RNA from a single animal was used for each microarray. We used the Agilent microarray (20,000 genes) for the cultured neuron studies. Here, cDNA products from pairs of animals (exposed and control) were synthesized and labeled with Cy5 and Cy3 and directly hybridized to the array. The differential fluorescence signals were collected and analyzed.

2.1.3 Data Analysis

Data from each chip should be normalized (e.g., to 50th percentile or intensity-dependent normalization), then filtered (by absent calls or missing values in all microarray chips). In both studies (animal vs cell exposure), data from control groups were compared to that from treatment groups to produce characteristic scatterplots (e.g., Fig. 1) (Pan et al., 2006). Outliers become apparent and can be subsequently validated and studied. In Affymetrix array, to examine the treatment effect, the 5 and 10 animal exposure groups were compared to controls. In Agilent array, differential expression between red and green channels (drug exposure vs control on each chip) is expressed as log ratio, and statistical cutoffs are established (sometimes arbitrarily, e.g., two-fold) for subsequent evaluation.

With enormous numbers of variables being measured per sample (thousands of genes) coupled with a relatively small sample size, the chance of measured differences (gene expression ratios) being significant is very small, if using standard post hoc multiple comparison tests (e.g., Bonferroni). It is widely accepted to apply a less conservative, yet appropriate approach, namely, Benjamini–Hochberg false discovery rate (BHFDR), for identification of a small number differences for subsequent validation. Using the Affymetrix platform (animal exposure), two-way ANOVA with BHFDR is used to differentiate the changes while eliminating batch effects from one group of samples/chips run at a different time from the others. In our Agilent experiments (cell exposure), one-way ANOVA with BHFDR was applied and a small, manageable list of differentially expressed genes was revealed.

2.1.4 Quantitative RT-PCR

Depending on the list size, microarray results should be confirmed by comparison with mRNA levels obtained by the qRT-PCR, using selected gene-specific primer pairs and the same RNA samples as used in the microarray

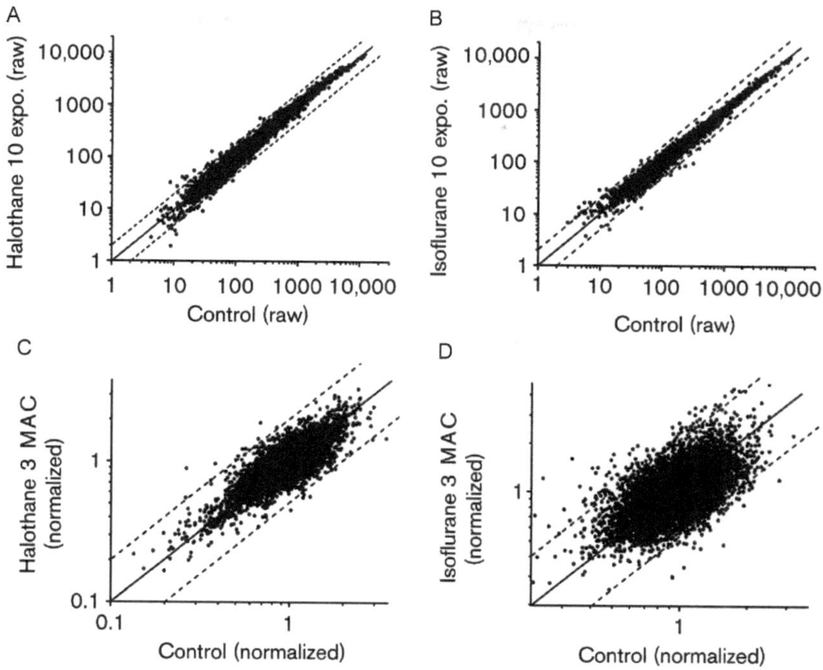

Fig. 1 Scatterplot of microarray data. Average control raw data vs average (animal study using Affymetrix chips) (A) halothane 10 exposure data and (B) isoflurane 10 exposure data. The lines of identity and the two-fold lines are shown. Note the overall low level of variation and the slightly greater scatter for the halothane data compared to isoflurane. The analogous results for the cell exposure data for (C) halothane and (D) isoflurane, except expressed as normalized data because of the array type (Agilent microarray). In this case, note the slightly larger variation for the isoflurane results.

experiments. Methods for qRT-PCR are readily available, but briefly, RNA is reverse transcribed with SuperScript II with random primers (Invitrogen). The PCR reactions are then carried out using ~10 ng of cDNA, 50 nmol of each primer, and SYBR Green master mix (Applied Biosystems) in 10 μL reactions. Levels of qRT-PCR product are measured using SYBR Green fluorescence; collected during real-time PCR on an Applied Biosystems 7900HT system. A control cDNA dilution series is created for each gene to establish a standard curve. Each reaction is subjected to melting point analysis to confirm single amplified products. The expression value of each gene should be normalized to a constitutively expressed "housekeeping" gene, such as β-actin or glyceraldehyde 3-phosphate dehydrogenase (GAPDH) cDNA, to calculate a relative amount of RNA present in each sample. The ratio of normalized expression values for each treated sample to the mean of the corresponding controls is then determined. The mean fold change is the geometric mean of those ratios within a group. Student's *t*-test was conducted between the ratios of normalized relative expression values for treated samples to controls and to determine statistical significance.

2.1.5 Limitations

It is important to realize that the gene chip batch effect can be significant. If not all the samples are run on the same batch of chips, the sample/chip batch effect could contribute to the overall variance in the data, and reduce an ability to detect changes. In our experiments, we had to perform two-way ANOVA (experimental condition vs sample/chip batch) to eliminate variation from samples collected at different time or using different batch of chips. Because of the expense of gene chips, some investigators choose to pool biological samples, but this obviously eliminates any measure of biological variability and demands post hoc validation.

2.2 Proteomics: Binding

The direct binding proteome for the volatile anesthetics can only be accessed via photolabeling, since the low affinity of all these drugs for their targets indicates very fast off-rates. Please see chapter on "Photolabeling Methodology" by Woll et al., as well as reviews on the subject for a more complete description and background (Eckenhoff, Xi, & Dailey, 2010; Weiser et al., 2014). The basic methodology is to first select the tissue of interest, typically some component of the central nervous system (CNS), and then to produce a homogenate at no more than 1 mg/mL of total protein. This is incubated with any of three current volatile anesthetic photolabels: halothane, azi-isoflurane, or azi-sevoflurane, at around a 10th of the clinical EC_{50}, for 10–15 min. This mixture is then placed in gas-tight quartz cuvettes of various volumes, and exposed to UV light of the appropriate wavelength, and for appropriate durations, both of which depend on the particular photolabel. For example, halothane requires 254 nm light, azi-isoflurane and azi-sevoflurane about 300 nm. All require durations on the order of 10–15 min. UV lamps have varying power output, so titration of distance and time may be necessary to achieve reasonable binding efficiencies. Some lamps have significant radiation outside of the listed peak, and filters are occasionally required to limit short UV exposure, as it can damage proteins significantly. After light exposure, the proteins are separated from unbound ligand or photoproducts with a combination of precipitation and centrifugation, and then typically separated on SDS-PAGE or 2D-PAGE. Detecting bound proteins require either a radiolabel or mass spectrometry.

2.2.1 Radiolabel Detection

Initial experiments used halothane; this anesthetic required no modification because photolysis at the C–Br bond was easily achieved with 254 nm light, producing a fairly reactive carbon-centered free radical. A ^{14}C version was made, thus any bound protein should have excess radioactivity, which is easily quantitated with scintillation counting, or autoradiography. Fig. 2 shows a schematic workflow of a human brain proteomic experiment using ^{14}C-halothane (Pan et al., 2007). After brain samples are labeled, they are separated on 2D-PAGE, and an autoradiogram prepared. Spots demonstrating radioactivity are excised, digested, and the peptides mass-profiled using mass spectrometry, and the proteins thereby identified. These data are then fed into network and pathway analyses (see Sections 2.4.1 and 2.4.2).

2.2.2 Mass Adduct Detection

A photolabeled protein has additional mass adducted to one or more residues, and of a very precise and known amount. Combined with the fact that it is exceedingly expensive to radiolabel azi-isoflurane and azi-sevoflurane, mass spectrometry has improved to the point where it is the technique of choice for detecting protein modifications. Thus, after samples are photolabeled, they can be either digested and submitted directly to liquid chromatography/mass spectrometry (LC/MS) to detect modified proteins, or first separated or enriched, and then digested and submitted for LC/MS. In both cases, the output is the identification of anesthetic-binding proteome, which can then be submitted as above for network and pathway analysis.

2.3 Proteomics: Expression

2.3.1 Volatile Anesthetic Exposure

2.3.1.1 Animal Exposure

As mentioned earlier, it is important to rule out physiological effects of the anesthetics as being responsible for any omics changes. Thus, pilot studies are often carried out to measure and report the magnitude of these effects and reduce their magnitude if possible. For example, in our experiment, we studied physiology in a group of 12 rats exposed to 1 MAC of different inhaled anesthetics (isoflurane (1.5% v/v) or halothane (1% v/v)), or to only 30% oxygen in nitrogen (control), for 5 h ($n = 4$ per condition). Blood gases,

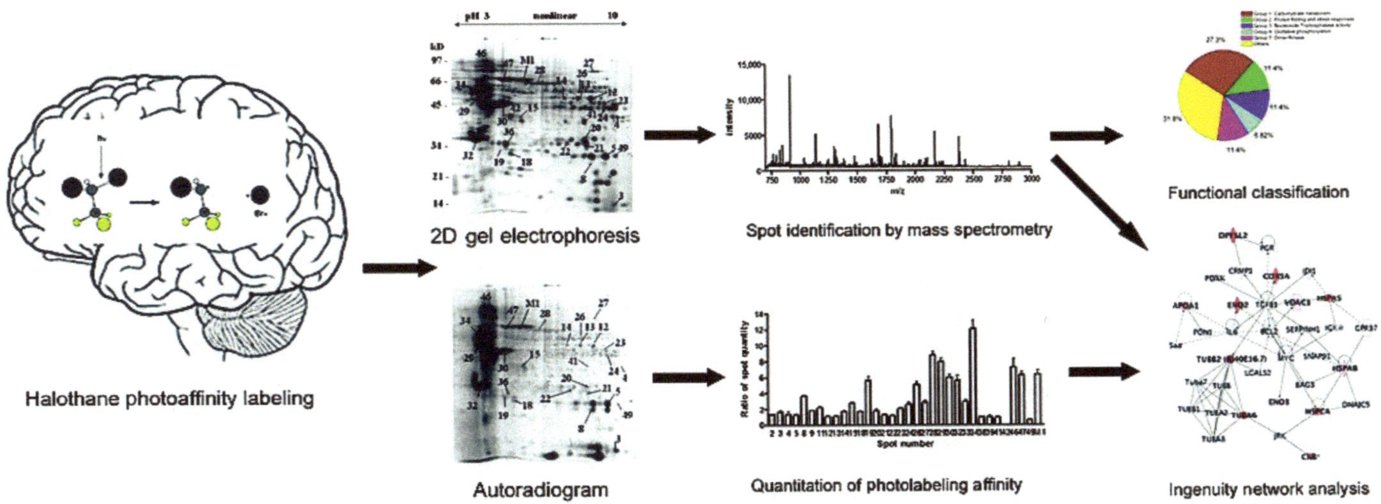

Fig. 2 Schematic flow chart of a human brain proteomic binding experiment. Both membrane and soluble proteins were prepared from brain tissue and separated by 2D gel electrophoresis. Those protein spots, photolabeled by halothane, were then submitted for identification by mass spectrometry. Further analyses of these proteins were performed to reveal the potential networks and pathways involved in their actions.

arterial blood pressure, respiratory rate, and rectal temperature were monitored and recorded every hour during the 5 h exposure. We found only modest changes, giving us confidence that physiology per se was not contributing significantly to the proteomic results. Such a pilot group should not be included in the proteomics study, because of the interventions necessary to collect the physiological data.

For the DIGE/protein study, a larger group of rats (in our case, 32) are exposed to 1 MAC anesthetics (or 30% oxygen only) for 5 h, followed by recovery. A variable recovery (0, 2, 24 h) was introduced in our experiments as it is clear that some proteomic responses can take time. At each time point, rats were rapidly anesthetized and perfused with ice-cold saline via the left ventricle, the brains removed, and dissected into regions (hippocampus and cerebral cortex), followed by rapid freezing in liquid nitrogen, and storage at −80°C.

2.3.1.2 Cell Exposure

The methods used to prepare and expose primary cultured cortical neurons to inhaled anesthetics are the same as described earlier (Pan et al., 2006). At DIV 16, three of the nine plates from each mother were exposed to 3 MAC halothane (3% v/v), 3 MAC isoflurane (4.5% v/v), or control gas for 6 h (5% CO_2/21% O_2/balance N2).

Various approaches for cell anesthetic exposure can be used. In Section 2.1, gas flow into the cell culture incubator passed through a calibrated vaporizer first. While simple, this requires a separate incubator for the control samples and equilibrates slowly due to the very high volume. In these proteomic studies, we fabricated low volume gas-tight stainless chambers into which a calculated amount of liquid anesthetic could be injected. Several such chambers could then be loaded into a single cell culture chamber. Pilot studies were necessary to assure that neither oxygen nor CO_2 varied during the exposure. The advantage here is rapid equilibration and conservation of the liquid anesthetic.

After exposures, the cells could be allowed to recover for variable times (we used 2 h), then harvested, frozen in liquid nitrogen, and stored at −80° C. Pooling of material is inevitable as it is difficult to culture cells from each fetus. We elected to pool cells from one plate of the same exposure condition from each of the three pregnant rats as one sample for protein preparation. Thus, there were three replicates for each drug treatment and three controls, in each of which cells from all three litters were represented.

2.3.2 Two-Dimensional Gel Electrophoresis

By interacting with their various targets, inhaled anesthetics may induce translational and posttranslational regulations up- or downstream in the CNS tissue; such targets might give clues as to important pathways influenced by the anesthetic. 2D-DIGE technology allows one to map and measure differential protein expression as a function of drug, region, and time.

Samples are first incubated with specific fluorophores that allow proteins to become uniquely covalently labeled. Then different samples are mixed together so that they can be run under identical conditions in the same gel, but retain their identity via the label. For the first dimension, the combined samples are added and rehydrated into 18 cm pH 3–10 IPG gel, followed by isoelectric focusing at 120 kVh. For the second dimension, these strips are placed at the top of large precast 9%–16% gradient PAGE gels using low fluorescence glass plates with one plate presilanized to preferentially affix the gel. After 15 h electrophoresis, gels are scanned immediately at high (100 μm) resolution separately by dye-specific excitation and emission wavelengths using a Typhoon 9400 variable mode imager. The gel intensities were kept in the linear range of 60,000–90,000 pixels/gel for image capture.

2.3.3 Data Processing

Tagged image file format (.tiff) images are then cropped and exported for analysis using the DeCyder v5.0 software. In our studies, the differential in-gel analysis module was used to identify protein spots and quantitatively compare expression ratio of each individual protein spot (ratio of Cy5 to Cy3) on a gel after normalization to the Cy2 signal. The unique feature of in-gel normalization using Cy2 channel reduces the need for technical replicates and increases the quantitative sensitivity. Protein spots are first manually matched, to provide landmarks for DIGE decyder software, which then automatically matches the rest of the spots across the gels. In the biological variation analysis module, Cy2 channel across gels are matched and spot volume values of Cy2, Cy3, and Cy5 channels are reported (spot volume = spot intensity/gel intensity). Image and spot intensity data are then reported.

2.3.4 Data Analysis

Data should be further processed as the ratio of Cy5 and Cy3 after normalization by the Cy2 channel (pooled internal standard). We used Partek software (Genome Suite Software) for one-way ANOVA with multiple correction testing (BHFDR). Protein spots with significant changes were selected (BHFDR, $P < 0.05$). Finally, it is necessary to run a separate 2D gel and stain with Coomassie blue. Protein spots with significant regulation of expression are localized on this gel, and then excised, digested, and submitted for mass spectroscopic identification (example shown in Fig. 3) (Pan et al., 2008). In our data, a total of 39 different proteins (combining regions, anesthetics, and cultured neurons) out of an average of 500 detectable protein spots in each gel. In this study, we were surprised to find that halothane and isoflurane had brain region-specific effects (Pan et al., 2008). There were more protein changes following halothane exposure in the cortex, whereas isoflurane altered the expression more in the hippocampus. As might be expected, though, about half of the altered proteins are the same.

2.4 Network and Pathway Analyses

The typical genomic or proteomic experiment yields a large number of changes that, on the surface, are not easily interpreted. After obtaining large-scale genomic and proteomic data, the appropriate statistical analysis method is first selected. Then, after identification of the changes, protein/gene ontology analyses (network and pathway) are performed. To aid interpretation, both free and commercial platforms are available to analyze the list of target genes or proteins, such as DAVID (database for annotation, visualization, and integrated discovery) (Fig. 4) and IPA (ingenuity pathway analysis) (Fig. 5).

2.4.1 Database for Annotation, Visualization, and Integrated Discovery

DAVID groups species-specific gene/protein identifiers to agglomerate a diverse array of functional and sequence annotation from a variety of public genomic resources including NCBI, PIR, and Uniprot/Swiss-Prot. Briefly, the algorithm uses kappa statistics to quantitatively measure agreement of how genes/proteins share similar annotation terms. It then generates a gene-to-gene similarity matrix based on shared function using multiple functional annotation sources. With the use of this approach, an object (gene/protein) can participate in more than one cluster, which better reflects biology in that a given gene may be associated with more than one functional group of genes. The algorithm also allows automatic determination of the optimal numbers of clusters (K) and exclusion of members (genes/proteins) that have weak

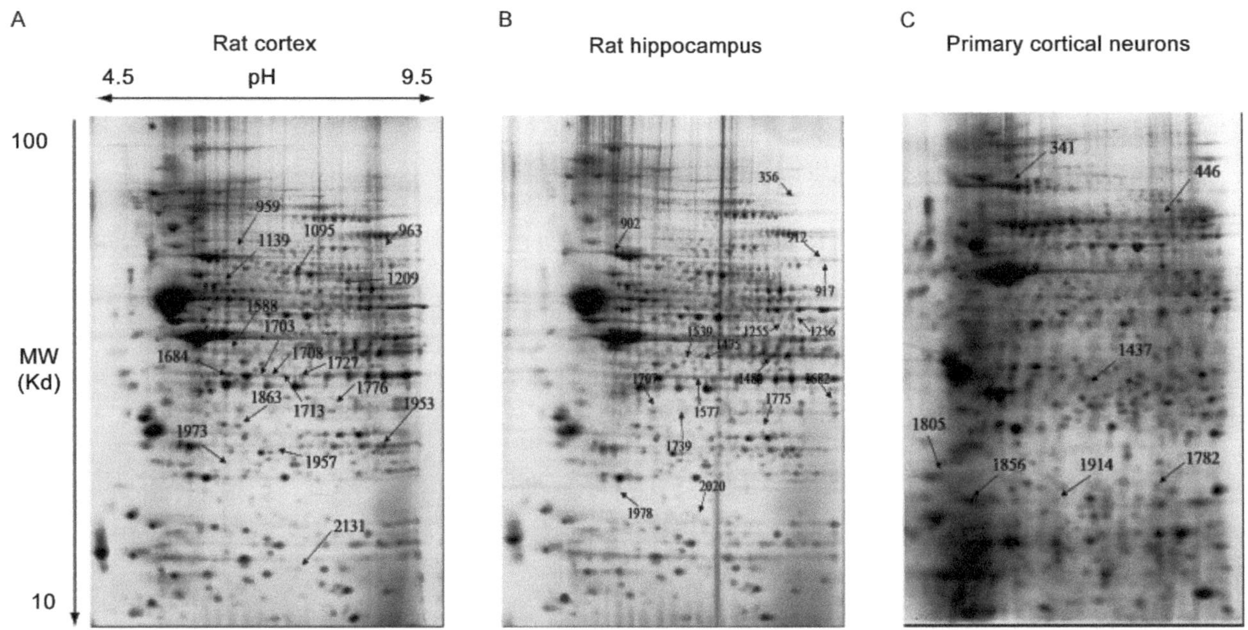

Fig. 3 Representative Coomassie Blue-stained 2D gels. (A) Rat cortex. (B) Rat hippocampus. (C) Rat primary cortical neurons. Spots were marked by their assigned numbers from DIGE study. The *horizontal arrows* show the pH gradient, which increases from 4.5 to 9.5 (nonlinear). The *vertical arrows* show protein spot separation at different molecular weight (MW) starting at 100 KDa (*top*) to 10 KDa (*bottom*) (nonlinear).

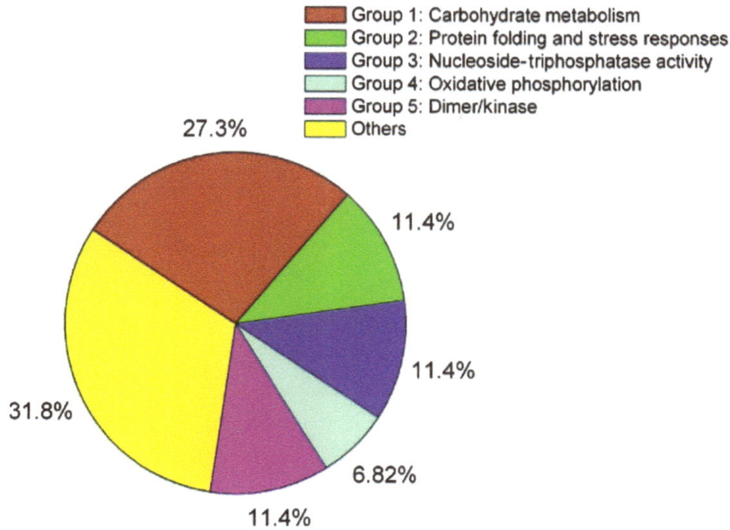

Fig. 4 Protein functions identified by DAVID functional classification.

relationships to other members. Thus, DAVID may allow clues not only for the functional pathways affected but also direct binding candidates for anesthetics.

2.4.2 Ingenuity Pathway Analysis

This Web-based program uses the ingenuity pathways knowledge base (IPKB), a database with large amounts of individually modeled relationships between proteins, to construct biological networks and pathways. The submitted proteins, called "focus proteins," are used as starting points for generating biological networks, which rely on IPKB for interactions with all the other proteins stored in the IPKB. A set of networks with a maximum size of 35 proteins each is generated, and a *P*-value for each is calculated according to the fit of our set of labeled proteins. The *P*-value compares the number of focus proteins that participate in a given network or pathway, to the total number of occurrences of the same proteins in all IPKB networks or pathways. The score of each network (given as the negative log) indicates the likelihood of the focus proteins occupying a network due to random chance. Therefore, a score of 2 has a 1% chance of being generated by chance alone. Importantly, this analysis allows labeling intensity (or change ratio) to be included, since the affinity and stoichiometry of binding, together with the change in expression, enhances the likelihood that the protein's function is altered.

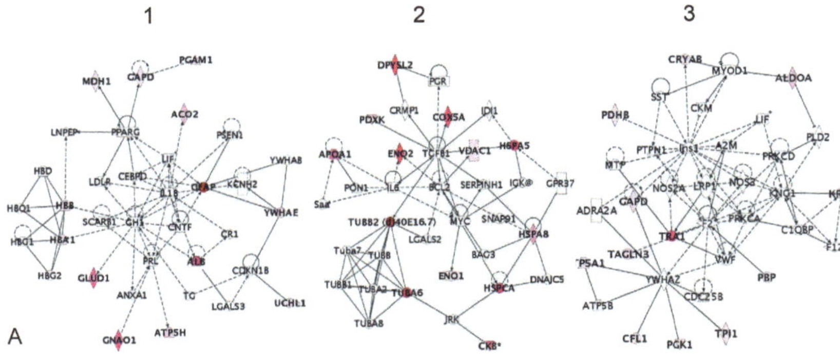

Network ID	Score of significance	# of focus proteins	Top functions
1	23	13	Cellular growth and proliferation, carbohydrate metabolism, Endocrine system development and function
2	23	13	Cell cycle, cell death, cancer
3	21	12	Cell-to-cell signaling and interaction, drug metabolism, cardiovascular system development and function

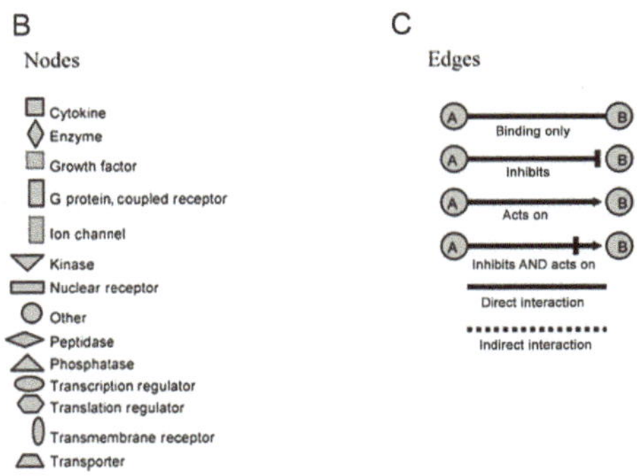

Fig. 5 Ingenuity pathway analysis (IPA) biological networks (summarized in (A); network 1: cell growth and proliferation; network 2: cell cycle and death; network 3: cell–cell signaling and interaction). Proteins are represented as nodes (B), and the biological relationship between two nodes is represented as an edge (line) (C). All edges are supported by at least one reference stored in the IPKB. The intensity of the node color indicates the labeling intensity. Nodes are displayed using *various shapes* that represent the functional class of the proteins.

3. SUMMARY AND CONCLUSION

Both inhaled anesthetics (halothane and isoflurane) induce only small degrees of changes in global gene expression. There are more gene responses to anesthetics in cultured neurons than rat whole brain, probably due to diluted drug effect in heterogeneous brain tissue and higher concentration of anesthetics in cell culture. The study by DIGE revealed inhaled anesthetics also only elicits limited degree of response at translational levels.

These detected protein responses are most likely caused by the physiological side effect from anesthesia, yet less likely from its primary effect. Halothane and isoflurane induce different proteomic responses in different brain regions. Statistical and network analyses are important parts of the studies, which give us the power to interpret the data.

REFERENCES

Culley, D. J., et al. (2006). Altered hippocampal gene expression 2 days after general anesthesia in rats. *European Journal of Pharmacology*, *549*(1–3), 71–78.

Eckenhoff, R. G. (2001). Promiscuous ligands and attractive cavities: How do the inhaled anesthetics work? *Molecular Interventions*, *1*(5), 258–268.

Eckenhoff, M. F., Chan, K., & Eckenhoff, R. G. (2002). Multiple specific binding targets for inhaled anesthetics in the mammalian brain. *The Journal of Pharmacology and Experimental Therapeutics*, *300*(1), 172–179.

Eckenhoff, R. G., Xi, J., & Dailey, W. P. (2010). Inhalational anesthetic photolabeling. *Methods in Molecular Biology*, *617*, 437–443.

Herold, K. F., Andersen, O. S., & Hemmings, H. C., Jr. (2017). Divergent effects of anesthetics on lipid bilayer properties and sodium channel function. *European Biophysics Journal*, *46*(7), 617–626.

Hochberg, Y., & Benjamini, Y. (1990). More powerful procedures for multiple significance testing. *Statistics in Medicine*, *9*(7), 811–818.

Howard, R. J., Trudell, J. R., & Harris, R. A. (2014). Seeking structural specificity: Direct modulation of pentameric ligand-gated ion channels by alcohols and general anesthetics. *Pharmacological Reviews*, *66*(2), 396–412.

Marota, J. J., Crosby, G., & Uhl, G. R. (1992). Selective effects of pentobarbital and halothane on c-fos and jun-B gene expression in rat brain. *Anesthesiology*, 77(2), 365–371.

Pan, J. Z., et al. (2006). Rat brain DNA transcript profile of halothane and isoflurane exposure. *Pharmacogenetics and Genomics*, *16*(3), 171–182.

Pan, J. Z., et al. (2007). Halothane binding proteome in human brain cortex. *Journal of Proteome Research*, *6*(2), 582–592.

Pan, J. Z., et al. (2008). Inhaled anesthetics elicit region-specific changes in protein expression in mammalian brain. *Proteomics*, *8*(14), 2983–2992.

Weiser, B. P., et al. (2014). Mechanisms revealed through general anesthetic photolabeling. *Current Anesthesiology Reports*, *4*(1), 57–66.

CHAPTER FOURTEEN

Neurochemistry of Anesthetic States

Ralph Lydic*,†,1, Helen A. Baghdoyan*,†, Amanda L. May*
*University of Tennessee, Knoxville, TN, United States
†Oak Ridge National Laboratory, Oak Ridge, TN, United States
[1]Corresponding author: e-mail address: rlydic@utk.edu

Contents

Abstract

Anesthetic mechanisms that eliminate consciousness and perception of pain are products of the nervous system. Chemical approaches to the study of anesthetic mechanisms have the potential to serve as an ideal interface between basic and clinical neuroscience. There are disproportionately more basic neurochemical studies than clinical studies of anesthetic mechanisms. Even within neuroscience, the study of anesthetic mechanisms is sparse. The Society for Neuroscience hosts one of the world's largest and most vibrant scientific meetings, yet the content themes of that meeting do not include anesthesia. One goal of this chapter is to facilitate neurochemical studies of anesthetic mechanisms by outlining user-friendly descriptions of existing and emerging techniques. The introduction provides a context for chapter goals. The second portion of this chapter focuses on microdialysis methods that enable the humane acquisition of neurochemical samples from intact, behaving animals during anesthetic induction, maintenance, and emergence. No single neurotransmitter and no single brain region regulate the physiological and behavioral traits characteristic of any anesthetic state.

Methods in Enzymology, Volume 603
ISSN 0076-6879
https://doi.org/10.1016/bs.mie.2018.01.024

This limitation is being addressed via application of new instrumentation and techniques in analytic chemistry. The final third of this chapter highlights selected omics approaches that are now being applied to the neurochemical study of anesthetic mechanisms. We hope that this brief chapter can stimulate basic and clinical metabolomic approaches aiming to elucidate the mechanisms of anesthetic action.

1. INTRODUCTION

Studies of anesthetic mechanisms confront a unique duality relative to the perspective that "It is the impulse of science to try to understand nature, and the impulse of technology to try to manipulate it" (Mukherjee, 2016). As an independent variable, chemistry enables the technology of manipulating states of consciousness. As a dependent variable, measures of anesthesia-induced changes in brain chemistry are essential for a scientific understanding of anesthetic mechanisms. Chemical data also gave rise to the theory that states of anesthesia are caused by molecular distortion of lipid membranes. Only since the 1980s has the lipid theory of anesthesia been supplanted by evidence that anesthesia is caused, in part, by binding of anesthetic molecules to cell surface proteins (Grasshoff, Drexler, Rudolh, & Antkowiak, 2006; Perouansky, 2012). In mammals, anesthesia is defined by a constellation of traits (unconsciousness, analgesia, amnesia, muscle relaxation, and blunted autonomic responses) that are actively generated by anatomically distributed neuronal networks. A unifying theory of anesthetic function must be able to explain complex sets of questions. First, what are the neuronal networks that regulate traits that define states of consciousness and how are these networks altered by anesthetics? Second, how do higher-level network phenotypes (electroencephalogram (EEG), motor hypotonia, autonomic dysregulation) vary as a function of administered drug class, dose, and route of administration, as well as recipient species, age, sex, and health status?

The shared circuits hypothesis provides a conceptual framework for the methods outlined in this chapter by postulating that brain mechanisms that evolved to generate naturally occurring states of consciousness are preferentially involved in generating traits characterizing anesthetic states (Lydic, 1996). Anesthesia and sleep are distinctly different states of consciousness sharing some similar traits (Hillman, Kaw, & Lydic, 2017). Evolutionary perspectives provide unbiased support for the shared circuits hypothesis. Regular light–dark cycles over the course of animal evolution promoted development of rest-activity cycles that in some species led to the development of neuronal networks generating sleep–wake cycles. In contrast to the influence of light cycles, exposure to anesthetics was not an evolutionary pressure. Thus, a logical postulate of the shared circuits hypothesis is that states of anesthesia are caused by drug actions on neural networks that evolved to generate states of sleep and wakefulness. The forgoing also is supported by convergent evolutionary comparisons. For example, volatile anesthetics eliminate wakefulness in species such as fruit flies (Allada & Nash, 1993; Weber et al., 2009), which have existed for several hundred million years, as well as in humans, which emerged as recently as 300,000 years ago (Schlebusch, Malmström, Gunther, et al., 2017). Many of the neural networks connecting brain regions that evolved to generate sleep (Fig. 1) are now recognized to contribute to the loss of wakefulness caused by anesthetic drugs (Baghdoyan & Lydic, 2012).

The extent to which state-dependent brain activity is correlatively or causally involved in the loss of normal wakefulness relative to sleep or

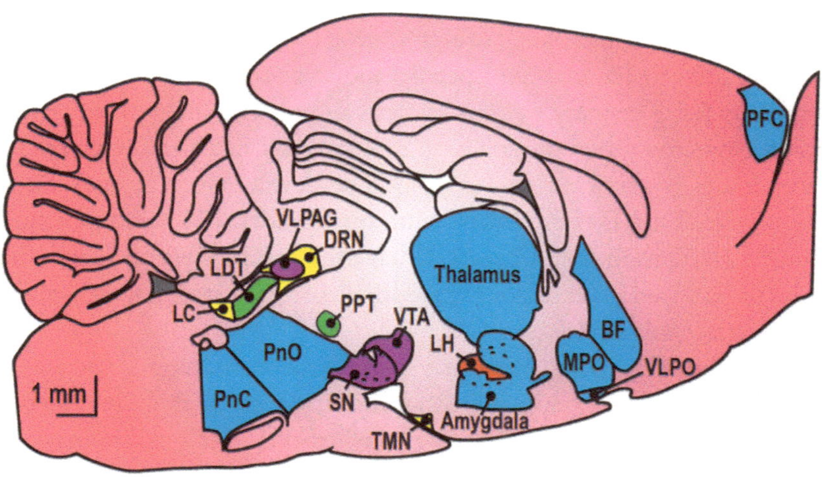

Fig. 1 Drawing of the rat brain in sagittal view illustrates brain regions with state-dependent activity profiles during sleep and/or anesthesia that differ from activity during wakefulness. The figure, and the chapter from which it is reproduced (Baghdoyan & Lydic, 2012) also emphasize how many brain regions and molecules remain to be studied across states of consciousness. *BF*, basal forebrain; *DRN*, dorsal raphé nucleus; *LC*, locus coeruleus; *LDT*, laterodorsal tegmental nucleus; *LH*, lateral hypothalamus; *MPO*, medial preoptic area; *PFC*, prefrontal cortex; *PnC*, pontine reticular formation, caudal part; *PnO*, pontine reticular formation, oral part; *PPT*, pedunculopontine tegmental nucleus; *SN*, substantia nigra; *TMN*, tuberomammillary nucleus; *VLPAG*, ventrolateral periaqueductal gray; *VLPO*, ventrolateral preoptic area; *VTA*, ventral tegmental area.

anesthesia is an active research area. The neural networks generating the multiple traits defining states of wakefulness, sleep, and anesthesia are anatomically distributed. Few studies continue to propose that a single brain region or single mechanism generates states of conscious. The conceptual model guiding the methods described later postulates the need to develop a chemical cartography of brain function across multiple states of consciousness. Section 2 describes the use of in vivo microdialysis for central drug administration and collection of endogenous molecules. Section 3 outlines the use of liquid chromatography-tandem mass spectrometry equipped with orbitrap mass analyzer for the detection of neurotransmitters and metabolites. Considered together, these two approaches make it possible to quantify the effects of anesthetics on many molecules measured simultaneously in specific brain regions.

2. MICRODIALYSIS FOR ANALYTE ACQUISITION AND DRUG DELIVERY

Analogous to the advantages of site localization that are conferred by regional vs general anesthesia, central nervous system drug administration is uniquely able to identify site-specific and state-specific changes in brain chemistry. Drug delivery by microdialysis also permits inferences regarding receptor systems, signal transduction cascades, and neuronal networks regulating states of consciousness (Fig. 2). The development of microdialysis 50 years ago (Bito, Davson, Levin, Murray, & Snider, 1966) and subsequent advances (Robinson & Justice, 1991) have made it possible to deliver molecules to localized brain regions while simultaneously sampling constituents of the extracellular space.

Drug delivery by microdialysis permits concentration response and antagonist-blocking experiments that can provide insight into pre- and postsynaptic receptor mechanisms. For experiments using microdialysis to deliver a molecule, it should be clear that the kilodalton (kDa) pore size of the dialysis membrane, enabling relatively selective analyte recovery, also limits the size and amount of a molecule that can be delivered by microdialysis. This feature explains why many studies report using high drug concentrations in anticipation of $<10\%$ crossing the membrane into the brain. The brain has no pain receptors and this makes it possible to use chronically implanted dialysis probes to obtain analyte measures from the same brain region before, during, and after systemic drug administration. Combining systemic drug administration with microdialysis drug delivery has

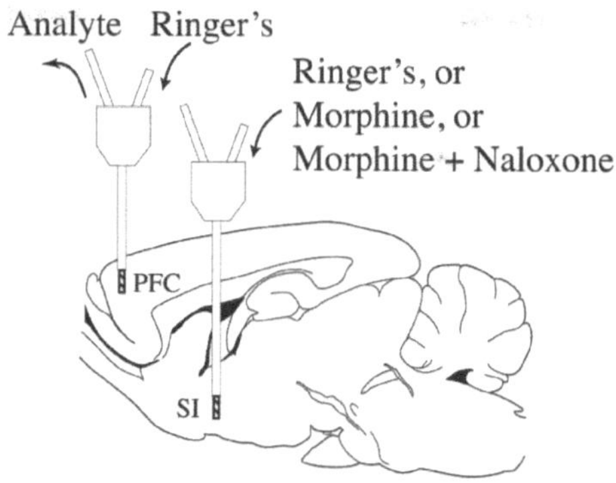

Fig. 2 Illustrates how simultaneous microdialysis in different brain regions can be used to identify drug action on neuronal network function. This dual dialysis method made it possible to show that dialysis delivery of morphine to the substantia innominata (SI) caused a concentration dependent, naloxone-sensitive decrease in acetylcholine (ACh) release within the prefrontal cortex (PFC) (Osman, Baghdoyan, & Lydic, 2005). Morphine also was administered systemically (i.v.), while measuring ACh release in the PFC. Combining dual dialysis and systemic drug delivery lead to the interpretation that opiates impair cognitive functions, in part, by inhibiting cholinergic neurons in the SI, resulting in decreased ACh release in the PFC.

been shown to unmask brain regions through which systemically administered drugs alter brain chemistry (Hambrecht-Wiedbusch, Mitchell, Firn, Baghdoyan, & Lydic, 2014; Osman et al., 2005).

2.1 Anesthesia-Induced Changes in Brain Chemistry

There are two major approaches to preclinical studies of brain chemistry using microdialysis. For studies of anesthetized subjects, after performing a craniotomy a microdialysis probe is stereotaxically positioned in the brain region of interest. Using anesthesia to hold arousal state of the animal at a chosen level, one can prevent the cycling between sleep–wake states and corresponding variations in measured analytes. This approach is especially useful for studies that quantify analytes across levels of anesthesia. Additionally, holding arousal level constant allows one to perform microdialysis drug delivery involving concentration response and antagonist-blocking experiments. When arousal state is clamped at a given level by anesthesia, one can infer that changes in analytes are caused by the concentration or type of drug delivered.

Microdialysis studies that aim to compare analytes during anesthesia with analytes during wakefulness are ideally performed using a within-subjects design. Such a design decreases variability associated with different microdialysis probes, exact brain site of probe placement, instrument sensitivity, etc. The first step for all studies of this type is to stereotaxically implant a guide tube above the brain region of interest. After a week or more to recover from surgery, a microdialysis probe can be inserted into the chronically implanted guide tube without the need for stereotaxic surgery. This experimental approach allows one to measure brain chemicals during wakefulness, during anesthetic induction, and during emergence from anesthesia. The within-subjects design also has the advantage of decreasing variability across groups of animals or caused by variation in microdialysis probe placement. Whether the experiments involve between- or within-subjects designs outlined earlier, a second part of any microdialysis protocol focuses on conditioning of the microdialysis probe. For most commercially available microdialysis probes the first task is to clear glycerol from the packaged probe and to insure that the dialysis probe membrane pores are opened.

2.2 Microdialysis Protocols

2.2.1 Microdialysis Probe Conditioning on the Day Before an Experiment

1. Check all tubing connecting the dialysis probe to the liquid switch (if using) to the perfusion syringe in order to ensure secure fittings and lack of air bubbles.
2. Attach the probe to the arm of the stereotaxic device, being careful not to touch the probe membrane.
3. Lower the probe membrane into a vial of 70% EtOH. Activate the perfusion pump and perfuse the probe with 70% EtOH for 10 min at 10 μL/min.
4. Lower the probe membrane in a vial of double distilled (dd) H_2O and perfuse with H_2O for 10 min at 10 μL/min.
5. Visually examine the probe membrane using a low-power microscope to ensure that there are no air bubbles and/or cracks.
6. If air bubbles are present, remove the probe from the stereotaxic arm. Hold the probe between your thumb and forefinger, and place your hand on an electronic vortex device and vibrate for 10 s. Reexam the probe under the microscope. If bubbles are present, repeat the probe vibration procedure using the vortex.
7. Perfuse the probe with dd H_2O at 0.3 μL/min overnight and collect the output into a graduated cylinder.

8. Performing this opening procedure on two probes increases the probability that the planned experiment will not be canceled due to failure of one probe. If both probes are functioning well, the second probe can be used within 2 or 3 days.

2.2.2 Microdialysis Probe Handling on the Day of Experiment

1. Use the volume of dd H_2O in the graduated cylinder to verify that, during the overnight perfusion, the probe has collected the amount of fluid anticipated from the probe perfusion rate.
2. Turn the flow rate up to 2 μL/min.
3. Place each probe into a vial containing Ringer's and perfuse with Ringer's for 10 min.
4. Check for air bubbles under the microscope during perfusion with Ringer's.
5. If bubbles are detected, vortex as described earlier.
6. Reduce the probe perfusion rate to the μL/min that will be used for the in vivo experiment.

2.2.3 Generating a Standard Curve and Measuring Probe Recovery

1. When using HPLC for measuring single analytes, one must generate a standard curve on the morning of the experiment for the analyte of interest.
2. The standard curve should have an *r*-value of ≥ 0.98
3. Place the dialysis probe into a known concentration of the analyte that is within the midrange of the standard curve.
4. Obtain three or four dialysis samples from the midrange standards. Convert chromatograms to units/dialysis time (e.g., pmol/10 min) in order to quantify the percent analyte recovery achieved by the microdialysis probe. Results will vary by analyte and analytic instrument, but typical microdialysis probe recoveries are commonly in the range of 5%–8%.
5. Place the probe in a vial of Ringer's and flush the probe with Ringer's to clear the standard solution.
6. Single analyte detectors employing ultraviolet or electrochemical sensors can exhibit changes in sensitivity over the course of a 10–12 h experiment. To determine whether there have been changes in instrument sensitivity, create a standard curve for the instrument before and again after running the biological samples. Compare the two standard curves using correlation and slope analyses.

7. In long-duration experiments, another potential problem is decreased analyte recovery by a microdialysis probe membrane. Impaired probe recovery can be caused by adhesion molecules or blood clots and is referred to as biofouling. It is important, therefore, to know how much of the change in measured analyte can be attributed to a drug being studied, and how much can be attributed to changes in the microdialysis membrane. For this reason, we recommend performing post-experiment probe recoveries using the same analyte standard that was used at the beginning of the experiment. This makes it possible to use inferential statistics to determine if the postexperiment probe recovery was statistically different from the preexperiment probe recovery. Without this approach, one cannot know the extent to which changes in measured analyte reflect the drug administered vs changes in recovery or measurement.
8. After the preexperiment probe recovery has been determined, the dialysis probe can be inserted into the brain under stereotaxic guidance.

2.3 Limitations of Microdialysis

Microdialysis measures have an inverse relationship between probe perfusion rate (volume/time) and the concentration of analytes in the dialysate. Depending on the analyte and the brain region studied, microdialysis studies have variable temporal resolution. That limitation also will vary as a function of the lower limits of detection (LOD) by the analytic instrument and the required microdialysis sample volume. Microdialysis studies of small brain regions require smaller probes that, of course, exacerbate the problem of temporal resolution. This can be visualized in Fig. 1 by comparing the size of the very small ventrolateral preoptic (VLPO) region to other brain areas known to contribute to arousal state control. We are unaware of any microdialysis measures across states of consciousness that have been obtained exclusively from the VLPO region.

2.4 Future Directions for Microdialysis Methods

Studies quantifying drug effects on neurotransmitters have, until recently, focused on a single analyte. Use of multiple analytic instruments and/or cross-study comparisons demonstrate the potential for novel insights from simultaneous measures of more than one transmitter across states of consciousness.

As a trait characterizing states of anesthesia and sleep, the GABA/ACh ratio (Fig. 3) indicates that NREM sleep is more similar to isoflurane anesthesia than to REM sleep, and that REM sleep is more similar to wakefulness than to NREM sleep or anesthesia. The data in Fig. 3 show this to be true for

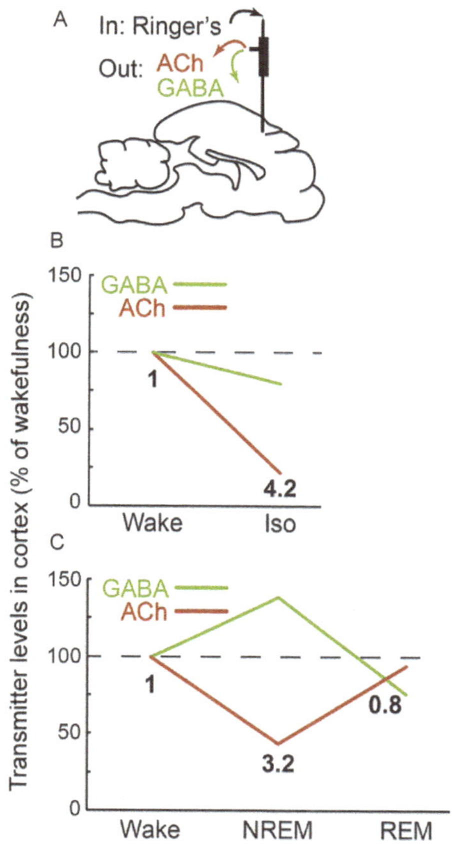

Fig. 3 Plots data used to calculate the GABA/ACh ratio in the cortex during isoflurane anesthesia and sleep. (A) The probe is perfused with Ringer's and used to collect samples containing GABA and ACh (Vanini, Lydic, & Baghdoyan, 2012). (B) Shows levels of GABA and ACh measured in cortical dialysis samples collected during electrographically confirmed states of wakefulness and 1 MAC isoflurane anesthesia. Values are plotted as a percent change from waking levels. Numbers within the graph indicate the GABA/ACh ratio. (C) Plots GABA/ACh ratios of cortical dialysis samples collected during electrographically confirmed states of wakefulness, nonrapid eye movement (NREM) sleep, and REM sleep (Marrosu et al., 1995; Vanini et al., 2012). *Panel (B): Data from Dong, H.-l., Fukuda, S., Murata, E., & Higuchi, T. (2006). Excitatory and inhibitory actions of isoflurane on the cholinergic ascending arousal system of the rat.* Anesthesiology, 104, *122–132.*

some regions of cortex and additional studies are needed for other brain regions. These state-dependent differences in transmitter profiles are consistent with higher-level phenotypes such as the EEG, which is similar between states of anesthesia and NREM sleep, and between REM sleep and wakefulness. Relative to the EEG of wakefulness and REM sleep, states of isoflurane anesthesia and NREM sleep exhibit an EEG of slower frequency and increased amplitude. EEG is a higher-level phenotype that is known to be the product of chemically regulated cell excitability. Thus, inferences about mechanisms of anesthetic action based on a higher-level phenotype such as EEG may reach different conclusions than inferences based on lower-level neurochemical phenotypes. The next section outlines advances in analytic chemistry that can now be applied to studies of anesthetic mechanisms. These newer methods make it possible to obtain simultaneous measures of many small molecule neurotransmitters (Zhang, Baer, Harp, Baghdoyan, & Lydic, 2017) and components of the brain metabolome (Baer, Bourdon, Spano, et al., 2017).

3. APPLICATION OF METABOLOMICS FOR ANESTHESIOLOGY RESEARCH

3.1 Anesthetic Pharmacology Alters the Entire Metabolome

Systemically administered anesthetics eliminate wakefulness and block pain perception because of the ability to alter neuronal excitability. Drug-induced changes in cell excitability likewise cause widespread alterations in metabolism, blood flow, and the plethora the small molecules regulating these responses. The profiling of the brain metabolome already has been shown by preclinical (Makaryus et al., 2011) and clinical (Jacob et al., 2012) studies to vary by anesthetic agent. In the United States about 60,000 patients per day receive general anesthesia, yet the mechanisms by which these molecules eliminate wakefulness and block pain perception are unknown (Brown, Lydic, & Schiff, 2010). The potential for metabolomic profiling to elucidate anesthetic mechanisms justifies the perspective of metabolomics as the "apogee of omics" approaches to neuroscience (Patti, Yanes, & Siuzdak, 2012).

3.2 Metabolomics Definition, Background, and Uses

Metabolomics uses a variety of techniques (Baghdoyan & Lydic, 2012; Keun et al., 2003; Zhang et al., 2011) including liquid chromatography-mass spectrometry (LC-MS) (Crews et al., 2009; Dulka et al., 2017; Lu et al., 2010; Stough et al., 2016) to measure relative concentrations of metabolites from cells. By quenching the metabolism quickly, a global picture of the current state of the system can be measured. For anesthesiology, the analytic tools described herein can be applied to any organ system from which samples can be obtained. Recent advances in technology for measuring metabolism such as the orbitrap mass analyzer (Hu et al., 2005; Lu et al., 2010) have made metabolomics a more routine analysis for all fields. The high mass resolution and accuracy of the orbitrap allow for the instrument to run in full scan mode, and the identification of metabolites relies on retention time and exact mass (Lu et al., 2010). Using metabolomics, coupled with other cooperative techniques, allows identification of potential new therapeutic targets (Gendelman, Zhang, Santamaria, et al., 2017; Stough et al., 2016) or biomarkers (Ippolito et al., 2005; Soga, Baran, Suematsu, et al., 2006; Yang et al., 2004) that can be confirmed with further testing.

3.3 Metabolomics Protocol (Lu, Bennett, & Rabinowitz, 2008; Yuan, Bennett, & Rabinowitz, 2008)

3.3.1 Equipment

- Thermo Fisher Scientific LC/MS
  - Ultimate 3000 LC pump equipped with Synergi 2.5 μm Hydro-RP 100, size 100 × 2.0 mm LC column (Phenomenex)
  - Exactive Plus Orbitrap Mass Spectrometer (Thermo Fischer Scientific)
  - XCalibur software
- Peak Processing Software, e.g., Metabolomic Analysis and Visualization Engine (MAVEN) (Clasquin, Melamud, & Rabinowitz, 2012; Lu et al., 2010) and XCMS (Benton, Wong, Trauger, & Siuzdak, 2008; Huan, Forsberg, Rinehart, et al., 2017; Smith, Want, O'Maille, Abagyan, & Siuzdak, 2006)
- Data Analysis Software (commercial software such as Graph Pad Prism or freeware such as R; Hu et al., 2005)

3.3.2 Reagents

- Metabolite Extraction Solvents (if necessary)
  - 40:40:20 Acetonitrile:methanol:water with 0.1 *M* formic acid or
  - 80% Methanol in water
- LC-MS Mobile Phases
  - (A) 97:3 Water:methanol containing 10 m*M* tributylamine and 15 m*M* acetic acid
  - (B) Methanol

3.3.3 Metabolite Extraction *(Fig. 4A)*

- All samples are kept at −80°C until analysis
- Extraction solvent, kept at −80°C, is added to sample and mixed
- Extraction is allowed to continue for 20 min at −20°C
- Sample is centrifuged to remove particulates (5 min, 13,000 rpm, 4°C) and the resulting supernatant is collected
- The remaining pellet is then washed twice with extraction solvent and combined with the previous supernatant
- Combined supernatant is dried under nitrogen
- Once dry, the sample is resuspended in 150 μL Milli-Q water

Note: Samples collected via microdialysis are normalized to equal volumes and each diluted to at least 50 μL using Ringer's, and placed directly on the instrument; therefore, does not require a metabolite extraction.

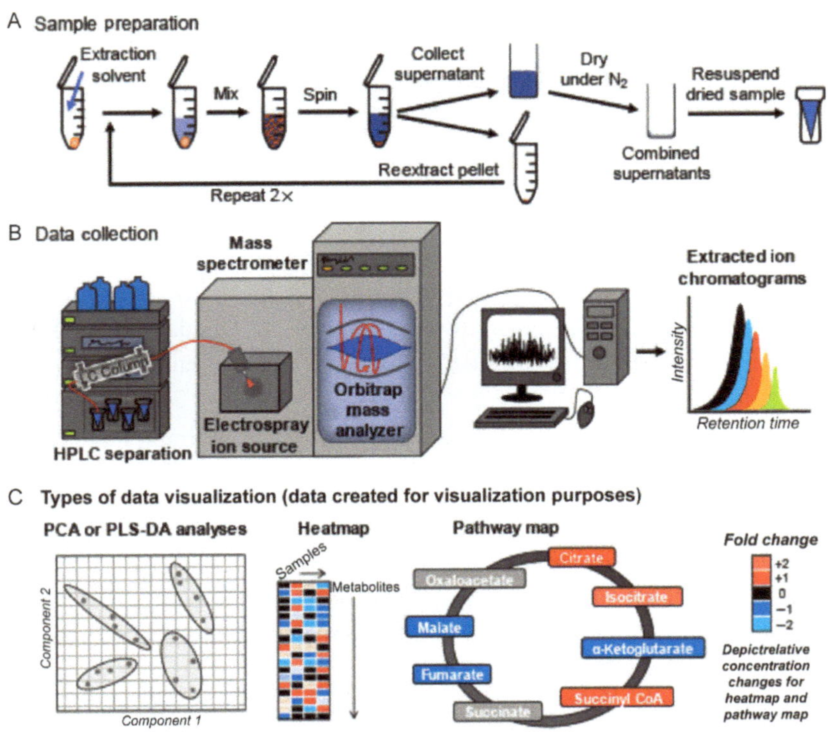

Fig. 4 Schematic of (A) sample preparation, (B) data collection, and (C) types of data visualization used for metabolomics.

3.3.4 *Instrument Method* (*Fig. 4B*)

- A 10 μL aliquot is injected onto the column kept at 25°C
- Metabolites are separated on column using the mobile phases above with following multistep gradient: $t=0$ min, 0% B; $t=5$ min, 0% B; $t=13$ min, 20% B; $t=15.5$ min, 55% B; $t=19$ min, 95% B; $t=25$ min, 0% B; at a flow rate of 200 μL/min
- Metabolites are introduced into the mass spectrometer via electrospray ionization and analyzed in negative mode using a resolving power of 140,000 and the scan window shifting from 72–800 m/z at 0–9 min to 100–1000 m/z at 9–25 min

Note: It is important to randomize the sample order so as not to introduce instrument bias/error into the sample analysis.

3.3.5 *Data Processing*

- Raw data files are first converted to a universal format, mzML (Martens, Chambers, Sturm, et al., 2011), using MSConvert software, an open source program offered with the ProteoWizard package (Chambers, Maclean, Burke, et al., 2012; Kessner, Chambers, Burke, Agus, & Mallick, 2008).
- Peaks can be identified using a peak-picking program such as XCMS (Benton et al., 2008; Huan et al., 2017; Smith et al., 2006) or MAVEN (Clasquin et al., 2012; Lu et al., 2010).
- Once samples are successfully converted, they are uploaded to MAVEN and peaks are manually selected for a set of ~300 metabolites by matching previously determined m/z and retention times (Lu et al., 2010).
- Lastly, MAVEN will run an algorithm to identify all spectral features within a sample.
- All known metabolites and unknown spectral features are exported from MAVEN as area under the curve to represent the relative concentrations of each metabolite between samples.

3.4 Analysis of Selected Peaks

Once the collection of data and processing of peaks are completed, the data are normalized and fold changes are calculated to determine concentrations differences in metabolites between conditions. These changes are visualized to help glean biological meaning from the data. Data normalization is performed to remove instrument and sample variability (Hochrein et al., 2015; Kind, Tolstikov, Fiehn, & Weiss, 2007; van den Berg, Hoefsloot, Westerhuis, Smilde, & van der Werf, 2006; Warrack et al., 2009;

Xia, Psychogios, Young, & Wishart, 2009; Xia & Wishart, 2002). This assures the changes observed in relative metabolite concentrations are due to the condition(s) tested. Eliminating sample variation can be accomplished by using a sample-specific normalization and can be in a variety of units (i.e., mass, volume, cell count, protein, etc.). Standards with unique mass to charge ratios (*m/z*) can be added to the sample prior to extraction in order to remove the instrument variability (Bijlsma et al., 2006). Preferably before further data analysis is done, an evaluation of how to handle any missing values is completed (Gromski et al., 2014). Based on the evaluation, the so-called blanks or NA values are replaced by a best estimation (Armitage, Godzien, Alonso-Herranz, López-Gonzálvez, & Barbas, 2015), typically a very small number, i.e., ½ minimum, if thought to be below the LOD. Many helpful online tools (Huan et al., 2017; Xia & Wishart, 2002) offer additional methods of replacement such as average/median values, *k*-nearest neighbor, and probabilistic principal component analysis (PCA). Alternatively, the analyte can be removed from the analysis if too many "NAs" are found. If the variable distribution is skewed, data transformation (van den Berg et al., 2006; Xia & Wishart, 2002) (e.g., log and cube root) or scaling (van den Berg et al., 2006; Xia & Wishart, 2002) (auto, pareto, range, etc.) may be appropriate in an effort to obtain a more Gaussian-like distribution. At present, no particular method is considered better or worse and the ideal normalization is one in which the distribution resembles a bell curve in order to make meaningful biological comparisons.

Fold changes of metabolites are then calculated from the normalized peak areas to determine an increase or decrease in concentration as compared to a control. These changes can be visualized in a multitude of ways (Huan et al., 2017; Katajamaa, Miettinen, & Oresic, 2006; Lu et al., 2010; Xia & Wishart, 2002) (Fig. 4C) with the most common being a heat map to get a global picture of metabolic changes. A heat map uses color variation to indicate the amount of change from the control either in a positive or negative direction. Some software from which heat maps are generated will cluster metabolites and sample groups to give a picture based on similarities of the resulting fold change (Eisen, Spellman, Brown, & Botstein, 1998; Xia & Wishart, 2002). However, grouping metabolites based on pathways can lead to a different picture focused on specific pathways. No matter how a heat map is organized, it can be helpful as a first pass elucidation of the data to help focus further analyses.

In combination with a heat map, performing either PCA or partial least squares discriminant analyses (PLS-DA) of the samples can help extrapolate the factors driving the changes observed between sample groups (Crews et al., 2009; Keun et al., 2003; Kind et al., 2007; Xia & Wishart, 2002). By using either of these methods, samples are grouped by scores driven by the variation within the sample. PCA uses linear principle components to perform the analysis to help recognize trends or patterns within the samples. PLS-DA performs a regression analysis on dependent variables within a sample to maximize the covariance to then help recognize the relevant variables of the group separation (Gromski et al., 2015).

3.5 Quality Control and Data Interpretation

A first pass through using univariate *t*-tests and ANOVA helps to focus the analysis on single metabolite changes such as biomarker discovery (Kind et al., 2007). Multivariate tests such as PCA, clustering, and PLS-DA help identify patterns and trends between groups. No single analytic technique, however, is considered as correct. Current evidence discussed in this chapter supports the view that the combination from start to finish of the experimental setup to data processing and analysis helps metabolomic approaches provide insight into biological mechanisms. Mapping the metabolic changes to specific pathways (Fig. 4C), and understanding the connectivity of the metabolites measured, as well as metabolite interactions with one another within or between specific metabolic pathways, allows for a broader understanding of the global changes within the system.

3.6 Metabolomics Advances and Future Endeavors

Metabolomics has been a field of interest for many years and novel applications make the interpretation of data a challenge. Further, the lack of an ideal statistical analysis to determine which metabolite concentrations are biologically significant prevents real insights into the metabolism of a particular system. Currently, new analytic programs are being developed to help interpret chemical changes. Additional analytic approaches aim to identify unknown spectral features that exhibit significant changes in relative concentration. Flux analyses can also be performed using stable isotopes to help in the understanding of pool size changes due to increased/decreased synthesis/usage (Yuan et al., 2008). Finally, combining this metabolomics approach with other "omics" techniques and/or other cooperative techniques would allow for a systems approach and help to understand the interactions between the different backbones of biology.

4. CONCLUSION

The anesthetic functions most valued by patients and clinical providers alike are the elimination of waking consciousness and surgical pain. These two functions are products of chemical actions on the nervous system. This fact leads us to consider anesthesia as a branch of clinical neuroscience. We welcome the Brain Health Initiative promoted by the American Society of Anesthesiologists (Berger & Eckenhoff, 2016) as validation of a neurocentric perspective with the potential to advance the neurochemical study of anesthetic mechanisms.

REFERENCES

Allada, R., & Nash, H. A. (1993). Drosophila melanogaster as a model for study of general anesthesia: The quantitative response to clinical anesthetics and alkanes. *Anesthesia and Analgesia*, 77, 19–26.

Armitage, E. G., Godzien, J., Alonso-Herranz, V., López-Gonzálvez, Á., & Barbas, C. (2015). Missing value imputation strategies for metabolomics data. *Electrophoresis*, *36*, 3050–3060.

Baer, A., Bourdon, A. K., Spano, G., et al. (2017). Metabolomic analysis of microdialysis samples from the frontal association cortex (FrA) of C57BL/6J (B6) mouse during isoflurane anesthesia. *The FASEB Journal*, *31*(1058), 1012.

Baghdoyan, H. A., & Lydic, R. (2012). The neurochemistry of sleep and wakefulness. In S. T. Brady, R. W. Albers, D. L. Price, & G. J. Siegel (Eds.), *Basic neurochemistry* (pp. 982–999). New York: Elsevier.

Benton, H. P., Wong, D. M., Trauger, S. A., & Siuzdak, G. (2008). XCMS2: Processing tandem mass spectrometry data for metabolite identification and structural characterization. *Analytical Chemistry*, *80*, 6382–6389.

Berger, M., & Eckenhoff, R. (2016). ASA brain health initiative: The science behind the scene. *ASA Monitor*, *80*, 12–14.

Bijlsma, S., Bobeldijk, L., Verheij, E. R., Ramaker, R., Kochhar, S., Macdonald, I. A., et al. (2006). Large-scale human metabolomics studies: A strategy for data (pre-) processing and validation. *Analytical Chemistry*, *78*, 567–574.

Bito, H., Davson, H., Levin, E., Murray, M., & Snider, N. (1966). Determination of brain interstitial concentrations by microdialysis. *Journal of Neurochemistry*, *52*, 1741–1750.

Brown, E. N., Lydic, R., & Schiff, N. D. (2010). General anesthesia, sleep, and coma. *The New England Journal of Medicine*, *363*, 2638–2650.

Chambers, M. C., Maclean, B., Burke, R., et al. (2012). A cross-platform toolkit for mass spectrometry and proteomics. *Nature Biotechnology*, *30*, 918–920.

Clasquin, M. F., Melamud, E., & Rabinowitz, J. D. (2012). LC-MS data processing with MAVEN: A metabolomic analysis and visualization engine. *Current Protocols in Bioinformatics*, *37*. 14.11.11-14.11.23.

Crews, B., Wikoff, W. R., Patti, G. J., Woo, H. K., Kalisiak, E., Heideker, J., et al. (2009). Variability analysis of human plasma and cerebral spinal fluid reveals statistical significance of changes in mass spectrometry-based metabolomics data. *Analytical Chemistry*, *81*, 8538–8544.

Dulka, B. N., Bourdon, A. K., Clinard, C. T., Muvvala, M. B. K., Campagna, S. R., & Cooper, M. A. (2017). Metabolomics reveals distinct neurochemical profiles associated with stress resilience. *Neurobiology of Stress*, 7, 103–112.

Eisen, M. B., Spellman, P. T., Brown, P. O., & Botstein, D. (1998). Cluster analysis and display of genome-wide expression patterns. *Proceedings of the National Academy of Sciences of the United States of America, 95*, 14863–14868.

Gendelman, H. E., Zhang, Y. N., Santamaria, P., et al. (2017). Evaluation of the safety and immunomodulatory effects of sargramostim in a randomized, double-blind phase 1 clinical Parkinson's disease trial. *NPJ Parkinson's Disease, 3*, 10.

Grasshoff, C., Drexler, B., Rudolh, U., & Antkowiak, B. (2006). Anaesthetic drugs: Linking molecular actions to clinical effects. *Current Pharmaceutical Design, 12*, 3665–3679.

Gromski, P. S., Muhamadali, H., Ellis, D. I., Xu, Y., Correa, E., Turner, M. L., et al. (2015). A tutorial review: Metabolomics and partial least squares-discriminant analysis—A marriage of convenience or a shotgun wedding. *Analytica Chimica Acta, 879*, 10–23.

Gromski, P. S., Xu, Y., Kotze, H. L., Correa, E., Ellis, D. I., Armitage, E. G., et al. (2014). Influence of missing values substitutes on multivariate analysis of metabolomics data. *Metabolites, 4*, 433–452.

Hambrecht-Wiedbusch, V. S., Mitchell, M. F., Firn, K. A., Baghdoyan, H. A., & Lydic, R. (2014). Benzodiazepine site agonists differentially alter acetylcholine release in rat amygdala. *Anesthesia and Analgesia, 118*, 1293–1300.

Hillman, D., Kaw, R., & Lydic, R. (2017). Sleep and anesthesia: Different states with shared pathophysiological traits. In S. Chokroverty (Ed.), *Sleep disorders medicine* (pp. 143–151). New York, NY: Springer.

Hochrein, J., Zacharias, H. U., Taruttis, F., Samol, C., Engelmann, J. C., Spang, R., et al. (2015). Data normalization of 1H NMR metabolite fingerprinting data sets in the presence of unbalanced metabolite regulation. *Journal of Proteome Research, 14*, 3217–3228.

Hu, Q. Z., Noll, R. J., Li, H. Y., Makarov, A., Hardman, M., & Cooks, R. G. (2005). The Orbitrap: A new mass spectrometer. *Journal of Mass Spectrometry, 40*, 430–443.

Huan, T., Forsberg, E. M., Rinehart, D., et al. (2017). Systems biology guided by XCMS online metabolomics. *Nature Methods, 14*, 461–462.

Ippolito, J. E., Xu, J., Jain, S., Moulder, K., Mennerick, S., Crowley, J. R., et al. (2005). An integrated functional genomics and metabolomics approach for defining poor prognosis in human neuroendocrine cancers. *Proceedings of the National Academy of Sciences of the United States of America, 102*, 9901–9906.

Jacob, Z., Li, H., Makaryus, R., Zhang, S., Reinsel, R., Lee, H., et al. (2012). Metabolomic profiling of children's brains undergoing general anesthesia with sevoflurane and propofol. *Anesthesiology, 117*, 1062–1071.

Katajamaa, M., Miettinen, J., & Oresic, M. (2006). MZmine: Toolbox for processing and visualization of mass spectrometry based molecular profile data. *Bioinformatics, 22*, 634–636.

Kessner, D., Chambers, M., Burke, R., Agus, D., & Mallick, P. (2008). ProteoWizard: Open source software for rapid proteomics tools development. *Bioinformatics, 24*, 2534–2536.

Keun, H. C., Ebbels, T. M. D., Antti, H., Bollard, M. E., Beckonert, O., Holmes, E., et al. (2003). Improved analysis of multivariate data by variable stability scaling: Application to NMR-based metabolic profiling. *Analytica Chimica Acta, 490*, 265–276.

Kind, T., Tolstikov, V., Fiehn, O., & Weiss, R. H. (2007). A comprehensive urinary metabolomic approach for identifying kidney cancer. *Analytical Biochemistry, 363*, 185–195.

Lu, W., Bennett, B. D., & Rabinowitz, J. D. (2008). Analytical strategies for LC-MS-based targeted metabolomics. *Journal of Chromatography B, 871*, 236–242.

Lu, W., Clasquin, M. F., Melamud, E., Amador-Noguez, D., Caudy, A. A., & Rabinowitz, J. D. (2010). Metabolomic analysis via reversed-phase ion-pairing liquid chromatography coupled to a stand alone orbitrap mass spectrometer. *Analytical Chemistry, 82*, 3212–3221.

Lydic, R. (1996). Reticular modulation of breathing during sleep and anesthesia. *Current Opinion in Pulmonary Medicine, 2*, 474–481.

Makaryus, R., Lee, H., Yu, M., Xhang, S., Smith, S. D., Rebecchi, M., et al. (2011). The metabolomic profile during isoflurane anesthesia differs from propofol anesthesia in the live rodent brain. *Journal of Cerebral Blood Flow and Metabolism*, *31*, 1422–1442.

Marrosu, F., Portas, C., Mascia, M. S., Casu, M. A., Fa, M., Giagheddu, M., et al. (1995). Microdialysis measurement of cortical and hippocampal acetylcholine release during sleep-wake cycle in freely moving cats. *Brain Research*, *671*, 329–332.

Martens, L., Chambers, M., Sturm, M., et al. (2011). mzML—A community standard for mass spectrometry data. *Molecular & Cellular Proteomics*, *10*. R110.000133. https://doi.org/10.1074/mcp.R110.000133.

Mukherjee, S. (2016). *The gene*. New York: Scribner.

Osman, N. I., Baghdoyan, H. A., & Lydic, R. (2005). Morphine inhibits acetylcholine release in rat prefrontal cortex when delivered systemically or by microdialysis to basal forebrain. *Anesthesiology*, *103*, 779–787.

Patti, G. J., Yanes, O., & Siuzdak, G. (2012). Innovation: Metabolomics: The apogee of the omics trilogy. *Nature Reviews. Molecular Cell Biology*, *13*, 263–269.

Perouansky, M. (2012). A quest for a unified model of anesthetic action: A century in Claude Bernard's shadow. *Anesthesiology*, *117*, 465–474.

Robinson, T. E., & Justice, J. B. (1991). *Microdialysis in the neurosciences* (pp. 1–476). New York: Elsevier.

Schlebusch, C. M., Malmström, H., Gunther, T., et al. (2017). Southern African ancient genomes estimate modern human divergence to 350,000 to 260,000 years ago. *Science*, *357*, eaao6266.

Smith, C. A., Want, E. J., O'Maille, G., Abagyan, R., & Siuzdak, G. (2006). XCMS: Processing mass spectrometry data for metabolite profiling using nonlinear peak alignment, matching, and identification. *Analytical Chemistry*, *78*, 779–787.

Soga, T., Baran, R., Suematsu, M., et al. (2006). Differential metabolomics reveals ophthalmic acid as an oxidative stress biomarker indicating hepatic glutathione consumption. *The Journal of Biological Chemistry*, *281*, 16768–16776.

Stough, J. M. A., Dearth, S. P., Denny, J. E., LeCleir, G. R., Schmidt, N. W., Campagna, S. R., et al. (2016). Functional characteristics of the gut microbiome in C57BL/6 mice differentially dusceptible to Plasmodium yoelii. *Frontiers in Microbiology*, 7, 1520.

van den Berg, R. A., Hoefsloot, H. C., Westerhuis, J. A., Smilde, A. K., & van der Werf, M. J. (2006). Centering, scaling, and transformations: Improving the biological information content of metabolomics data. *BMC Genomics*, 7, 1–15.

Vanini, G., Lydic, R., & Baghdoyan, H. A. (2012). GABA-to-ACh ratio in basal forebrain and cerebral cortex varies significantly during sleep. *Sleep*, *35*, 1325–1334.

Warrack, B. M., Hnatyshyn, S., Ott, K.-H., Reily, M. D., Sanders, M., Zhang, H., et al. (2009). Normalization strategies for metabonomic analysis of urine samples. *Journal of Chromatography B*, *877*, 547–552.

Weber, B., Schaper, C., Bushey, D., Rohlfs, M., Steinfath, M., Tononi, G., et al. (2009). Increased volatile anesthetic requirement in short-sleeping drosophila mutants. *Anesthesiology*, *110*, 313–316.

Xia, J. G., Psychogios, N., Young, N., & Wishart, D. S. (2009). MetaboAnalyst: A web server for metabolomic data analysis and interpretation. *Nucleic Acids Research*, *37*, W652–W660.

Xia, J., & Wishart, D. S. (2002). Using metaboanalyst 3.0 for comprehensive metabolomics data nalysis. *Current Protocols in Bioinformatics*, *55*, 14.10.11–14.10.91.

Yang, J., Xu, G., Zheng, Y., Kong, H., Pang, T., Lv, S., et al. (2004). Diagnosis of liver cancer using HPLC-based metabonomics avoiding false-positive result from hepatitis and hepatocirrhosis diseases. *Journal of Chromatography B*, *813*, 59–65.

Yuan, J., Bennett, B. D., & Rabinowitz, J. D. (2008). Kinetic flux profiling for quantitation of cellular metabolic fluxes. *Nature Protocols*, *3*, 1328–1340.

Zhang, X., Baer, A., Harp, N. R., Baghdoyan, H. A., & Lydic, R. (2017). *Isoflurane anesthesia alters monoamine levels in prefrontal cortex (PFC) of C57BL/6J (B6) mouse*. Washington, DC: Society for Neuroscience. (Program, 286.21).

Zhang, B., Halouska, S., Schiaffo, C. E., Sadykov, M. R., Somerville, G. A., & Powers, R. (2011). NMR analysis of a stress response metabolic signaling network. *Journal of Proteome Research*, *10*, 3743–3754.

CHAPTER FIFTEEN

Brief Introduction to Electroencephalography

Alex Proekt[1]
Perelman School of Medicine, Department of Anesthesiology and Critical Care, University of Pennsylvania, Philadelphia, PA, United States
[1]Corresponding author: e-mail address: proekta@uphs.upenn.edu

Contents

Abstract

Electroencephalography (EEG) has a long history in neuroscience starting with its original description in humans by Hans Berger in 1929 (Berger, 1932). Investigations of EEG under anesthesia started soon after in the mid-1930s (Gibbs, 1937). No single methodology paper can credibly cover all of the issues relating to this rich field. The purpose of this chapter is to introduce some caveats that complicate and inform analysis of the EEG. Special emphasis will be given to common issues such as choice of reference electrode, filtering, artifact rejection, and spectral analysis. We will specifically emphasize high-density EEG recordings that have become the norm due to technological improvement in electrode and data acquisition design methods. In the last section we will discuss some applications of EEG analysis techniques to the study of the effects of anesthetics on the nervous system.

1. WHAT DOES EEG MEASURE?

The neurophysiological basis of Electroencephalography (EEG) signals will not be covered here in any significant detail. Please consult Niedermeyer and da Silva (2005) and Nunez and Srinivasan (2006) for a

Methods in Enzymology, Volume 603
ISSN 0076-6879
https://doi.org/10.1016/bs.mie.2018.02.009

detailed discussion of the biophysical origins of EEG signals. In what follows we will briefly introduce the key issues that inform and complicate the interpretation of the EEG.

Electrical signals are fundamental to the functioning of the nervous system. Neurons maintain a negative transmembrane potential (typically in the neighborhood of ~-70 mV) relative to the extracellular space and can fire action potentials (brief voltage impulses) depending on the synaptic inputs (postsynaptic potentials). This transmembrane voltage (V_m) can be measured as the potential difference between the inside of the cell and ground. Recordings of V_m provide the most direct measure of neuronal activity. Yet, these measurements are not easily obtained in the intact vertebrate brain and are almost never available in human subjects.

Any ion movement across an excitable membrane be that of a soma, dendrite, or an axon must be accompanied by movement of ions in the extracellular space. Thus, transmembrane currents contribute to the electric field that can be recorded extracellularly. The strength of this electric field (V_e) can be recorded as a potential difference between two electrodes one of which is defined as reference. EEG is a specific case of V_e that involves placing electrodes onto the skull. Many different electrode configurations exist and are to various degrees used in clinical and research settings. Here we will focus on the discussion of high-density EEG. Currently high density refers to >100 channels of EEG recorded simultaneously from different points on the skull.

There is a fundamental difference between V_m and V_e: while V_m reflects the voltage state of a single neuron, V_e necessarily reflects a complex mixture of contributions from myriad neuronal and glial sources (Buzsáki, Anastassiou, & Koch, 2012; Dietzel, Heinemann, & Lux, 1989). Thus, for all extracellular signals including the EEG (see below) it is crucial to distinguish between the sensors (i.e., electrodes where the potential is recorded) and the sources (neuronal substrates and processes that give rise to the observed extracellular potential).

If one knew the sources of neuronal activity, i.e., currents flowing through each patch of membrane in each neuron, one could in principle predict the voltage recorded by an extracellular electrode (Holt & Koch, 1999; Rall & Shepherd, 1968). This approach has recently been incorporated into a simulation environment (Lindén et al., 2013). Going from the microscopic variables such as current fluxes across patches of membrane to macroscopic variables such as extracellular field is known as the forward problem. The opposite—i.e., going from the macroscopic variables such as the local field or EEG to microscopic variables—is known as the "inverse problem." Unlike the forward problem, the inverse problem does not have a unique solution—many different microscopic current fluxes can give rise to the same macroscopic potential. Thus, the EEG cannot inform us directly about activity of individual neurons (Nunez & Srinivasan, 2006). That being said, there is significant research effort aimed at constraining the likely distribution of sources that could give rise to the observed EEG signals (Grech et al., 2008).

While V_e ultimately reflects many components of neuronal activity such as action and postsynaptic potentials, under most conditions V_e in general and EEG in particular is dominated by the contribution of postsynaptic potentials. There are several reasons for this. The primary reason is that in order to be discernible in the extracellular recording, transmembrane currents in many sources must overlap in time. Action potentials last approximately 1 ms and are thus extremely unlikely to have significant temporal overlap between different neurons in a small volume of brain tissue. In contrast postsynaptic potentials can last 10s of milliseconds and are much more synchronous across neurons (Buzsáki et al., 2012; Niedermeyer & da Silva, 2005). Furthermore, because the amplitude of the electric fields decays rapidly as a function of distance from the source, extracellular voltage will be dominated by sources closest to it. Because EEG is placed onto the surface of the head, the closest electrical sources are located in the superficial cortical layers which contain apical dendrites of pyramidal neurons. Apical dendrites of pyramidal neurons have a preferred orientation normal to the cortical surface. Consequently many different dendrites will have current flow in approximately the same direction. Thus, contributions from multiple dendrites add to the extracellular field rather than cancel each other out (Kirschstein & Köhling, 2009). Synaptic inputs into apical dendrites come from the so-called matrix projections originating in the thalamus (Jones, 1998, 2001) with some contribution from corticocortical synapses. Thus, EEG signal is dominated by the fluctuations of the synaptic input into apical dendrites of cortical pyramidal cells which reflect reverberating activity of thalamocortical loops.

Last important issue to mention in the context of origins of EEG signals is the issue of spatial scale. A small electrode placed into the brain parenchyma will faithfully record extracellular fields dominated by the neuronal sources in its immediate vicinity. For instance, using silicone multielectrode arrays spacing less than 1 mm between electrodes can be readily attained in experimental settings (Buzsáki et al., 2015). This allows recording of activity of single neurons. In contrast, even with high-density EEG recordings each channel of the EEG records electrical activity arising from 10 cm^2 patch of cortex or more. In order to be recorded in the EEG despite being far from the source of the signal and separated by cerebrospinal fluid, skull, muscle, and scalp, the local fields across areas of cortex many orders of magnitude larger than a single neuron have to be correlated (Nunez & Srinivasan, 2010). EEG is therefore primarily suited for investigating fluctuations in this macroscopic-scale brain activity. Thus, because spatial correlations between local fields depend on the global brain states (Destexhe, Contreras, & Steriade, 1999), it is not surprising that changes in global brain states such as loss and recovery of consciousness induced with general anesthetics are readily observed in the EEG (Gibbs, 1937).

2. TECHNICAL CONSIDERATIONS FOR RECORDING EEG SIGNALS

2.1 Reference Electrode

EEG amplifier records potential difference between each electrode and reference. Clearly, the choice of reference will influence the recorded signals. Ideally, reference should be electrically stable and have a zero potential. Thus, it may seem appropriate to place the reference electrode far away from the subject. In practice, however, the long wires that would be required for this would serve as antennae that would pick up ambient electrical noise. Because EEG signals are small (microvolts) recording noise can readily overwhelm the signal. Thus, traditionally, reference is placed on the subject. Many different reference choices have been used in the literature such as the tip of the nose (Andrew & Pfurtscheller, 1996), vertex (Lehmann, Strik, Henggeler, Koenig, & Koukkou, 1998), mastoid or ear (Başar, Rahn, Demiralp, & Schürmann, 1998), ring of electrodes around the neck (Kuklinski, 1983; Nunez & Srinivasan, 2006), bilateral mastoids or ears (Gevins & Smith, 2000), or average of all recorded electrodes (Offner, 1950). Unfortunately, none of these points on the body have a true zero electrical potential, and thus, reference choice has a strong influence on the signal recorded in each electrode. There is a theoretical advantage to the average reference in that the surface integral of the potential of a conductor is zero under most circumstances. This advantage is only well realized in situations where a large number of electrodes are used (high-density EEG). If the EEG montage has a small number of electrodes, the average reference can introduce complex artifacts into the recordings. Furthermore, care must be taken to remove all of the noisy electrodes before averaging. For this reason, we suggest first recording the data with a single-electrode reference (e.g., mastoid or vertex). If conversion to an average or other (see below) reference is desired, this can be done in postprocessing.

While physically recording the signal with respect to a distant reference is technically not feasible because of electrical noise, recently a mathematical algorithm has been proposed to simulate this recording configuration (Chella, Marzetti, Pizzella, Zappasodi, & Nolte, 2014; Yao et al., 2005). This reference electrode standardization technique (REST) model is based on an assumption of a "three concentric spheres" model of head conductance (Nunez & Srinivasan, 2006; Rush & Driscoll, 1969). Some simulation studies (Lei & Liao, 2017) have demonstrated advantages of REST reference, and this reference configuration has been used in a number of experimental settings (Chella et al., 2014; Chella, Pizzella, Zappasodi, & Marzetti, 2016; Liu et al., 2015; Tian & Yao, 2013; Yao et al., 2005). The advantage of REST reference relative to average is moderate for a high-density EEG recording in simulation studies (Lei & Liao, 2017). The algorithm for computing REST reference has been recently published in an open-source MATLAB® toolbox (Dong et al., 2017). The best choice of reference may depend on the specifics of the analysis performed, but it most strongly affects analytic techniques that look at the interrelations between different EEG signals (e.g., coherence, connectivity, etc.). Thus, it seems prudent to us to verify that the overall conclusions are not strongly affected by the choice of reference scheme. The only way to currently accomplish this is to build in different reference schemes into the analysis pipeline.

2.2 Filtering and Sampling Frequency

All modern applications of the analysis of electrophysiological signals such as the EEG are performed digitally. That is signals are sampled with a certain sampling rate onto a digital computer where analysis is subsequently performed. Thus, an experimenter needs to decide how frequently to sample the data. If the data are sampled too frequently, the resulting datasets will be unnecessarily large and this would complicate subsequent analysis. Conversely, if the data are sampled too sparsely in time, then information will be lost. Decisions concerning sampling frequency should be based on Nyquist theorem which states that if the highest frequency of interest in the data is f then the sampling interval dt between consecutive samples should be $1/(2f)$.

Most power in the EEG signals is contained in the relatively low frequencies (Niedermeyer & da Silva, 2005; Nunez & Srinivasan, 2006), thus sampling of EEG beyond 1 kHz is probably unnecessary. Downsampling from the original sampling rate can be performed post hoc if desired for speeding up the analysis code. To avoid introducing sharp discontinuities in the signal we suggest using a low-pass filter prior to downsampling rather than simply removing some data points.

Most forms of EEG analysis require some filtering of the data. Filtering generally refers to selectively attenuating some frequencies for the purposes of enhancing others. Typically filters fall into several categories, such as high pass, low pass, band pass, notch, based on the kind of frequencies they select for. For instance, high-pass filter selectively attenuates lower frequencies, while low-pass filter attenuates higher frequencies. Filters can differ by their design such as for instance finite impulse response and infinite impulse response (IIR) filters. Much like the choice of reference, choice of filter type and filter settings can have dramatic consequences on the data. For instance, consider a simple signal that is constructed by adding two sine waves (Fig. 1). In order to extract just the higher frequency oscillation a sixth order Butterworth (IIR) high-pass filter was applied to the signal. In the filtered waveform (dashed red trace in the bottom panel), the slow oscillation is indeed suppressed well. This is because the two frequencies were chosen to be quite different from each other for the purposes of illustration. Yet, the filtered waveform does not recover the original sine wave that went into the construction of the signal (compare the dashed and the solid line). In this example, the filtered waveform is phase shifted relative to

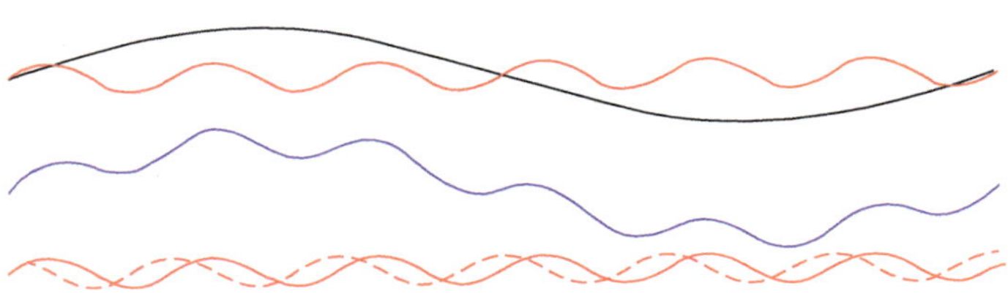

Fig. 1 Filtering introduces phase shifts. The signal of interest consists of two sine waves: slow (*black*) and fast (*red*) in the *top panel*. The overall signal is the sum of the two oscillations (*blue*) in the *middle panel*. A sixth order high-pass Butterworth filter was constructed in MATLAB® using *butter* command and applied to the data. The cut-off frequency was set in the middle of the frequency interval separating the fast and the slow oscillation. The filtered signal (*dashed bottom*) does indeed recover the fast oscillation. However, this oscillation is phase shifted with respect to the original signal (*red solid line*).

the original sine wave. This is just one of many distortions in the signal that filters can induce (Percival & Walden, 1993). Thus, it is our opinion that filtering should be minimized. This is especially relevant for applications where phases of oscillations are important such as phase coupling which constitutes the basis of the bispectrum—a mathematical technique at the core of the BIS index of anesthetic depth (Liu, Singh, & White, 1996). Some of the distortions induced by filtering such as phase delays can be minimized by using post hoc filtering of the data rather than using the filters built into the EEG amplifier. One simple method, for instance, is to run the same filter forward and backward along the same strip of data. This way the phase delays cancel each other out, and the overall signal phase is better preserved. This method is implemented in the MATLAB® *filtfilt* command that we find rather useful in analysis of EEG data.

Despite best recording practices it is not always possible to completely remove the 60 Hz (or 50 Hz outside the US) oscillation from the original recording. Thus, a common practice is to apply a notch filter. A notch (also known as band stop) filter selectively attenuates oscillations in a narrow frequency band. Yet, band stop filters are most problematic. Conventional notch filters are known to produce band holes in the spectrum, significant distortions around the notched frequencies, as well as phase distortions and Gibbs rippling in the time domain. An alternative approach to conventional notch filter has been proposed in Mitra (2007). The basic ideas are that the signal is first subjected to a sliding window multitaper spectral estimation (see below). Then a Thomson spectral F-test is computed for the regression coefficient fit to each complex-valued element in the spectrum. This detects statistically significant deviations of a particular narrow-band signal from the expected spectrum. The complex-valued spectral coefficients identified using this method can be used to construct a sine wave which is then subtracted from the data in the time domain. This avoids many of the pitfalls of conventional notch filtering. The code for performing this type of filtering can be found in the Chronux toolbox for MATLAB® or in CleanLine plugin for EEGLAB (Bokil, Andrews, Kulkarni, Mehta, & Mitra, 2010).

2.3 Artifact Rejection

The importance of proper artifact rejection cannot be overstated for processing any EEG recording. Because EEG electrodes are placed on the skull rather than directly onto the brain, the electrical signals recorded by the electrodes are not solely due to electrical fields generated in the brain.

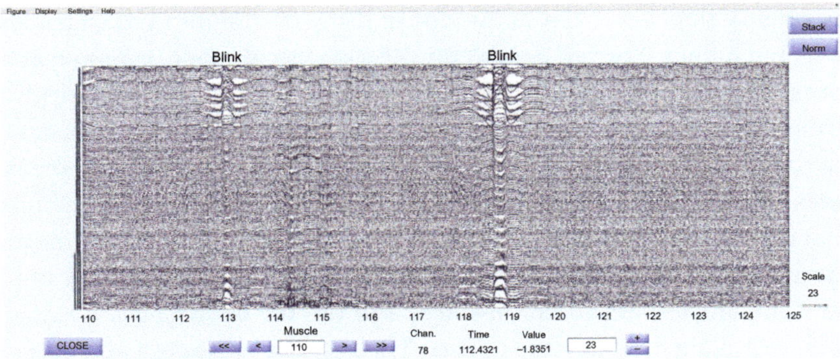

Fig. 2 Artifacts in EEG. Snapshot of high density (128 channels) of EEG during the awake state in a human subject. The signals are displayed using the *eegplot* function. This function is a part of the EEGLAB toolbox and is quite useful in visually inspecting the signals. Blink artifacts are much larger in amplitude than the EEG signals. Muscle artifact appears as characteristic high-frequency oscillations. Removing these artifacts is critical for further analysis. Using *eegplot* function, artifacts can be marked in the signal. Regions that contain artifacts can be excluded from subsequent analysis.

For instance, electrical activity in the muscles can be much larger in amplitude than electrical activity generated by the brain. Other very common sources of artifacts are blinks and movements (e.g., Fig. 2). There are two classes of approaches to artifact removal: remove the segments of data containing the artifact, and remove the artifact from the data. Analysis of most datasets involves some combination of both of these approaches. Obviously it is desirable to keep as much of the data as possible. Thus, we will briefly mention here some techniques that can be used to remove the artifacts from the data. One way to remove the muscle artifact is to simply low-pass filter the data below 30 Hz or so. Most EMG activity occurs at faster frequencies and filtering will therefore attenuate activity due to muscle. That being said, some neuronal activity of interest such as gamma oscillations, for instance, does occur above 30 Hz, and this way of removing muscle artifact will also remove the signal of interest.

Another popular approach that has been successfully used to remove artifacts from EEG data is based on an assumption that the spatial location of sources of noise (and neuronal activity) is constant. For instance, the electrical fields generated by eye blinks ought to be consistently stronger in the frontal electrodes. A class of methods known as the independent component analysis (ICA) is a particularly useful tool with multiple publications in EEG literature (Delorme & Makeig, 2004; Delorme, Sejnowski, & Makeig, 2007; Jung, Makeig, Bell, & Sejnowski, 1998; Vigário, 1997;

Vigario, Sarela, Jousmiki, Hamalainen, & Oja, 2000, to name a few). ICA is at the core of a popular open-source MATLAB®-based EEG analysis toolbox called EEGLAB (Delorme & Makeig, 2004) and is also utilized in another MATLAB®-based EEG toolbox called Fieldtrip (Oostenveld, Fries, Maris, & Schoffelen, 2011). There are several fast algorithms that compute ICA (e.g., Bell & Sejnowski, 1997; Oja & Yuan, 2006), but they all seek to accomplish the same goal—to breakdown the multivariate distribution of observed data (multiple channels of EEG recordings in this case) into multiple independent and non-Gaussian sources that can be added together to reconstruct the original signal. In the case of ICA, "independence" is accomplished by attempting to minimize the mutual information between signals (Bell & Sejnowski, 1997). Deviation from Gaussian distribution is attained by maximizing kurtosis. At the completion of an ICA run, one ends up with a mixing and unmixing matrix. These matrices quantify how each individual channel of EEG contributes to the activation of each independent component. Rather than studying statistical properties of the raw signals (e.g., spectrum) recorded in each EEG channel, one can instead study statistical properties of each independent component (e.g., Fig. 3) (Delorme & Makeig, 2004). This figure illustrates a typical application of ICA to identify and remove blink artifacts from EEG data. Once an independent component corresponding to blinks for instance is identified, it can be removed from the unmixing matrix without totally removing all of the data that occurs during the blink. While there are many published methods for automatic EEG artifact rejection (Krishnaveni, Jayaraman, Anitha, & Ramadoss, 2006; LeVan, Urrestarazu, & Gotman, 2006; Li, Ma, Lu, & Li, 2006; Mognon, Jovicich, Bruzzone, & Buiatti, 2011; Shao, Shen, Ong, Wilder-Smith, & Li, 2009, to name a few), we strongly advocate using semiautomated methods where the data are processed using one of the above-mentioned algorithms but is subsequently inspected by an expert observer.

3. BRIEF INTRODUCTION TO SPECTRAL ANALYSIS OF THE EEG

It is believed that oscillations play a significant role in neuronal processing (Buzsaki, 2004). Thus, many analytic techniques for EEG and other forms of neuronal activity recordings focused on identifying and characterizing the different oscillations. One very broad class of methods that has been used extensively in neuroscience in general and in analysis of EEG in

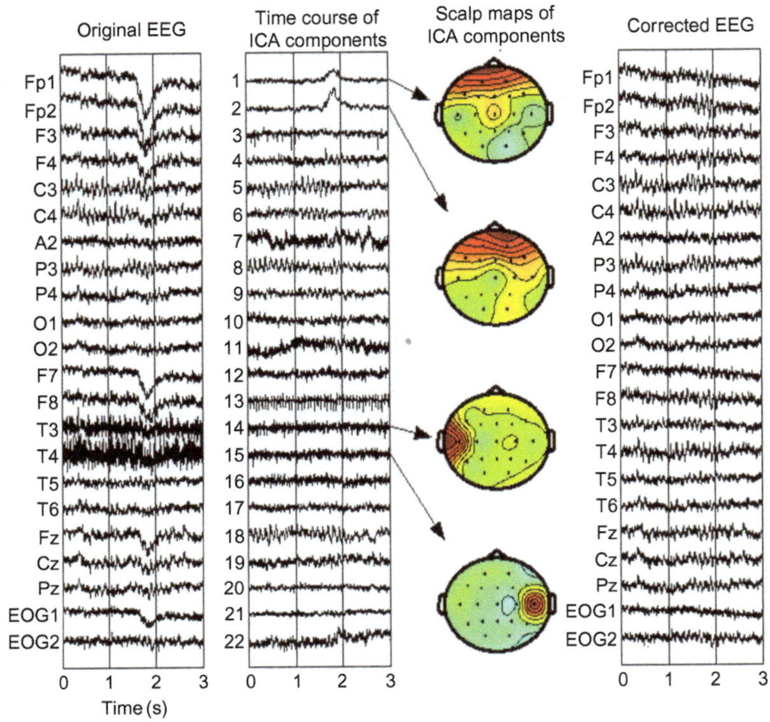

Fig. 3 Artifact rejection using independent component analysis. Image from the EEGLAB tutorial. Close-up of blink and muscle EEG artifact is shown on the *left*. Note the similarity to those shown in Fig. 2. ICA decomposition of the EEG converts the data from channels × time matrix to component × time matrix. Note that activation of components 1 and 2 closely corresponds to the blink artifacts, while the high frequency of activity in components 14 and 15 appears similar to the muscle activity. This intuition is corroborated by the fact that the majority of the blink artifact is found in the frontal leads (*third column*), while muscle artifacts are most closely related to temporalis activity. Components that contain artifacts (Berger, 1932; Buzsáki et al., 2015; Gibbs, 1937; Nunez & Srinivasan, 2010) can be removed from the mixing matrix. This matrix can be inverted to then recover the EEG data without the artifact.

the context of anesthesia specifically is spectral analysis. The mathematical basis of spectral analysis is the Fourier transform which transforms a function in time domain (e.g., EEG) into a complex-valued function of frequency. This corresponds to expressing a function as a sum of sines and cosines. Yet, the proper application of spectral analysis to complex and nonstationary signals such as those recorded in the EEG cannot be accomplished by simply applying the fast Fourier transform algorithm (Press, 2007) to the recorded signals. This section will only introduce the very basics of spectral analysis and make suggestions concerning a good choice of analytic techniques. A much more detailed account of spectral analysis can be found in Bokil et al. (2010), Mitra (2007), and Percival and Walden (1993). Several very useful and pragmatic tutorials can be found at http://chronux.org/.

All recorded signals are evaluated at discrete time points set by the sampling frequency. If we had an infinitely long stretch of data we would be able to properly evaluate its discrete Fourier transform. Yet, all real signals are finite. Furthermore, when the signals are nonstationary (see below), even smaller chunks of data have to be chosen for spectral analysis. Taking a chunk of data is equivalent to multiplying the signal by a function that is equal to 1 within the desired window and is 0 elsewhere. Thus, taking a chunk of data for spectral analysis is equivalent to convolving the Fourier transform of the signal with a Fourier transform of a rectangle. The Fourier transform of the rectangle, however, distorts the signal by introducing both narrow and broadband bias and profoundly distorts the spectrum. Other functions can be convolved with the signal to extract a segment of data (e.g., Hanning, Tukey, Exponential, B-Spline, Boxcar, Parzen, Welch, etc.). The choice of the window has significant implications for spectral analysis. Yet, one specific choice of a spectral window—discrete prolate spheroidal (Slepian) sequences—deserves special attention. These sequences constitute an optimal choice of a windowing function in a sense that concentrate power in the desired frequency range while minimizing power outside of the range. Remarkably, in order to find such windowing function, one needs to solve an eigenvalue problem where the desired functions emerge as eigenvectors. Because the matrix is symmetric, the resultant eigenvectors are orthogonal. As a consequence, the same short segment of data can be convolved with each one of such eigenvectors in turn to get an independent estimate of the spectrum of the signal within the data. This allows for estimation of the confidence intervals around the spectrum for a given window of data. Confidence interval estimation can be performed using a variety of bootstrapping methods such as Jacknife for instance. These discrete prolate spheroidal sequences are also known as tapers and are the basis for what is referred to as multitaper spectral analysis (Bokil et al., 2010; Mitra, 2007; Percival & Walden, 1993). While many alternative choices are possible, our opinion is that for most applications, multitaper spectral analysis is a very good choice of technique for spectral estimation. Some code for performing multitaper spectral estimation and computing Slepian sequences is native to MATLAB®. Additional spectral estimation routines for multitaper spectra and coherences are found in the Chronux toolbox for MATLAB®. Python packages for multitaper spectral estimation are also freely available at http://www.thomas-cokelaer.info/software/spectrum/html/contents.html.

When performing multitaper spectral estimation, one needs to settle on a choice of several key parameters that have a strong influence on results. The first crucial choice is the number of tapers. This choice is related to the degree of desired smoothing in the frequency domain. Recall that the tapers maximize power in the frequency range W. It turns out that the maximum number of tapers is $K=2NW\Delta t$ where N is the number of points in a window of data and Δt is the inverse of the sampling frequency. Thus, the choice of the number of tapers and the choice of the window length is closely related. Choosing an appropriate spectral window plays a very strong role in the resulting spectral analysis. The choice of an appropriate window length depends on the specifics of experimental design and the question the analysis is trying to answer. Choosing a particular window length is equivalent to assuming that the signal is approximately stationary and linear over the course of this window (Mitra, 2007; Percival & Walden, 1993). In our experience some applications (e.g., Alonso et al., 2014; Solovey et al., 2015) require extremely short windows, while other (e.g., Hudson, Calderon, Pfaff, & Proekt, 2014) require exceptionally long spectral windows. As is the case with most other parameters the safest strategy is to explore the range of parameter values to demonstrate that the specific choice does not have a very strong influence on the results.

4. MULTIVARIATE SPECTRAL ANALYSIS: COHERENCE AND PHASE LAG INDEX

Spectrum of a signal is very closely related to its autocorrelation. By analogy crossspectrum is closely related to the crosscorrelations between signals. The major advantage of spectra over correlations is that statistically robust methodologies exist for computing spectral quantities for short segments of data. Techniques similar to that used for estimation of the spectrum of a single-time series (i.e., one channel of the EEG) can be generalized to a pair of signals. One common way of representing the results is to plot the appropriately normalized absolute value of the crossspectrum between a pair of signals as a function of frequency. This quantity [0,1] is typically referred to as coherence and measures frequency-dependent correlations between the signals. If the coherence value is close to 1 at a particular frequency, this means that the phase difference between two signals is approximately constant at that frequency. Conversely, if the number is close to 0, then the phase differences are more uniformly distributed. A version of coherence analysis referred to as phase lag index (PLI) is essentially coherence where the phase lag between two signals is nonzero (i.e., the signals are asynchronous) (Stam, Nolte, & Daffertshofer, 2007). In this way, coherence and PLI are forms of "functional connectivity" (Bastos & Schoffelen, 2015).

There has been very significant interest in quantifying connectivity between brain regions and brain activity signals. It is not possible to review the various approaches here in any significant detail. Here, we will only address a single caveat that complicates all forms of this analysis known as the common reference problem. Recall that EEG signals measure potential difference between a pair of points. One such point is called reference. Thus, if two different electrodes are recording activity with respect to the same reference, fluctuations in voltage at the reference electrode will be reflected in both electrodes and would yield spurious connectivity results. One solution to this problem is to use a different reference for each channel. This can, for instance, be accomplished by using locally referenced bipolar derivation (Bosman et al., 2012). This is not assumption free (Bastos & Schoffelen, 2015) and decreases the number of total electrodes by a factor of ½. Closely related to the common reference problem is the problem of field spread/ volume conduction. As we alluded to earlier for extracellular recordings such as the EEG, there is a fundamental distinction between the sensors (i.e., electrodes) and the sources of electric activity. Because electrical fields travel readily through conductive medium, each sensor will record mixture of activity from different sources. As discussed earlier, it is not strictly speaking possible to unambiguously extract the different sources from EEG recordings. Thus, coherence observed in two electrodes may simply reflect spreading of the electric fields rather than true phase locking between different sources of neuronal activity (Nunez et al., 1997, 1999; Schoffelen & Gross, 2009; Srinivasan, Winter, Ding, & Nunez, 2007; Winter, Nunez, Ding, & Srinivasan, 2007). The solution to the problem of volume conduction is conceptually similar to that of the common reference. The choice of referencing significantly affects the spatial scale on which connectivity is considered (Nunez et al., 1997). Arguably the most parsimonious solution is to compute a surface Laplacian reference (Srinivasan, Nunez, & Silberstein, 1998). The simplest form of the Laplacian is to reference each electrode to the average of all of its nearest neighbors (Cimenser et al., 2011). This technique has the advantage of mathematical simplicity but at the cost of potentially introducing artifacts into the data. More sophisticated approaches based on spline approximations of the surface potentials are reviewed in

Carvalhaes and de Barros (2015). Code for generating Laplacians is readily available through Fieldtrip (Oostenveld et al., 2011). It has been argued that PLI is less sensitive to the issues of common reference and field spread because it only considers non-zero phase lag (Stam et al., 2007). That being said, synchrony may play an important role in neuronal processing.

5. EFFECTS OF ANESTHETICS ON THE EEG

Changes in the amplitude and frequency of EEG induced with general anesthesia became readily apparent to early encephalographers (Gibbs, 1937). Indeed, Gibbs (1937) suggested that "Electroencephalography may be of value in controlling depth of anesthesia and sedation." Indeed there are now many different "depth of anesthesia" monitors in clinical use. Furthermore, there is increasing interest in using the EEG to modify the incidence of postoperative cognitive disturbances (Whitlock et al., 2014).

Oscillations in the EEG observed during the awake and anesthetized states are typically grouped into several categories based on their frequencies.

Band	Frequency (Hz)
δ	<4
θ	4–7
α	8–15
β	16–30
γ	>30

It should be noted, however, that by and large the power spectrum of the EEG is $1/f$ noise. This means that $P \propto f^{\beta}$ where P is the power and f is the frequency. This conspicuous relationship between power and frequency has been a subject of vigorous debate (e.g., Bédard, Kröger, & Destexhe, 2006; Beggs & Plenz, 2003; He, Zempel, Snyder, & Raichle, 2010) which we will not cover here.

In the case of the EEG one significant frequency component that deviates from the usual relationship between power and frequency is the alpha wave. Indeed, alpha wave was first described by Hans Berger in his original manuscript on human EEG and it has been referred to as the Berger wave in the early literature. Alpha wave has been of significant interest in anesthesia research. In the awake brain alpha wave is most prominent in the occipital leads when the subject is relaxed and their eyes are closed. It has been noted by Michenfelder and colleagues (Tinker, Sharbrough, & Michenfelder, 1977) that loss of consciousness is associated with the spatial shift in the alpha oscillations such that just around the point where responsiveness is lost, alpha wave assumes an anterior dominance. This anteriorization of the alpha rhythm has been systematically investigated by John et al. (2001) and was suggested to be the hallmark of loss of consciousness. Recent robust computational techniques by Brown and colleagues revealed that frontal alpha oscillations are coherent during loss and recovery of consciousness induced with propofol (Cimenser et al., 2011). Yet there are some important differences among the anesthetics with respect to the alpha oscillations. For instance, dexmedetomidine produces a spindle-like pattern consisting of intermittent oscillations (12–16 Hz) (Akeju et al., 2016a). Furthermore, Mashour and colleagues (Blain-Moraes et al., 2015) found that the anteriorization of alpha rhythm is not consistently found in the EEG of sevoflurane-anesthetized human subjects. The same group demonstrated that ketamine—a dissociative anesthetic—does not elicit an increase in the frontal alpha-band power (Blain-Moraes, Lee, Ku, Noh, & Mashour, 2014).

Deepening of the anesthetic state is associated with the increase in the slow oscillations referred to as the delta waves (Gibbs, 1937). These slow oscillations arise as a result of the interplay between ion channels and synaptic transmission in the thalamocortical circuits (Steriade, McCormick, & Sejnowski, 1993). In increasing doses many, but not all, anesthetics elicit delta oscillations (Brown, Lydic, & Schiff, 2010). A notable exception to this pattern is ketamine. Under ketamine anesthesia delta and gamma oscillations alternate in a burst-like pattern (Akeju et al., 2016b; Maksimow et al., 2006). Another interesting exception to the delta wave pattern associated with deeper stages of anesthesia is nitrous oxide. Avramov and colleagues first found that delta waves appear transiently (Avramov, Shingu, & Mori, 1990) during the administration of nitrous oxide. In a paradoxical reaction Avramov et al. also demonstrated that painful stimulation, typically associated with desynchronization of the EEG, under nitrous oxide can be associated with emergence of delta oscillations (Avramov et al., 1990). There has been a recent resurgence of interest in using delta oscillations under anesthesia to quantify anesthetic depth. For instance, Tracey and colleagues observe that loss of consciousness induced with propofol (Ní Mhuircheartaigh, Warnaby, Rogers, Jbabdi, & Tracey, 2013). Warnaby and colleagues extended these findings to show that slow wave activity saturation is a hallmark of loss of consciousness for the volatile anesthetics in addition to propofol (Warnaby, Sleigh, Hight, Jbabdi, & Tracey, 2017).

If the anesthetic concentration is increased further, one typically observes a characteristic pattern of the EEG referred to as burst suppression. Burst suppression, as the name suggests corresponds to alterations between period of high-amplitude fluctuations in the EEG signals and periods of suppressed EEG (Steriade, Amzica, & Contreras, 1994). First demonstrated by Derbyshire, Rempel, Forbes, and Lambert (1936), burst suppression is a feature of neuronal activity induced by many anesthetics. Interestingly halothane, in the clinically useful concentration range, does not seem to reliably elicit burst suppression (Murrell, Waters, & Johnson, 2008; Yli-Hankala, Eskola, & Kaukinen, 1989).

In addition to the traditional spectral analysis of the EEG, recent studies turned to the quantification of "functional connectivity" in the anesthetized states. Functional connectivity can be loosely defined as the statistical interrelationship between recordings obtained in different parts of the brain. As alluded to earlier, care must be taken to ensure that this relationship is not spuriously observed because of effects of common ground and volume conductance. Several measures of connectivity have been applied to the EEG under anesthesia. The measure most closely associated with spectral analysis is coherence. Coherence is the frequency domain counterpart of correlation. The advantage of the coherence measure is that it can be robustly estimated in short segments of data using, for instance, multitaper spectral estimates. An example of this approach is (Cimenser et al., 2011) that demonstrate that the alpha waves found over anterior leads of the EEG are coherent. A closely related technique is PLI. PLI seeks to minimize effects of volume conduction by only focusing at coherence observed with a nonzero phase lag. An example of application of PLI to analyzing the EEG under ketamine anesthesia is by Blain-Moraes et al. (2014, 2015). Yet, other measures of functional connectivity rely on information-theoretic quantities such as transfer entropy a symbolic transfer entropy (Staniek & Lehnertz, 2008). As the mathematics of these measures is rather involved, they would not be discussed in significant detail here. Some useful code for computing these quantities can be found in TRENTOOL—a MATLAB®-based toolbox (Wibral, Vicente, Priesemann, & Lindner, 2011) and MuTE (Montalto, Faes, & Marinazzo, 2014). An exciting development in the field of functional connectivity as it relates to the study of anesthesia is the proposal that mechanistically distinct anesthetic agents that induce different patterns of oscillations in the EEG may nonetheless uniformly disrupt connectivity between the frontal and the parietal cortices. This work has been pioneered by Lee and Mashour (e.g., Ku, Lee, Noh, Jun, & Mashour, 2011; Lee et al., 2009, 2013).

6. CONCLUDING REMARKS

Here we introduced the very basics of EEG and pointed toward some useful references that address some of the fundamental issues that complicate analysis of the EEG in much more detail. On the one hand proliferation of software tools that allow complex processing of EEG signals has made EEG more accessible to scientists interested in neuronal activity. Thus, it may no longer be necessary for all researchers involved with EEG recordings to do very sophisticated mathematical calculations de novo. On the other hand, this accessibility may give the false impression that EEG analysis can be performed in an "out of the box" unsupervised fashion. The major purpose of this chapter was to introduce some caveats and assumptions involved in nearly all forms of EEG analysis without getting into the mathematical details. The history of the investigation of the effects of anesthetics on the EEG is almost as long as the history of the EEG itself. Thus, here we just highlighted some of the classic papers and some of the recent findings that characterize the effects of anesthetics on the EEG.

REFERENCES

Akeju, O., et al. (2016a). Spatiotemporal dynamics of dexmedetomidine-induced electroencephalogram oscillations. *PLoS One*, *11*, e0163431.

Akeju, O., et al. (2016b). Electroencephalogram signatures of ketamine anesthesia-induced unconsciousness. *Clinical Neurophysiology*, *127*, 2414–2422.

Alonso, L. M., et al. (2014). Dynamical criticality during induction of anesthesia in human ECoG recordings. *Frontiers Neural Circuits*, *8*, 20.

Andrew, C., & Pfurtscheller, G. (1996). Dependence of coherence measurements on EEG derivation type. *Medical & Biological Engineering & Computing*, *34*, 232–238.

Avramov, M. N., Shingu, K., & Mori, K. (1990). Progressive changes in electroencephalographic responses to nitrous oxide in humans: A possible acute drug tolerance. *Anesthesia and Analgesia*, *70*, 369–374.

Başar, E., Rahn, E., Demiralp, T., & Schürmann, M. (1998). Spontaneous EEG theta activity controls frontal visual evoked potential amplitudes. *Electroencephalography and Clinical Neurophysiology*, *108*, 101–109.

Bastos, A. M., & Schoffelen, J.-M. (2015). A tutorial review of functional connectivity analysis methods and their interpretational pitfalls. *Frontiers in Systems Neuroscience*, *9*, 175.

Bédard, C., Kröger, H., & Destexhe, A. (2006). Does the 1/f frequency scaling of brain signals reflect self-organized critical states? *Physical Review Letters*, *97*, 118102.

Beggs, J. M., & Plenz, D. (2003). Neuronal avalanches in neocortical circuits. *The Journal of Neuroscience*, *23*, 11167–11177.

Bell, A. J., & Sejnowski, T. J. (1997). The 'independent components' of natural scenes are edge filters. *Vision Research*, *37*, 3327–3338.

Berger, H. (1932). Über das Elektrenkephalogramm des Menschen. *Archiv für Psychiatrie und Nervenkrankheiten*, *97*, 6–26.

Blain-Moraes, S., Lee, U., Ku, S., Noh, G., & Mashour, G. A. (2014). Electroencephalographic effects of ketamine on power, cross-frequency coupling, and connectivity in the alpha bandwidth. *Frontiers in Systems Neuroscience*, *8*, 114.

Blain-Moraes, S., et al. (2015). Neurophysiological correlates of sevoflurane-induced unconsciousness. *Anesthesiology*, *122*, 307–316.

Bokil, H., Andrews, P., Kulkarni, J. E., Mehta, S., & Mitra, P. P. (2010). Chronux: A platform for analyzing neural signals. *Journal of Neuroscience Methods*, *192*, 146–151.

Bosman, C. A., et al. (2012). Attentional stimulus selection through selective synchronization between monkey visual areas. *Neuron*, *75*, 875–888.

Brown, E. N., Lydic, R., & Schiff, N. D. (2010). General anesthesia, sleep, and coma. *The New England Journal of Medicine*, *363*, 2638–2650.

Buzsaki, G. (2004). Neuronal oscillations in cortical networks. *Science*, *304*, 1926–1929.

Buzsáki, G., Anastassiou, C. A., & Koch, C. (2012). The origin of extracellular fields and currents—EEG, ECoG, LFP and spikes. *Nature Reviews. Neuroscience*, *13*, 407–420.

Buzsáki, G., et al. (2015). Tools for probing local circuits: High-density silicon probes combined with optogenetics. *Neuron*, *86*, 92–105.

Carvalhaes, C., & de Barros, J. A. (2015). The surface Laplacian technique in EEG: Theory and methods. *International Journal of Psychophysiology*, *97*, 174–188.

Chella, F., Marzetti, L., Pizzella, V., Zappasodi, F., & Nolte, G. (2014). Third order spectral analysis robust to mixing artifacts for mapping cross-frequency interactions in EEG/MEG. *NeuroImage*, *91*, 146–161.

Chella, F., Pizzella, V., Zappasodi, F., & Marzetti, L. (2016). Impact of the reference choice on scalp EEG connectivity estimation. *Journal of Neural Engineering*, *13*, 036016.

Cimenser, A., et al. (2011). Tracking brain states under general anesthesia by using global coherence analysis. *Proceedings of the National Academy of Sciences of the United States of America*, *108*, 8832–8837.

Delorme, A., & Makeig, S. (2004). EEGLAB: An open source toolbox for analysis of single-trial EEG dynamics including independent component analysis. *Journal of Neuroscience Methods*, *134*, 9–21.

Delorme, A., Sejnowski, T., & Makeig, S. (2007). Enhanced detection of artifacts in EEG data using higher-order statistics and independent component analysis. *NeuroImage*, *34*, 1443–1449.

Derbyshire, A. J., Rempel, B., Forbes, A., & Lambert, E. F. (1936). The effects of anesthetics on action potentials in the cerebral cortex of the cat. *American Journal of Physiology-Legacy Content*, *116*, 577–596.

Destexhe, A., Contreras, D., & Steriade, M. (1999). Spatiotemporal analysis of local field potentials and unit discharges in cat cerebral cortex during natural wake and sleep states. *The Journal of Neuroscience*, *19*, 4595–4608.

Dietzel, I., Heinemann, U., & Lux, H. D. (1989). Relations between slow extracellular potential changes, glial potassium buffering, and electrolyte and cellular volume changes during neuronal hyperactivity in cat brain. *Glia*, *2*, 25–44.

Dong, L., et al. (2017). MATLAB toolboxes for reference electrode standardization technique (REST) of scalp EEG. *Frontiers in Neuroscience*, *11*, 601.

Gevins, A., & Smith, M. E. (2000). Neurophysiological measures of working memory and individual differences in cognitive ability and cognitive style. *Cerebral Cortex*, *10*, 829–839.

Gibbs, F. A. (1937). Effect on the electro-encephalogram of certain drugs which influence nervous activity. *Archives of Internal Medicine*, *60*, 154.

Grech, R., et al. (2008). Review on solving the inverse problem in EEG source analysis. *Journal of Neuroengineering and Rehabilitation, 5*, 25.

He, B. J., Zempel, J. M., Snyder, A. Z., & Raichle, M. E. (2010). The temporal structures and functional significance of scale-free brain activity. *Neuron, 66*, 353–369.

Holt, G. R., & Koch, C. (1999). Electrical interactions via the extracellular potential near cell bodies. *Journal of Computational Neuroscience, 6*, 169–184.

Hudson, A. E., Calderon, D. P., Pfaff, D. W., & Proekt, A. (2014). Recovery of consciousness is mediated by a network of discrete metastable activity states. *Proceedings of the National Academy of Sciences of the United States of America, 111*, 9283–9288.

John, E. R., et al. (2001). Invariant reversible QEEG effects of anesthetics. *Consciousness and Cognition, 10*, 165–183.

Jones, E. G. (1998). Viewpoint: The core and matrix of thalamic organization. *Neuroscience, 85*, 331–345.

Jones, E. G. (2001). The thalamic matrix and thalamocortical synchrony. *Trends in Neurosciences, 24*, 595–601.

Jung, T.-P., Makeig, S., Bell, A. J., & Sejnowski, T. J. (1998). Independent component analysis of electroencephalographic and event-related potential data. In *Central auditory processing and neural modeling* (pp. 189–197). Springer.

Kirschstein, T., & Köhling, R. (2009). What is the source of the EEG? *Clinical EEG and Neuroscience, 40*, 146–149.

Krishnaveni, V., Jayaraman, S., Anitha, L., & Ramadoss, K. (2006). Removal of ocular artifacts from EEG using adaptive thresholding of wavelet coefficients. *Journal of Neural Engineering, 3*(4), 338.

Ku, S.-W., Lee, U., Noh, G.-J., Jun, I.-G., & Mashour, G. A. (2011). Preferential inhibition of frontal-to-parietal feedback connectivity is a neurophysiologic correlate of general anesthesia in surgical patients. *PLoS One, 6*, e25155.

Kuklinski, W. S. (1983). Electric fields of the brain. The Neurophysics of EEG by P. L. Nunez with contributions by R. D. Katznelson. *Medical Physics, 10*, 119–120.

Lee, U., et al. (2009). The directionality and functional organization of frontoparietal connectivity during consciousness and anesthesia in humans. *Consciousness and Cognition, 18*, 1069–1078.

Lee, U., et al. (2013). Disruption of frontal–parietal communication by ketamine, propofol, and sevoflurane. *Anesthesiology, 118*, 1264–1275.

Lehmann, D., Strik, W. K., Henggeler, B., Koenig, T., & Koukkou, M. (1998). Brain electric microstates and momentary conscious mind states as building blocks of spontaneous thinking: I. Visual imagery and abstract thoughts. *International Journal of Psychophysiology, 29*, 1–11.

Lei, X., & Liao, K. (2017). Understanding the influences of EEG reference: A large-scale brain network perspective. *Frontiers in Neuroscience, 11*, 205.

LeVan, P., Urrestarazu, E., & Gotman, J. (2006). A system for automatic artifact removal in ictal scalp EEG based on independent component analysis and Bayesian classification. *Clinical Neurophysiology, 117*, 912–927.

Li, Y., Ma, Z., Lu, W., & Li, Y. (2006). Automatic removal of the eye blink artifact from EEG using an ICA-based template matching approach. *Physiological Measurement, 27*, 425–436.

Lindén, H., et al. (2013). LFPy: A tool for biophysical simulation of extracellular potentials generated by detailed model neurons. *Frontiers in Neuroinformatics*, 7, 41.

Liu, J., Singh, H., & White, P. F. (1996). Electroencephalogram bispectral analysis predicts the depth of midazolam-induced sedation. *Anesthesiology, 84*, 64–69.

Liu, Q., et al. (2015). Estimating a neutral reference for electroencephalographic recordings: The importance of using a high-density montage and a realistic head model. *Journal of Neural Engineering, 12*, 056012.

Maksimow, A., et al. (2006). Increase in high frequency EEG activity explains the poor performance of EEG spectral entropy monitor during S-ketamine anesthesia. *Clinical Neurophysiology, 117*, 1660–1668.
Mitra, P. (2007). *Observed brain dynamics*. Oxford University Press.
Mognon, A., Jovicich, J., Bruzzone, L., & Buiatti, M. (2011). ADJUST: An automatic EEG artifact detector based on the joint use of spatial and temporal features. *Psychophysiology, 48*, 229–240.
Montalto, A., Faes, L., & Marinazzo, D. (2014). MuTE: A MATLAB toolbox to compare established and novel estimators of the multivariate transfer entropy. *PLoS One, 9*, e109462.
Murrell, J. C., Waters, D., & Johnson, C. B. (2008). Comparative effects of halothane, isoflurane, sevoflurane and desflurane on the electroencephalogram of the rat. *Laboratory Animals, 42*, 161–170.
Ní Mhuircheartaigh, R., Warnaby, C., Rogers, R., Jbabdi, S., & Tracey, I. (2013). Slow-wave activity saturation and thalamocortical isolation during propofol anesthesia in humans. *Science Translational Medicine, 5*, 208ra148.
Niedermeyer, E., & da Silva, F. H. L. (2005). *Electroencephalography: Basic principles, clinical applications, and related fields*. Lippincott Williams & Wilkins.
Nunez, P. L., & Srinivasan, R. (2006). *Electric fields of the brain: The neurophysics of EEG*. USA: Oxford University Press.
Nunez, P. L., & Srinivasan, R. (2010). Scale and frequency chauvinism in brain dynamics: Too much emphasis on gamma band oscillations. *Brain Structure & Function, 215*, 67–71.
Nunez, P. L., et al. (1997). EEG coherency. I: Statistics, reference electrode, volume conduction, Laplacians, cortical imaging, and interpretation at multiple scales. *Electroencephalography and Clinical Neurophysiology, 103*, 499–515.
Nunez, P. L., et al. (1999). EEG coherency II: Experimental comparisons of multiple measures. *Clinical Neurophysiology, 110*, 469–486.
Offner, F. F. (1950). The EEG as potential mapping: The value of the average monopolar reference. *Electroencephalography and Clinical Neurophysiology, 2*, 213–214.
Oja, E., & Yuan, Z. (2006). The fastICA algorithm revisited: Convergence analysis. *IEEE Transactions on Neural Networks, 17*, 1370–1381.
Oostenveld, R., Fries, P., Maris, E., & Schoffelen, J.-M. (2011). FieldTrip: Open source software for advanced analysis of MEG, EEG, and invasive electrophysiological data. *Computational Intelligence and Neuroscience, 2011*, 156869.
Percival, D. B., & Walden, A. T. (1993). *Spectral analysis for physical applications*. Cambridge, UK: Cambridge University Press.
Press, W. H. (2007). Numerical recipes. In *The art of scientific computing* (3rd ed.). Cambridge University Press.
Rall, W., & Shepherd, G. M. (1968). Theoretical reconstruction of field potentials and dendrodendritic synaptic interactions in olfactory bulb. *Journal of Neurophysiology, 31*, 884–915.
Rush, S., & Driscoll, D. A. (1969). EEG electrode sensitivity—an application of reciprocity. *IEEE Transactions on Biomedical Engineering, BME-16*, 204.
Schoffelen, J.-M., & Gross, J. (2009). Source connectivity analysis with MEG and EEG. *Human Brain Mapping, 30*, 1857–1865.
Shao, S.-Y., Shen, K.-Q., Ong, C. J., Wilder-Smith, E. P. V., & Li, X.-P. (2009). Automatic EEG artifact removal: A weighted support vector machine approach with error correction. *IEEE Transactions on Biomedical Engineering, 56*, 336–344.
Solovey, G., et al. (2015). Loss of consciousness is associated with stabilization of cortical activity. *The Journal of Neuroscience, 35*, 10866–10877.
Srinivasan, R., Nunez, P. L., & Silberstein, R. B. (1998). Spatial filtering and neocortical dynamics: Estimates of EEG coherence. *IEEE Transactions on Biomedical Engineering, 45*, 814–826.

Srinivasan, R., Winter, W. R., Ding, J., & Nunez, P. L. (2007). EEG and MEG coherence: Measures of functional connectivity at distinct spatial scales of neocortical dynamics. *Journal of Neuroscience Methods*, *166*, 41–52.

Stam, C. J., Nolte, G., & Daffertshofer, A. (2007). Phase lag index: Assessment of functional connectivity from multi channel EEG and MEG with diminished bias from common sources. *Human Brain Mapping*, *28*, 1178–1193.

Staniek, M., & Lehnertz, K. (2008). Symbolic transfer entropy. *Physical Review Letters*, *100*, 158101.

Steriade, M., Amzica, F., & Contreras, D. (1994). Cortical and thalamic cellular correlates of electroencephalographic burst-suppression. *Electroencephalography and Clinical Neurophysiology*, *90*, 1–16.

Steriade, M., McCormick, D., & Sejnowski, T. (1993). Thalamocortical oscillations in the sleeping and aroused brain. *Science*, *262*, 679–685.

Tian, Y., & Yao, D. (2013). Why do we need to use a zero reference? Reference influences on the ERPs of audiovisual effects. *Psychophysiology*, *50*, 1282–1290.

Tinker, J. H., Sharbrough, F. W., & Michenfelder, J. D. (1977). Anterior shift of the dominant EEG rhythm during anesthesia in the java monkey. *Anesthesiology*, *46*, 252–259.

Vigário, R. N. (1997). Extraction of ocular artefacts from EEG using independent component analysis. *Electroencephalography and Clinical Neurophysiology*, *103*, 395–404.

Vigario, R., Sarela, J., Jousmiki, V., Hamalainen, M., & Oja, E. (2000). Independent component approach to the analysis of EEG and MEG recordings. *IEEE Transactions on Biomedical Engineering*, *47*, 589–593.

Warnaby, C. E., Sleigh, J. W., Hight, D., Jbabdi, S., & Tracey, I. (2017). Investigation of slow-wave activity saturation during surgical anesthesia reveals a signature of neural inertia in humans. *Anesthesiology*, *127*, 645–657.

Whitlock, E. L., et al. (2014). Postoperative delirium in a substudy of cardiothoracic surgical patients in the BAG-RECALL clinical trial. *Anesthesia & Analgesia*, *118*, 809–817.

Wibral, M., Vicente, R., Priesemann, V., & Lindner, M. (2011). TRENTOOL: An open source toolbox to estimate neural directed interactions with transfer entropy. *BMC Neuroscience*, *12*, P200.

Winter, W. R., Nunez, P. L., Ding, J., & Srinivasan, R. (2007). Comparison of the effect of volume conduction on EEG coherence with the effect of field spread on MEG coherence. *Statistics in Medicine*, *26*, 3946–3957.

Yao, D., et al. (2005). A comparative study of different references for EEG spectral mapping: The issue of the neutral reference and the use of the infinity reference. *Physiological Measurement*, *26*, 173–184.

Yli-Hankala, A., Eskola, H., & Kaukinen, S. (1989). EEG spectral power during halothane anaesthesia. A comparison of spectral bands in the monitoring of anaesthesia level. *Acta Anaesthesiologica Scandinavica*, *33*, 304–308.

CHAPTER SIXTEEN

Using Positron Emission Tomography in Revealing the Mystery of General Anesthesia: Study Design Challenges and Opportunities

Harry Scheinin[*,†,‡,1], Emilee C. Alkire[§], Annalotta Scheinin[*,†], Michael T. Alkire[§,¶], Oskari Kantonen[†,§], Jaakko Långsjö[*,||]

[*]Turku PET Centre, University of Turku and the Hospital District of Southwest Finland, Turku, Finland
[†]Turku University Hospital, Turku, Finland
[‡]Integrative Physiology and Pharmacology, Institute of Biomedicine, University of Turku, Turku, Finland
[§]The Center for the Neurobiology of Learning and Memory, University of California, Irvine, Irvine, CA, United States
[¶]VA Long Beach Healthcare System, Long Beach, CA, United States
[||]Tampere University Hospital, Tampere, Finland
[1]Corresponding author: e-mail address: harry.scheinin@utu.fi

Contents

Abstract

Functional neuroimaging with positron emission tomography (PET) is one of the cornerstones for studying the central nervous system effects of general anesthetics and anesthesia mechanisms. General anesthesia offers a unique and safe way to directly

Methods in Enzymology, Volume 603
ISSN 0076-6879
https://doi.org/10.1016/bs.mie.2018.01.025

manipulate consciousness, and can thus be used as a powerful research tool to study the neurobiology of human consciousness. In this chapter, we will address the possibilities of PET imaging in revealing the mysteries of general anesthesia and anesthetic induced unconsciousness and summarize some of the recent advancements in the field. Importantly, we will discuss possible ways to separate brain activity changes associated with the changing level of consciousness from the concentration or dose-dependent direct or indirect drug effects on the brain. We will try to demonstrate how state-of-the-art clinical pharmacology, use of specific anesthetic drugs, and innovative study design solutions could be utilized.

1. INTRODUCTION

After the first positron emission tomography (PET) study performed on anesthetized humans by Dr. Alkire in the mid-1990s (Alkire et al., 1995), functional neuroimaging has become an important tool for the study of general anesthesia and the effects of anesthetics on the central nervous system. The first human brain imaging studies of general anesthesia focused on the cerebral metabolic rate of glucose (CMR_{glu}) utilization (Alkire, Haier, Shah, & Anderson, 1997; Alkire et al., 1999). These studies confirmed much of the findings from previous animal studies investigating the local CMR_{glu} utilization (Cavazzuti, Porro, Barbieri, & Galetti, 1991; Dam et al., 1990; Maekawa, Tommasino, Shapiro, Keifer-Goodman, & Kohlenberger, 1986; Ori, Dam, Pizzolato, Battistin, & Giron, 1986), but also helped to establish the magnitude of the global metabolic suppression seen for each human when given a dose of anesthetic sufficient to induce unconsciousness. Early studies primarily used a region of interest analysis technique for comparison with the previous animal work. Later, more sophisticated voxel-based analysis methods were applied to find critical brain structures and network interactions related to anesthetic effects (Alkire, 2008; Alkire, Hudetz, & Tononi, 2008; Alkire & Miller, 2005). Also, specific receptor ligands have been applied in studying anesthesia mechanisms in the living human brain.

Because general anesthesia offers a unique and safe way to directly manipulate consciousness, it can also be used as a powerful research tool to study neurobiology of human consciousness (Beecher, 1947). With neuroimaging, such as PET, it is possible to visualize the state-related changes and try to pinpoint the brain structures or neural mechanisms related to changes in consciousness. The search for the neural correlates of consciousness (NCC) is based on a simple principle: identify two conditions that differ only with respect to consciousness and measure and compare brain activity in some way. The difference between the conditions should reflect those brain processes that are specifically related to the conscious experience. The NCC paradigm has especially been applied in the study of visual consciousness focusing on studying the contents of consciousness rather than consciousness as a state. It is important to understand, that "consciousness" is an extremely complex superordinate heading covering a number of different phenomena, such as the contents of consciousness, the states of consciousness (and unconsciousness), different types of consciousness (phenomenal consciousness, reflective consciousness, access consciousness, and self-awareness), and altered states of consciousness such as dreaming and hypnosis (Långsjö, Revonsuo, & Scheinin, 2014). One interesting simplification, but readily applicable in anesthesiology and clinical medicine, is the division of consciousness to its two fundamental components of *arousal* (or level) and *awareness* (or contents) as suggested by Laureys (2005) (Fig. 1).

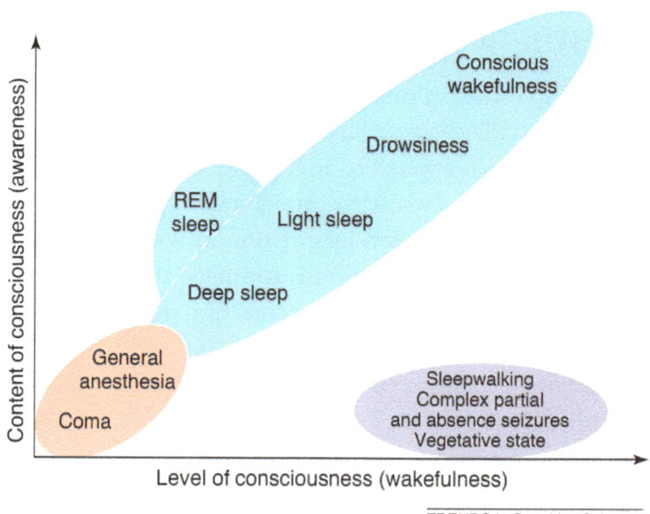

Fig. 1 Simplified illustration of the two major components of consciousness: the level (i.e., wakefulness or arousal) and the content (i.e., awareness or experience) of consciousness. In normal physiological states, the level and content are positively correlated with the exception of dream activity during REM-sleep (*blue–green*), in which the "content > level" of consciousness. Patients in pathological or pharmacological coma (i.e., general anesthesia!) are unconscious (*red*). In dissociated states of consciousness, such as the vegetative state, absence and complex partial seizures, and sleepwalking (*purple*), the "level > contents" of consciousness. *Reproduced with permission from Laureys, S. (2005). The neural correlate of (un)awareness: Lessons from the vegetative state.* Trends in Cognitive Sciences, 9, *556–59.*

Mashour and Hudetz have recently suggested how "bottom-up" and "top-down" paradigms can be synthetized in describing anesthetic induced unconsciousness (Mashour & Hudetz, 2017). Anesthetics suppress both the level (through "bottom-up" mechanisms) and contents (through "top-down" mechanisms) of consciousness partly explaining certain empirical observations regarding the asymmetry of induction and emergence from general anesthesia, as well as the diversity of anesthetic drugs.

In this chapter, we will address the possibilities of using modern PET imaging in revealing the mysteries of general anesthesia and summarize some of the recent advancements in the field. The separation of brain activity changes associated with the changing level of consciousness from the concentration or dose-dependent direct drug effects on the brain is a special challenge. We will try to demonstrate how state-of-the-art clinical pharmacology and innovative study design solutions could be utilized.

2. ANESTHETIC INDUCED UNCONSCIOUSNESS

The crucial question in studying anesthesia mechanisms is how different anesthetic agents affect consciousness. The NCC paradigm introduced earlier has classically been applied to investigate anesthetic induced unconsciousness in the following two ways: brain activity is first measured awake and then after anesthetic induced loss of responsiveness (LOR) or loss of consciousness (LOC). Another option is to image the subjects after terminating drug administration to let the concentration decrease until the subject regains consciousness (return of consciousness, ROC). In both approaches, the two conditions are contrasted to find the neural processes that associate with consciousness.

But how can we really know that the subject is unconscious? Since consciousness cannot be directly measured, the researchers have to use indirect ways to differentiate the conscious and unconscious states. Typically, the difference between these states is defined by using *responsiveness*, which in this context is a goal-directed behavioral response performed by the subject typically after researcher's command like "squeeze my hand if you can hear this." There are, however, serious issues related to using responsiveness as an indicator of LOC and ROC, as recent evidence suggests that conscious experiences are possible in several unresponsive conditions previously considered to represent unconsciousness (Leslie et al., 2009;

Monti et al., 2010; Noreika et al., 2011; Owen et al., 2006; Sanders, Tononi, Laureys, & Sleigh, 2012; Siclari et al., 2017).

The commonly accepted definition for conscious state is to experience at least some inner, subjective processes of sentience or awareness (Sanders, Raz, Banks, Boly, & Tononi, 2016; Searle, 2000). Using this proper definition for consciousness widens the spectrum of possible conscious states beyond those that can be detected by measuring responsiveness, as subjects can have covert inner experiences in a form of dreaming during anesthesia and normal sleep. Dream reports have been acquired frequently after experimental anesthesia (Noreika et al., 2011) and, albeit less commonly, from subjects awakened from deep N3 sleep (Siclari, Larocque, Postle, & Tononi, 2013) and after emergence from general anesthesia in surgical patients (Leslie et al., 2009). These are examples of *disconnected consciousness*, which can, as of now, only be detected by collecting reports of covert experiences during unresponsiveness.

Connectedness refers to the relationship between consciousness and the environment so that the conscious experience or its qualitative content is affected by the environmental stimuli (Sanders et al., 2012). Anesthesia awareness is an example of *connected consciousness* where *responsiveness* is usually absent either because of drug-induced muscle paralysis or impaired ability to react to or understand the stimuli because of the anesthetic drug that is used. The incidence of anesthesia awareness with explicit recall is relatively low at 0.1%–0.2%, but studies using the isolated forearm technique (where one arm is left unaffected by muscle relaxant because of the use of a tourniquet) suggest that connected consciousness under general anesthesia, albeit without recall, might be much higher at reported frequencies ranging from 4.7% to 37% (Sanders et al., 2017). Anesthetics are potent agents for blocking memory (Alkire & Gorski, 2004). In addition to posing serious ethical questions, these findings demonstrate that the amnesic effects of anesthetic agents make the lack of postanesthesia reports unreliable in proving that there were no connected or disconnected consciousness present, even during anesthetic depth generally regarded as sufficient for surgical procedures. A simplified schematic of differences in consciousness, connectedness, and responsiveness is illustrated in Fig. 2.

In spite of the ambiguity inherent in the terminology of anesthetic induced unconsciousness vs unresponsiveness, the terms are used as synonyms in this review in order to follow the terminology used in the original manuscripts.

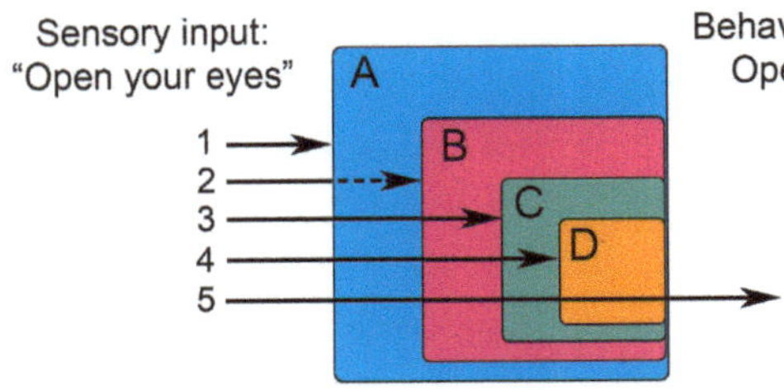

Fig. 2 Schematic of differences in consciousness, connectedness, and responsiveness. Sensory input results in purposeful behavioral output only when all aspects of conscious information processing (A = conscious state, B = awareness and comprehension of the stimulus) and motor readiness (C = will and intention to respond, D = ability to respond) are functional. *Arrows* represent signal processing in different states of consciousness: 1 = all stimuli blocked from reaching consciousness (deep anesthesia or brain death), 2 = dreaming during sleep or anesthesia; surrounding stimuli may (but not necessarily: *dashed arrow*) influence dream content, 3 = deficient will to act (e.g., anterior cingulate lesion), 4 = inability to respond regardless of total awareness (e.g., awareness during anesthesia), and 5 = commands lead to purposeful responses (normal waking consciousness). *Reproduced with permission from Långsjö, J. W., Alkire, M. T., Kaskinoro, K., Hayama, H., Maksimow, A., Kaisti, K. K., et al. (2012). Returning from oblivion: Imaging the neural core of consciousness.* Journal of Neuroscience, 32, *4935–43.*

3. PRINCIPLES OF PET

PET is based on the use of short-lived positron emitting radioisotopes like ^{11}C (half-life 20 min), ^{13}N (10 min), ^{15}O (2 min), and ^{18}F (110 min). Drugs and naturally occurring substances can be labeled without changing their physicochemical or pharmacological properties. Most PET tracers have to be synthesized (necessitating onsite generation of isotopes in a cyclotron), purified, quality tested, and injected into the study subject or patient in less than an hour. After emission, a positron (i.e., a positively charged electron, antimatter) travels only an extremely short distance in the body before encountering an electron (i.e., negatively charged, matter) with consequent destruction of the particles (i.e., annihilation in a matter–antimatter collision). Two gamma rays of equal energy (511 keV) traveling in opposite directions (i.e., exactly 180 degree apart) are formed, and PET imaging is based on simultaneous detection of these paired gamma rays in an array of Geiger Counter-like detector crystals located in a ring outside the body. PET, by using computed tomography reconstruction of the decay events, can thus measure the anatomical distribution of the radiolabeled molecules in the body, and it can, for example, be used to study physiological phenomena and interactions of drugs with specific receptors or enzyme systems.

PET is an extremely sensitive and specific research tool. As drugs can be labeled to very high specific radioactivity, often an intravenous injection of less than a microgram is sufficient to perform a PET study. Even subpicomolar concentrations can be reliably detected. Thus, "tracer doses," i.e., doses without any pharmacological effects can be used which enables the use of this technology in humans even at the early stages of drug development. The advantages of PET over other isotope methods are the ability to quantitatively measure tracer concentrations in tissue, better spatial (2–6 mm) and temporal resolution, and superior sensitivity due to multiple detectors. The benefits also include avoidance of invasive procedures and localization of regions of interest in relatively small anatomical structures, typically 1–3 cm^3. Physiological and pharmacological phenomena and biological correlates can be estimated in vivo by mathematical modeling of the tissue and plasma time–activity curves obtained through dynamic imaging and blood sampling. PET is currently the only technique that permits noninvasive quantification of regional tissue perfusion, glucose utilization, fatty acid uptake and oxidation, and oxygen consumption in absolute units. For example, $[^{15}O]H_2O$, $[^{15}O]O_2$, and $[^{15}O]CO$ can be used to determine regional cerebral blood flow (CBF), cerebral metabolic rate of oxygen (CMR_{O_2}), and cerebral blood volume (CBV), respectively, and ^{18}F-labeled fluorodeoxyglucose ($[^{18}F]FDG$) to determine regional CMR_{glu}. Examples of PET tracers relevant to anesthesiology are shown in Table 1.

Because of the short half-life of ^{15}O, it is possible to perform several PET scans during a single experimental session. A $[^{18}F]FDG$ PET study may be needed to complement the metabolic assessment, but a clear disadvantage is that baseline and posttreatment PET studies have to be performed in separate sessions. As a rule of thumb, the time interval between two PET scans should be at least five half-lives of the radionuclide. A limiting factor in designing and performing PET studies in patients and healthy subjects is related to the radiation exposure inherent to the method.

4. IMAGING GENERAL ANESTHESIA WITH PET

4.1 Early Studies

PET has been crucial in discovering how different anesthetics affect cerebral metabolism and perfusion. Global suppression of brain activity is consistently observed during anesthesia with GABAergic drugs (Alkire et al., 1995, 1997, 1999; Kaisti et al., 2003, 2002), α_2-Adrenoceptor agonist dexmedetomidine (Prielipp et al., 2002) and with the NMDA-receptor antagonist xenon

Table 1 Examples of PET Tracers Used in Brain Research Relevant to Anesthesiology

Tracer	Biological Measure	Abbreviation	Administration	Half-Life
$[^{15}O]$ Tracers				2 min
$[^{15}O]H_2O$	Cerebral blood flow	CBF	Intravenous	
$[^{15}O]CO$	Cerebral blood volume	CBV	Inhalation	
$[^{15}O]O_2$	Cerebral metabolic rate of oxygen	CMR_{O_2}	Inhalation	
$[^{11}C]$ Tracers				20 min
$[^{11}C]$Carfentanil	Opiate receptors in the brain		Intravenous	
$[^{11}C]$Flumazenil	$GABA_A$ (Benzodiazepine) receptors in the brain		Intravenous	
$[^{11}C]$Ketamine	NMDA-receptors in the brain		Intravenous	
$[^{18}F]$ Tracers				110 min
$[^{18}F]$FDG	Cerebral metabolic rate of glucose	CMR_{glu}	Intravenous	
$[^{18}F]$Flumazenil	$GABA_A$ (Benzodiazepine) receptors in the brain		Intravenous	

(Laitio et al., 2007, 2009). Volatile anesthetics often induce clear global reduction in metabolism with less distinct changes in perfusion (Kaisti et al., 2003; Laitio et al., 2009; Schlunzen, Juul, Hansen, Gjedde, & Cold, 2010). During surgical levels of anesthesia with sevoflurane, oxygen and glucose metabolism are globally suppressed while perfusion is decreased significantly only in the frontoparietal and occipital cortices, the thalamus, the caudate, and in the cerebellum (Kaisti et al., 2003; Schlunzen et al., 2010). Propofol, on the other hand, tends to maintain the balance slightly better and reduces CBF and glucose metabolism down to approximately 50% of the awake values during surgical levels of anesthesia (Schlunzen, Juul, Hansen, & Cold, 2012). The dose-dependent reduction of CMR_{glu} along deepening levels of propofol or isoflurane anesthesia in a single individual is shown in Fig. 3.

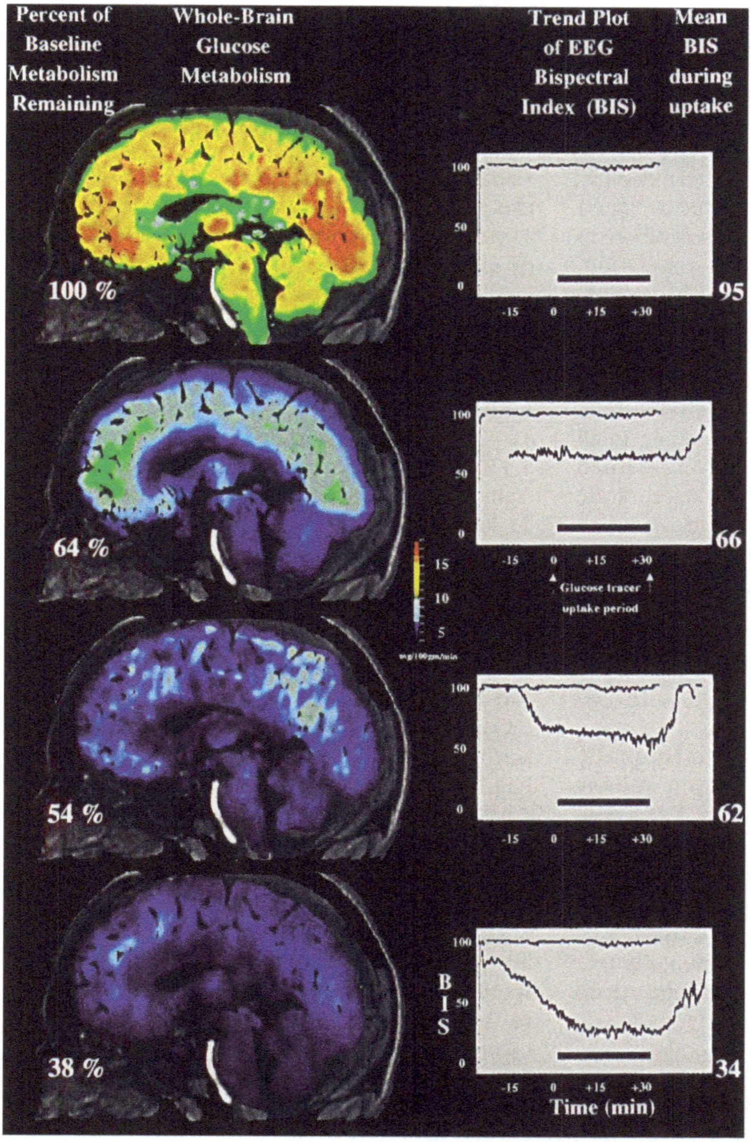

Fig. 3 Reduction of CMR_{glu} and the bispectral index (BIS) of frontal EEG along deepening propofol or isoflurane anesthesia in a single individual. From top to down: awake baseline, sedation, light anesthesia, and deep anesthesia. In each row, the percentage of baseline metabolism is shown on the *left* and mean BIS recorded during PET data acquisition (*black horizontal bar*) on the *right*. The awake baseline BIS plot is replicated in the plots of the other conditions for comparison. *Reproduced with permission from Alkire, M. T. (1998). Quantitative EEG correlations with brain glucose metabolic rate during anesthesia in volunteers.* Anesthesiology, 89, *323–33.*

The effects of the most classical NMDA-antagonist, ketamine, are nearly inverse compared to other anesthetics, since ketamine has been shown to induce a concentration-dependent increase in cerebral perfusion. CMR_{glu} is increased especially during subanesthetic (sedative) doses of racemic ketamine while *S*-ketamine anesthesia increases glucose metabolism significantly only in the thalamus (Långsjö et al., 2003, 2005, 2004). Surprisingly, the flow-metabolism pattern induced by xenon, another NMDA-receptor antagonist, resembles that associated with volatile anesthetics (Laitio et al., 2007, 2009).

In principle, the PET studies assessing the effects of anesthetics on the brain can be divided into two classes based on their study designs. The more clinically oriented studies applied an approach mimicking clinical practice where anesthesia is induced with a rapid administration of large dose of an anesthetic agent (Kaisti et al., 2003, 2002; Laitio et al., 2007, 2009; Långsjö et al., 2005; Schlunzen et al., 2012, 2010, 2004; Schlunzen, Cold, Rasmussen, & Vafaee, 2006). This approach is practical for the study of deep surgical levels of anesthesia and was meant to reveal the metabolic perfusion profiles of different anesthetics on clinically relevant doses from the safety point of view. Although many of these studies were performed on healthy subjects, the obtained results are important in selecting optimal anesthetic regimen especially for the neurosurgical or neurointensive patient population.

While being relevant for clinical anesthesiology, little attention was paid to the actual transition point from consciousness to unconsciousness in these earlier studies. Induction of anesthesia being almost instantaneous, unconsciousness was reached far too quickly for the transition to be studied with most imaging techniques that have a minimum temporal resolution required for obtaining a sufficient sample of data. A slightly different approach was used in the early studies by Alkire and coworkers. Alkire et al. applied a protocol of anesthetic titration until the subjects just became unresponsive (Alkire et al., 1995, 1997, 1999). These studies laid the foundation for research trying to capture the changes related to the actual LOC. With slow and gradual administration of an anesthetic agent, it became possible to approach the transition point of consciousness and unconsciousness in a delicate and slow enough manner for it to be scientifically experimented. This was important also because PET imaging has relatively poor time resolution. With PET perfusion imaging, obtained scans represent a summation of cerebral perfusion fluctuations taking place in approximately minutes. Owing to differences in tracer kinetics, scans measuring CMR_{glu} represent a summation of at least 30–60 min of cerebral glucose metabolism fluctuation. Because it is not possible to make the time resolution of PET imaging better, it was necessary to slow down the process of anesthetic induction. At the same time, the risk of dose overshoot in drug administration, which is evident in fast anesthesia induction adopted from the operating room environment, was minimized. One of the first imaging studies to truly focus on the transition between consciousness and unconsciousness was performed by Fiset et al. (1999). In their study, propofol was administered as target-controlled infusion (TCI) to reach steady endpoints of mild and deep levels of sedation and finally a state of unresponsiveness. PET–CBF measurements performed upon these conditions revealed that the decrease in cerebral perfusion was particularly pronounced in the medial thalamus, the cuneus, precuneus, and the posterior cingulate cortex (PCC) as well as in the orbitofrontal cortex and the right angular gyrus. In these regions, decreased CBF also correlated with the increasing concentrations of propofol (Fiset et al., 1999).

Based on the suppression patterns observed during anesthesia, deactivation of some brain regions could be more important for anesthetic induced unconsciousness (Alkire et al., 2008). Neuroimaging studies have shown that anesthesia is most consistently associated with decreased activity in the thalamus, prefrontal cortex, ACC, PCC, precuneus, and the temporo–parieto–occipital junction (Alkire et al., 1995, 1997, 1999; Bonhomme et al., 2001, 2008; Fiset et al., 1999; Kaisti et al., 2003, 2002; Veselis et al., 2004). Because these brain structures seem to be more active during resting wakefulness, they have been proposed to be involved in maintaining cerebral homeostasis (Cavanna, 2007; Gusnard, Raichle, & Raichle, 2001). This is supported by a finding that activity in these regions is decreased during goal-directed mental activity, i.e., when brain activity is required for something else than just "maintenance" (Cavanna, 2007; Gusnard et al., 2001). Owing to their higher activity level by *default* during wakefulness, these structures have been suggested to form the *default mode network* of the brain (Cavanna, 2007; Gusnard et al., 2001). It is not entirely clear whether these regions truly represent the NCC or simply just the regions most affected by the anesthetics. Nevertheless, it is of interest that these same regions are consistently deactivated during impairment of consciousness almost irrespective of the mechanism (Alkire & Miller, 2005; Baars, Ramsoy, & Laureys, 2003; Cavanna, 2007; Laureys, Boly, & Maquet, 2006; Laureys, Lemaire, Maquet, Phillips, & Franck, 1999; Maquet, 2000; Rainville et al., 1999; Thomas et al., 2000) (Fig. 4).

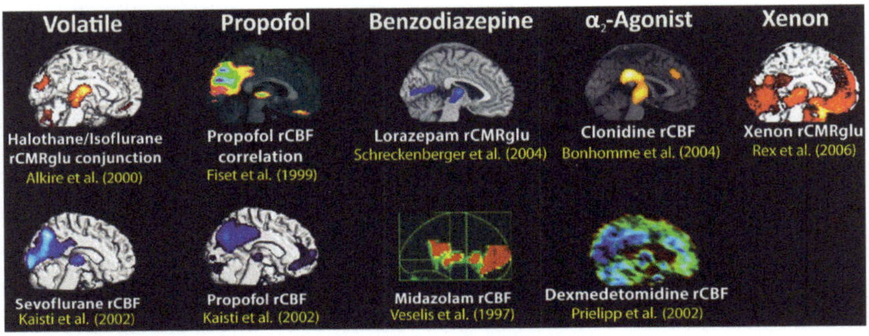

Fig. 4 The regional effects of anesthetics on the brain are shown in humans who were given nine different anesthetic agents representing five drug classes at doses that caused, or nearly caused, loss of responsiveness. The regional effects were measured using either blood flow or glucose metabolism-based imaging techniques by eight different groups of investigators. All images show regional decreases in activity caused by anesthesia compared to the awake state, except the propofol correlation image, which shows where increasing anesthetic dose correlates with decreasing blood flow. Suppression in the thalamus is seen with all anesthetics (Alkire et al., 2000; Bonhomme et al., 2008; Fiset et al., 1999; Kaisti et al., 2002; Prielipp et al., 2002; Rex et al., 2006; Schreckenberger et al., 2004; Veselis et al., 1997). *Modified with permission from Alkire, M. T., & Miller, J. (2005). General anesthesia and the neural correlates of consciousness.* Progress in Brain Research, 150, *229–44; Rex, S., Schaefer, W., Meyer, P. H., Rossaint, R., Boy, C., Setani, K., et al. (2006). Positron emission tomography study of regional cerebral metabolism during general anesthesia with xenon in humans.* Anesthesiology, 105: *936–43.*

Studies on deep surgical levels of anesthesia help to reveal the overall global direction of change in brain activity induced by different anesthetic agents: suppression with most drugs (Kaisti et al., 2002; Laitio et al., 2009; Schlunzen et al., 2012, 2010), increased with NMDA-antagonists ketamine and nitrous oxide (Kaisti et al., 2003; Långsjö et al., 2005). This background knowledge is important for the interpretation of more subtle changes observed at the transition of consciousness to unconsciousness usually visualized in respect to a change in global mean value in the brain (i.e., as *relative*). Thus, in a case of global activity suppression, the brain regions exhibiting significant *relative decreases* (often visualized by means of voxel analysis software, e.g., statistical parametric mapping or pattern analysis software, e.g., partial least squares) represent the structures where activity systematically decreased more than that seen in global mean values between the two conditions. On the other hand, *relative increases* would then represent the regions where activity decrease was systematically less than that seen in the global mean (Kaisti et al., 2003; Långsjö et al., 2003).

Conversely, the situation is reversed if anesthetic profile is dominated by, e.g., global increase in cerebral perfusion as seen with ketamine (Långsjö et al., 2003, 2005, 2004).

4.2 Studying Molecular Mechanisms

Most of the current anesthetics enhance inhibitory gamma-aminobutyric acid (GABA) neurotransmission in the brain through the $GABA_A$ receptor (activation of inhibition). Ketamine, xenon, and nitrous oxide are exceptions as their primary target has been proposed to be the inhibition of the excitatory glutamatergic system through *N*-methyl-D-aspartate (NMDA) receptor (inhibition of activation). However, there is very little in vivo human data of the receptor level mechanisms of different anesthetics.

By using ^{11}C labeled flumazenil and PET, it is possible to investigate the GABAergic system of the living human brain. Traditionally this method has been used in studying various neurological and psychiatric disorders. In anesthesia studies, the method is based on the idea that the enhanced binding of ^{11}C-flumazenil by allosteric modulation on the multisubunit $GABA_A$ receptor protein would indicate increased activity of the GABAergic system during anesthesia. Indeed, propofol and also volatile anesthetics isoflurane and sevoflurane have been demonstrated to enhance ^{11}C-flumazenil binding in the living human brain (Gyulai, Mintun, & Firestone, 2001; Salmi et al., 2004). The anesthetic state caused by volatile anesthetics likely involves other and probably multiple different molecular mechanisms. It has been shown that ex vivo benzodiazepine receptor density correlates with regional cerebral metabolic reduction studied with [^{18}F]FDG and PET induced by propofol but not isoflurane anesthesia (Alkire & Haier, 2001). Interestingly, neither xenon nor subanesthetic racemic ketamine affected ^{11}C-flumazenil binding in healthy subjects (Salmi et al., 2008, 2005) supporting the view that GABAergic mechanisms are not involved with these agents.

PET would be the method-of-choice for studying receptor mechanisms of anesthetic drugs in humans. Unfortunately, the number of suitable PET ligands is very limited at the moment. ^{11}C-ketamine has been used in experimental studies, but to our knowledge, it has not been used in studying NMDA mechanisms of anesthetics in the living human brain. Labeling an active drug with a PET radionuclide (e.g., ^{11}C) seldom yields an optimal PET ligand. Nevertheless, direct labeling of fluorinated anesthetics with ^{18}F might reveal important processes in the uptake and distribution of volatile anesthetics throughout the brain and body. An early attempt at this idea was performed in animals back in 1963 when halothane was labeled with ^{11}C. The study of various new ligands may also offer new clues as such ligands are developed. The radiochemical challenges related to the development and validation of novel PET ligands are beyond the scope of this review.

4.3 Disentangling the Mysteries of General Anesthesia and Human Consciousness

Voxel-based reanalysis of PET data collected in the first experiments in the 1990s has provided interesting results. Unified depression in the thalamus and the midbrain region, the cerebellum, cuneus, and parts of the frontal and temporal cortices at LOC (LOR) were demonstrated (Alkire, Haier, & Fallon, 2000). Later on, analysis of the same data revealed decreased thalamocortical and corticocortical functional and effective connectivity upon anesthetic induced unconsciousness (White & Alkire, 2003) (Fig. 5). Structural equation modeling of [^{18}F]FDG PET data revealed subtle network effects of anesthetic induced unresponsiveness or unconsciousness. Comparing equipotent doses of different anesthetics have also provided interesting insight about possible differences between the drugs (Kaisti et al., 2003; Veselis et al., 2004). Both propofol and thiopental deactivate posterior parts of the brain while the suppression of the frontal lobe seems to be more pronounced with propofol (Veselis et al., 2004).

In recent years, the focus of NCC studies has been on the point of transition to or from unresponsiveness. Instead of high, surgical doses administered rapidly, anesthetic drugs can be titrated in small increments until the threshold for unresponsiveness is just reached. As already discussed, Fiset and coworkers were among the first to apply gradual transition from the awake state to unconsciousness (Fiset et al., 1999). In a recent study of our own, dexmedetomidine or propofol were applied for the manipulation of consciousness while performing PET perfusion imaging measurements at different states of consciousness (Långsjö et al., 2012). At escalating drug concentrations and deepening anesthetic level, suppression of the PCC, thalamus, precuneus, brainstem, and the frontal, parietal, and temporal cortices were observed with both drugs. Conversely, the emergence of consciousness, as assessed with a motor response to a spoken command, was associated with the activation of a core network involving subcortical and limbic regions (i.e., the brain stem, thalamus, hypothalamus, and the ACC) with surprisingly little contribution from the neocortex (Fig. 6). The unique property of dexmedetomidine (see Section 4.5) was exploited

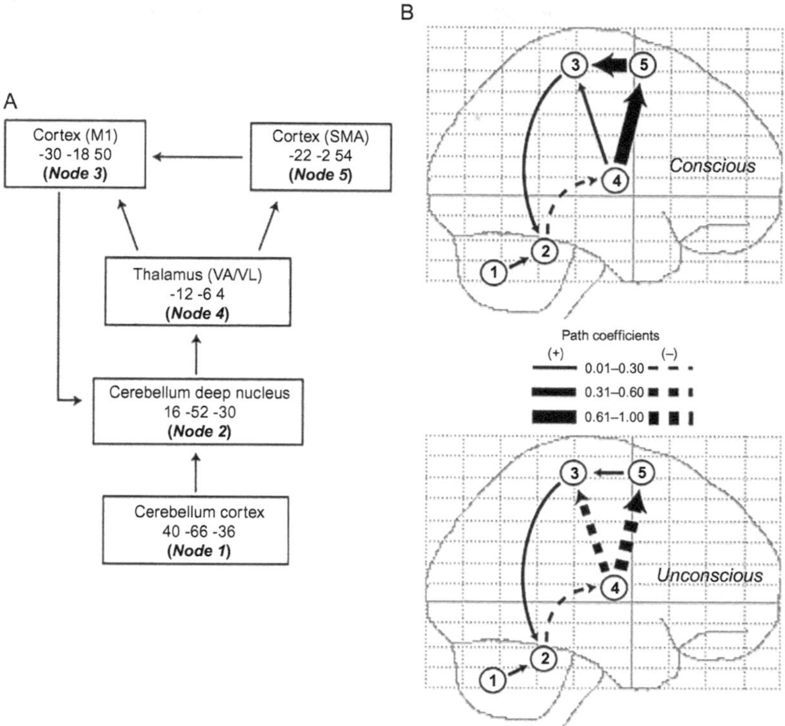

Fig. 5 Effective connectivity changes with anesthetic induced unresponsiveness in the human cerebellothalamocortical network. (A) The network nodes, with their Talairach coordinates, and their modeled interactions. (B) Structural equation modeling reveals that effective connectivity dramatically changes within this network, especially involving the thalamocortical and corticocortical interactions depending on the presence or absence of responsiveness (conscious or unconscious in the *graph*). The *type of arrow* indicates the sign of the influence (*solid* = positive, *dashed* = negative) and the *thickness of the arrows* the relative strength of the influence. Connectivity analyses can reveal network interactions and regional effects that might otherwise be missed with more traditional analysis techniques. *Reproduced with permission from White, N. S., & Alkire, M. T. (2003). Impaired thalamocortical connectivity in humans during general-anesthetic-induced unconsciousness.* NeuroImage, 19, *402–11.*

as it enabled us to separate brain activity changes associated with the changing level of consciousness from the drug-related effects on the brain. An active parietal region demonstrated greater functional connectivity with the ACC upon emergence from anesthesia during constant infusion and plasma concentration of dexmedetomidine. These findings revealed a functional network that activates with restored consciousness (Långsjö et al., 2012).

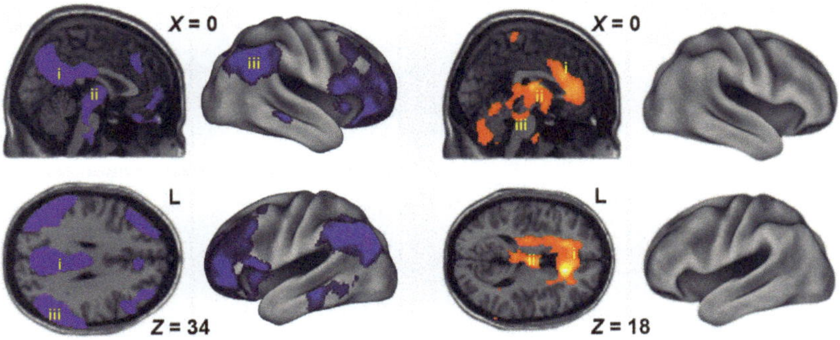

Fig. 6 $[^{15}O]H_2O$ PET findings showing the difference of the overall effects of anesthetic drugs (dexmedetomidine and propofol combined) from those related to consciousness (responsiveness). On the *left*, state-related activity suppressions (*blue color*) that follow the anesthetic dose in the posterior cingulate cortex (i), thalamus (ii), precuneus, brainstem, and the frontal, parietal (iii) and temporal cortices. These findings are well in line with earlier imaging studies. On the *right*, activations (*yellow* and *orange colors*) related to the emergence of responsiveness to verbal command during or after anesthetic infusion in the anterior cingulate cortex (i), thalamus (ii), and the brainstem locus coeruleus/parabrachial area (iii). The emergence from anesthesia appears to be associated with activation of deep, phylogenetically old brain structures rather than the neocortex. *Modified with permission from Långsjö, J. W., Alkire, M. T., Kaskinoro, K., Hayama, H., Maksimow, A., Kaisti, K. K., et al. (2012). Returning from oblivion: Imaging the neural core of consciousness.* Journal of Neuroscience, 32, *4935–43.*

More emphasis is now being placed in analyzing changes in connectivity, which refers to functional coupling and information transfer between different parts of the brain. The communication in both the thalamocortical and corticocortical circuits have been shown to diminish during unresponsiveness/unconsciousness, and the disruption of these connections has been proposed to prevent the formation of a coherent experience of the surroundings and self. In a recent PET study, dexmedetomidine-induced unconsciousness showed decreased CBF and CMR_{glu} in the thalamus and frontoparietal cortices, which agrees with previous findings (Akeju et al., 2014). However, emergence from this state showed sustained reduction in CBF, but restored thalamocortical functional connectivity.

4.4 Study Design Issues

As discussed earlier, most of the PET studies have compared the effects of anesthetics with awake baseline. To study the critical mechanisms related to anesthetic induced unresponsiveness and/or unconsciousness, deep sedation would be a more relevant reference. The concentration–effect relationships of general anesthetics are, however, extremely steep (Franks, 2008), and planning the optimal dosing is very challenging. Most of the intravenous anesthetics follow multicompartment pharmacokinetics, and maintaining steady-state or pseudo steady-state concentration during PET data acquisition requires computer-controlled infusions with a method called TCI. Slow zero order infusions have also been used in recent experiments, but with this approach, possible nonspecific, direct or indirect drug effects cannot be totally separated from those on consciousness. Furthermore, interindividual variability of general anesthetics, both due to pharmacokinetic and pharmacodynamic reasons, is large. Therefore, determining the individual dose or concentration for the selected behavioral end-point (e.g., LOR or LOC) before the PET scanning session should be considered (Långsjö et al., 2012).

4.5 How to Separate Nonspecific Drug Effects From Those on Consciousness?

This is a special challenge. We have exploited the unique property of a selective α_2-adrenoceptor agonist, dexmedetomidine, the effects of which can be reversed without changing the dosing or concentration of the drug (Maksimow et al., 2007). The subjects were aroused with loud voices and mild physical stimulation during constant drug infusion, titrated carefully to LOR (Långsjö et al., 2012). The subjects were scanned before and immediately after arousal, enabling us to separate brain activity changes associated with the changing state of consciousness from drug-related nonspecific effects on the brain. In our new study (ClinicalTrials.gov identifier NCT01889004), we are trying to take one step further. Traditionally, propofol-induced unresponsiveness has not been considered similarly reversible, but we recently discovered that with careful dose titration to achieve LOR, almost half of the subjects could be aroused also during constant propofol infusion (http://www.asaabstracts.com/strands/asaabstracts/abstract.htm?year=2016&index=10&absnum=4493). This finding is intriguing and may have implications in consciousness research.

Another possibility for not changing the concentration of the anesthetic is to antagonize its effects with another pharmacological agent. In previous experimental studies, physostigmine (Meuret, Backman, Bonhomme, Plourde, & Fiset, 2000; Xie et al., 2011), methylphenidate (Chemali, Van Dort, Brown, & Solt, 2012; Solt et al., 2011), and dextroamphetamine (Kenny, Taylor, Brown, & Solt, 2015) have been shown to activate emergence from propofol or isoflurane anesthesia. To our knowledge, human studies have only been published using physostigmine (Meuret et al., 2000; Röhm, Riechmann, Boldt, Schöllhorn, & Piper, 2007; Xie et al., 2011). Based on the results from an imaging study with PET and restoration of consciousness by physostigmine in healthy subjects, Xie et al. suggested that reductions in the activity in the thalamus and precuneus are functionally related to propofol-induced unconsciousness independently of any nonspecific effects of the drug (Xie et al., 2011). The inevitable problem when using pharmacological reversal agents is the fact that possible direct or indirect effects of the other intervention (e.g., physostigmine) cannot be excluded.

4.6 Choosing the Optimal Imaging Method: PET or fMRI?

Functional magnetic resonance imaging (fMRI) is another method for studying anesthetic induced changes in neural activity in the human brain. PET has traditionally been the method-of-choice for absolute quantification of biological measures, like regional CBF and CMR_{glu}. However, after the introduction of arterial spin labeling perfusion MRI, it may be that PET will soon be challenged as the dominant technique for measuring CBF (Dashdorj et al., 2013; Venkatraghavan et al., 2016; Wong, Buxton, & Frank, 1999). PET is, however, the gold standard for measuring cerebral metabolism. Furthermore, for studying receptor mechanisms in the living human brain, PET still remains the only available method.

Compared to PET, fMRI is not dependent on radioactive isotopes (no cyclotron needed), it is far more accessible (most hospitals have one), it is less expensive and has significantly better temporal resolution. However, fMRI suffers from signal-to-noise problems that must be overcome by using multiple behavioral transitions occurring in a relatively short timeframe. This can be challenging in capturing changes in consciousness caused by general anesthesia when the change in behavioral state takes minutes to develop.

Attempts to get rid of the noisy background signal on fMRI experiments led to an interesting observation. The background noise was found to fluctuate spontaneously at frequencies <0.1 Hz, giving rise to a completely new approach in fMRI signal analysis, resting state fMRI (rs-fMRI), which is based on the temporal coherence of BOLD signal fluctuation between different regions (Biswal, Yetkin, Haughton, & Hyde, 1995). Based on rs-fMRI, brain can be divided into several functional networks (Raichle, 2010). Subsequent rs-fMRI studies have demonstrated, e.g., that corticocortical and thalamocortical connectivity changes associate with propofol-induced unconsciousness and that connectivity is preserved within the visual and auditory networks. These findings suggest that sensory stimuli are passed through the thalamus and received by the primary sensory cortices even during unconsciousness (Boveroux et al., 2010). Although interesting, rs-fMRI methodology is strained with some open questions. It is not yet clear what biological phenomena BOLD fluctuations represent and whether or not these fluctuations involve significant or important informational transfer (Raichle, 2010).

How might one decide between designing an anesthesia study utilizing PET or utilizing fMRI? Does one method clearly triumph over the other? Each technique has its strengths and limitations. Generally, if state-related changes over time are the focus of one's hypothesis, then PET may be the better technique. For example, what is different in the brain when a dose of anesthetic is given that will put the brain in a lightly amnesic state vs a heavy amnesic state? (Veselis et al., 2004, 2008). On the other hand, if one is interested in the events associated with which items are able to be remembered when one is on a select amnesic dose of an agent, then fMRI may better show the brain regions associated with processing the specific events that lead to memory or subsequent forgetting (Hayama et al., 2012). Thus, the main advantage of PET imaging is that it gives a nice global overview of activity throughout the whole brain that is related to the experimental state a subject is in. Essentially, imagine being able to place an electrophysiology electrode in every neuron of the brain and then have subjects do an experiment, by recording all the activity in the brain the regions most related to the task should become evident. The task-related activity can thus be surveyed across the whole brain, noninvasively and provide the electrophysiologist with new targets for future studies, for which many may not have been obvious from other methods. However, as the field of cognitive neuroscience emerged, it became clear that much happens in the brain over the 30 min sampling window needed for a typical CMR_{glu} scan. The advantage of fMRI is the faster temporal resolution that takes experiments out of the realm of state-related changes into the realm of event-related neural activity.

The strongest design of an experiment utilizing fMRI and anesthesia would be an on/off design between conscious and unconscious states, alternating every 30 s for a number of minutes. It follows that a comparison between these states should reveal the associated cognitive changes underlying the change in consciousness state. Unfortunately, other than dexmedetomidine, there is no anesthetic that can be "turned on and off" so quickly as to allow fMRI signals to be reliably measured in the timeframe needed. Therefore, the question of which technique is better is not the correct question to ask. One should formulate a hypothesis and then use whichever technique is the best suited to answer the specific scientific question being asked.

5. SUMMARY AND CONCLUSION

Anesthetic drugs can be used as powerful tools to study conscious–unconscious state transition and human consciousness in general. A number of experimental neuroimaging studies with different anesthetic agents suggest that neuronal hyperpolarization block at the level of the thalamus, or thalamocortical and corticocortical reverberant loops, contributes to anesthetic induced unconsciousness. However, not all anesthetics suppress global cerebral metabolism and cause a regionally specific effect on the thalamus. Functional and effective connectivity analysis techniques can be used with neuroimaging to investigate anesthetic induced changes in network interactions. State-of-the-art clinical pharmacology and innovative study design solutions can be utilized to separate brain activity changes associated with the changing level of consciousness from the concentration or dose-dependent direct or indirect drug effects on the brain.

Considering the evidence for conscious experiences occurring regularly in behaviorally unresponsive subjects, it seems justified to suggest that the field of anesthesia mechanisms research needs to shift from considering consciousness as an either on or off state to seeing it as a continuum, where the richness of behavioral and experiential interactions with the environment can gradually fade and disconnect (supposedly) completely. Consciousness can also fluctuate, e.g., due to the change in the level of arousal during sudden intense surgical stimulation. If brain activity is contrasted between two conscious conditions separated only by capability for behavioral interactions, wrong inferences about consciousness can be made. A real challenge for consciousness research is to design study paradigms to find neural signatures for disconnected but conscious, and unconscious states. Furthermore, as evidence shows that preventing consciousness per se is not necessary to prevent experiences of a surgery (as long as the consciousness remains disconnected), it might be a time to start considering inducing *disconnectedness*, rather than unconsciousness, as the necessary component of successful general anesthesia.

FUNDING STATEMENT

Academy of Finland, Helsinki, Finland; Jane and Aatos Erkko Foundation, Helsinki, Finland; VSSHP-EVO, Turku, Finland; Doctoral Programme of Clinical Investigation, University of Turku Graduate School, Turku, Finland; The Finnish Medical Foundation, Helsinki, Finland; The Emil Aaltonen Foundation, Tampere, Finland.

CONFLICT OF INTEREST

None declared.

REFERENCES

Akeju, O., Loggia, M. L., Catana, C., Pavone, K. J., Vazquez, R., Rhee, J., et al. (2014). Disruption of thalamic functional connectivity is a neural correlate of dexmedetomidine-induced unconsciousness. *eLife, 4*, e04499.

Alkire, M. T. (2008). Probing the mind: Anesthesia and neuroimaging. *Clinical Pharmacology and Therapeutics, 84*, 149–152.

Alkire, M. T., & Gorski, L. A. (2004). Relative amnesic potency of five inhalational anesthetics follows the Meyer–Overton rule. *Anesthesiology, 101*, 417–429.

Alkire, M. T., & Haier, R. J. (2001). Correlating in vivo anaesthetic effects with ex vivo receptor density data supports a GABAergic mechanism of action for propofol, but not for isoflurane. *British Journal of Anaesthesia, 86*, 618–626.

Alkire, M. T., Haier, R. J., Barker, S. J., Shah, N. K., Wu, J. C., & Kao, Y. J. (1995). Cerebral metabolism during propofol anesthesia in humans studied with positron emission tomography. *Anesthesiology, 82*, 393–403.

Alkire, M. T., Haier, R. J., & Fallon, J. H. (2000). Toward a unified theory of narcosis: Brain imaging evidence for a thalamocortical switch as the neurophysiologic basis of anesthetic-induced unconsciousness. *Consciousness and Cognition, 9*, 370–386.

Alkire, M. T., Haier, R. J., Shah, N. K., & Anderson, C. T. (1997). Positron emission tomography study of regional cerebral metabolism in humans during isoflurane anesthesia. *Anesthesiology, 86*, 549–557.

Alkire, M. T., Hudetz, A. G., & Tononi, G. (2008). Consciousness and anesthesia. *Science, 322*, 876–880.

Alkire, M. T., & Miller, J. (2005). General anesthesia and the neural correlates of consciousness. *Progress in Brain Research, 150*, 229–244.

Alkire, M. T., Pomfrett, C. J., Haier, R. J., Gianzero, M. V., Chan, C. M., Jacobsen, B. P., et al. (1999). Functional brain imaging during anesthesia in humans: Effects of halothane on global and regional cerebral glucose metabolism. *Anesthesiology, 90*, 701–709.

Baars, B. J., Ramsoy, T. Z., & Laureys, S. (2003). Brain, conscious experience and the observing self. *Trends in Neurosciences, 26*, 671–675.

Beecher, H. K. (1947). Anesthesia's second power: Probing the mind. *Science, 105*, 164–166.

Biswal, B., Yetkin, F. Z., Haughton, V. M., & Hyde, J. S. (1995). Functional connectivity in the motor cortex of resting human brain using echo-planar MRI. *Magnetic Resonance in Medicine, 34*, 537–541.

Bonhomme, V., Fiset, P., Meuret, P., Backman, S., Plourde, G., Paus, T., et al. (2001). Propofol anesthesia and cerebral blood flow changes elicited by vibrotactile stimulation: A positron emission tomography study. *Journal of Neurophysiology, 85*, 1299–1308.

Bonhomme, V., Maquet, P., Phillips, C., Plenevaux, A., Hans, P., Luxen, A., et al. (2008). The effect of clonidine infusion on distribution of regional cerebral blood flow in volunteers. *Anesthesia and Analgesia, 106*, 899–909.

Boveroux, P., Vanhaudenhuyse, A., Bruno, M. A., Noirhomme, Q., Lauwick, S., Luxen, A., et al. (2010). Breakdown of within- and between-network resting state functional magnetic resonance imaging connectivity during propofol-induced loss of consciousness. *Anesthesiology, 113*, 1038–1053.

Cavanna, A. E. (2007). The precuneus and consciousness. *CNS Spectrums, 12*, 545–552.

Cavazzuti, M., Porro, C. A., Barbieri, A., & Galetti, A. (1991). Brain and spinal cord metabolic activity during propofol anaesthesia. *British Journal of Anaesthesia, 66*, 490–495.

Chemali, J. J., Van Dort, C. J., Brown, E. N., & Solt, K. (2012). Active emergence from propofol general anesthesia is induced by methylphenidate. *Anesthesiology, 116*, 998–1005.

Dam, M., Ori, C., Pizzolato, G., Ricchieri, G. L., Pellegrini, A., Giron, G. P., et al. (1990). The effects of propofol anesthesia on local cerebral glucose utilization in the rat. *Anesthesiology, 73*, 499–505.

Dashdorj, N., Corrie, K., Napolitano, A., Petersen, E., Mahajan, R. P., & Auer, D. P. (2013). Effects of subanesthetic dose of nitrous oxide on cerebral blood flow and metabolism: A multimodal magnetic resonance imaging study in healthy volunteers. *Anesthesiology, 118*, 577–586.

Fiset, P., Paus, T., Daloze, T., Plourde, G., Meuret, P., Bonhomme, V., et al. (1999). Brain mechanisms of propofol-induced loss of consciousness in humans: A positron emission tomographic study. *The Journal of Neuroscience, 19*, 5506–5513.

Franks, N. P. (2008). General anaesthesia: From molecular targets to neuronal pathways of sleep and arousal. *Nature Reviews. Neuroscience, 9*, 370–386.

Gusnard, D. A., Raichle, M. E., & Raichle, M. E. (2001). Searching for a baseline: Functional imaging and the resting human brain. *Nature Reviews Neuroscience, 2*, 685–694.

Gyulai, F. E., Mintun, M. A., & Firestone, L. L. (2001). Dose-dependent enhancement of in vivo GABA(A)-benzodiazepine receptor binding by isoflurane. *Anesthesiology, 95*, 585–593.

Hayama, H. R., Drumheller, K. M., Mastromonaco, M., Reist, C., Cahill, L. F., & Alkire, M. T. (2012). Event-related functional magnetic resonance imaging of a low dose of dexmedetomidine that impairs long-term memory. *Anesthesiology, 117*, 981–995.

Kaisti, K. K., Långsjö, J. W., Aalto, S., Oikonen, V., Sipilä, H., Teräs, M., et al. (2003). Effects of sevoflurane, propofol, and adjunct nitrous oxide on regional cerebral blood flow, oxygen consumption, and blood volume in humans. *Anesthesiology, 99*, 603–613.

Kaisti, K. K., Metsähonkala, L., Teräs, M., Oikonen, V., Aalto, S., Jääskeläinen, S., et al. (2002). Effects of surgical levels of propofol and sevoflurane anesthesia on cerebral blood flow in healthy subjects studied with positron emission tomography. *Anesthesiology, 96*, 1358–1370.

Kenny, J. D., Taylor, N. E., Brown, E. N., & Solt, K. (2015). Dextroamphetamine (but not atomoxetine) induces reanimation from general anesthesia: Implications for the roles of dopamine and norepinephrine in active emergence. *PLoS One, 10*, e0131914.

Laitio, R. M., Kaisti, K. K., Långsjö, J. W., Aalto, S., Salmi, E., Maksimow, A., et al. (2007). Effects of xenon anesthesia on cerebral blood flow in humans: A positron emission tomography study. *Anesthesiology, 106*, 1128–1133.

Laitio, R. M., Langsjo, J. W., Aalto, S., Kaisti, K. K., Salmi, E., Maksimow, A., et al. (2009). The effects of xenon anesthesia on the relationship between cerebral glucose metabolism and blood flow in healthy subjects: A positron emission tomography study. *Anesthesia and Analgesia, 108*, 593–600.

Långsjö, J. W., Alkire, M. T., Kaskinoro, K., Hayama, H., Maksimow, A., Kaisti, K. K., et al. (2012). Returning from oblivion: Imaging the neural core of consciousness. *Journal of Neuroscience, 32*, 4935–4943.

Långsjö, J. W., Kaisti, K. K., Aalto, S., Hinkka, S., Aantaa, R., Oikonen, V., et al. (2003). Effects of subanesthetic doses of ketamine on regional cerebral blood flow, oxygen consumption, and blood volume in humans. *Anesthesiology, 99*, 614–623.

Långsjö, J. W., Maksimow, A., Salmi, E., Kaisti, K., Aalto, S., Oikonen, V., et al. (2005). S-ketamine anesthesia increases cerebral blood flow in excess of the metabolic needs in humans. *Anesthesiology, 103*, 258–268.

Långsjö, J. W., Revonsuo, A., & Scheinin, H. (2014). Harnessing anesthesia and brain imaging for the study of human consciousness. *Current Pharmaceutical Design, 20*, 4211–4224.

Långsjö, J. W., Salmi, E., Kaisti, K. K., Aalto, S., Hinkka, S., Aantaa, R., et al. (2004). Effects of subanesthetic ketamine on regional cerebral glucose metabolism in humans. *Anesthesiology, 100*, 1065–1071.

Laureys, S. (2005). The neural correlate of (un)awareness: Lessons from the vegetative state. *Trends in Cognitive Sciences, 9*, 556–559.

Laureys, S., Boly, M., & Maquet, P. (2006). Tracking the recovery of consciousness from coma. *The Journal of Clinical Investigation, 116*, 1823–1825.

Laureys, S., Lemaire, C., Maquet, P., Phillips, C., & Franck, G. (1999). Cerebral metabolism during vegetative state and after recovery to consciousness. *Journal of Neurology, Neurosurgery, and Psychiatry, 67*, 121.

Leslie, K., Sleigh, J., Paech, M. J., Voss, L., Lim, C. W., & Sleigh, C. (2009). Dreaming and electroencephalographic changes during anesthesia maintained with propofol or desflurane. *Anesthesiology, 111*, 547–555.

Maekawa, T., Tommasino, C., Shapiro, H. M., Keifer-Goodman, J., & Kohlenberger, R. W. (1986). Local cerebral blood flow and glucose utilization during isoflurane anesthesia in the rat. *Anesthesiology, 65*, 144–151.

Maksimow, A., Snapir, A., Särkelä, M., Kentala, E., Koskenvuo, J., Posti, J., et al. (2007). Assessing the depth of dexmedetomidine-induced sedation with electroencephalogram (EEG)-based spectral entropy. *Acta Anaesthesiologica Scandinavica, 51*, 22–30.

Maquet, P. (2000). Functional neuroimaging of normal human sleep by positron emission tomography. *Journal of Sleep Research, 9*, 207–231.

Mashour, G. A., & Hudetz, A. G. (2017). Bottom-up and top-down mechanisms of general anesthetics modulate different dimensions of consciousness. *Frontiers in Neural Circuits, 11*, 44.

Meuret, P., Backman, S. B., Bonhomme, V., Plourde, G., & Fiset, P. (2000). Physostigmine reverses propofol-induced unconsciousness and attenuation of the auditory steady state response and bispectral index in human volunteers. *Anesthesiology, 93*, 708–717.

Monti, M. M., Vanhaudenhuyse, A., Coleman, M. R., Boly, M., Pickard, J. D., Tshibanda, L., et al. (2010). Willful modulation of brain activity in disorders of consciousness. *The New England Journal of Medicine, 362*, 579–589.

Noreika, V., Jylhänkangas, L., Moro, L., Valli, K., Kaskinoro, K., Aantaa, R., et al. (2011). Consciousness lost and found: Subjective experiences in an unresponsive state. *Brain and Cognition, 77*, 327–334.

Ori, C., Dam, M., Pizzolato, G., Battistin, L., & Giron, G. (1986). Effects of isoflurane anesthesia on local cerebral glucose utilization in the rat. *Anesthesiology, 65*, 152–156.

Owen, A. M., Coleman, M. R., Boly, M., Davis, M. H., Laureys, S., & Pickard, J. D. (2006). Detecting awareness in the vegetative state. *Science, 313*, 1402.

Prielipp, R. C., Wall, M. H., Tobin, J. R., Groban, L., Cannon, M. A., Fahey, F. H., et al. (2002). Dexmedetomidine-induced sedation in volunteers decreases regional and global cerebral blood flow. *Anesthesia and Analgesia, 95*, 1052–1059.

Raichle, M. E. (2010). Two views of brain function. *Trends in Cognitive Sciences, 14*, 180–190.

Rainville, P., Hofbauer, R. K., Paus, T., Duncan, G. H., Bushnell, M. C., & Price, D. D. (1999). Cerebral mechanisms of hypnotic induction and suggestion. *Journal of Cognitive Neuroscience, 11*, 110–125.

Rex, S., Schaefer, W., Meyer, P. H., Rossaint, R., Boy, C., Setani, K., et al. (2006). Positron emission tomography study of regional cerebral metabolism during general anesthesia with xenon in humans. *Anesthesiology, 105*, 936–943.

Röhm, K. D., Riechmann, J., Boldt, J., Schöllhorn, T., & Piper, S. N. (2007). Do patients profit from physostigmine in recovery from desflurane anaesthesia. *Acta Anaesthesiologica Scandinavica, 51*, 278–283.

Salmi, E., Kaisti, K. K., Metsähonkala, L., Oikonen, V., Aalto, S., Någren, K., et al. (2004). Sevoflurane and propofol increase 11c-flumazenil binding to gamma-aminobutyric acid A receptors in humans. *Anesthesia and Analgesia, 99*, 1420–1426.

Salmi, E., Laitio, R. M., Aalto, S., Maksimow, A. T., Langsjo, J. W., Kaisti, K. K., et al. (2008). Xenon does not affect gamma-aminobutyric acid type a receptor binding in humans. *Anesthesia and Analgesia, 106*, 129–134.

Salmi, E., Langsjo, J. W., Aalto, S., Nagren, K., Metsahonkala, L., Kaisti, K. K., et al. (2005). Subanesthetic ketamine does not affect 11c-flumazenil binding in humans. *Anesthesia and Analgesia, 101*, 722–725.

Sanders, R. D., Gaskell, A., Raz, A., Winders, J., Stevanovic, A., Rossaint, R., et al. (2017). Incidence of connected consciousness after tracheal intubation: A prospective, international, multicenter cohort study of the isolated forearm technique. *Anesthesiology, 126*, 214–222.

Sanders, R. D., Raz, A., Banks, M. I., Boly, M., & Tononi, G. (2016). Is consciousness fragile. *British Journal of Anaesthesia, 116*, 1–3.

Sanders, R. D., Tononi, G., Laureys, S., & Sleigh, J. W. (2012). Unresponsiveness not equal unconsciousness. *Anesthesiology, 116*, 946–959.

Schlunzen, L., Cold, G. E., Rasmussen, M., & Vafaee, M. S. (2006). Effects of dose-dependent levels of isoflurane on cerebral blood flow in healthy subjects studied using positron emission tomography. *Acta Anaesthesiologica Scandinavica, 50*, 306–312.

Schlunzen, L., Juul, N., Hansen, K. V., & Cold, G. E. (2012). Regional cerebral blood flow and glucose metabolism during propofol anaesthesia in healthy subjects studied with positron emission tomography. *Acta Anaesthesiologica Scandinavica, 56*, 248–255.

Schlunzen, L., Juul, N., Hansen, K. V., Gjedde, A., & Cold, G. E. (2010). Regional cerebral glucose metabolism during sevoflurane anaesthesia in healthy subjects studied with positron emission tomography. *Acta Anaesthesiologica Scandinavica, 54*, 603–609.

Schlunzen, L., Vafaee, M. S., Cold, G. E., Rasmussen, M., Nielsen, J. F., & Gjedde, A. (2004). Effects of subanaesthetic and anaesthetic doses of sevoflurane on regional cerebral blood flow in healthy volunteers. A positron emission tomographic study. *Acta Anaesthesiologica Scandinavica, 48*, 1268–1276.

Schreckenberger, M., Lange-Asschenfeld, C., Lochmann, M., Mann, K., Siessmeier, T., Buchholz, H. G., et al. (2004). The thalamus as the generator and modulator of EEG alpha rhythm: A combined PET/EEG study with lorazepam challenge in humans. *Neuroimage, 22*, 637–644.

Searle, J. R. (2000). Consciousness. *Annual Review of Neuroscience, 23*, 557–578.

Siclari, F., Baird, B., Perogamvros, L., Bernardi, G., LaRocque, J. J., Riedner, B., et al. (2017). The neural correlates of dreaming. *Nature Neuroscience, 20*, 872–878.

Siclari, F., Larocque, J. J., Postle, B. R., & Tononi, G. (2013). Assessing sleep consciousness within subjects using a serial awakening paradigm. *Frontiers in Psychology, 4*, 542.

Solt, K., Cotten, J. F., Cimenser, A., Wong, K. F., Chemali, J. J., & Brown, E. N. (2011). Methylphenidate actively induces emergence from general anesthesia. *Anesthesiology, 115*, 791–803.

Thomas, M., Sing, H., Belenky, G., Holcomb, H., Mayberg, H., Dannals, R., et al. (2000). Neural basis of alertness and cognitive performance impairments during sleepiness. I. Effects of 24 H of sleep deprivation on waking human regional brain activity. *Journal of Sleep Research, 9*, 335–352.

Venkatraghavan, L., Poublanc, J., Bharadwaj, S., Sobczyk, O., Crawley, A. P., Mandell, D. M., et al. (2016). Noninvasive measurement of cerebral blood flow under anesthesia using arterial spin labeling MRI: A pilot study. *Journal of Neurosurgical Anesthesiology, 28*, 331–336.

Veselis, R. A., Feshchenko, V. A., Reinsel, R. A., Dnistrian, A. M., Beattie, B., & Akhurst, T. J. (2004). Thiopental and propofol affect different regions of the brain at similar pharmacologic effects. *Anesthesia and Analgesia, 99*, 399–408.

Veselis, R. A., Reinsel, R. A., Beattie, B. J., Mawlawi, O. R., Feshchenko, V. A., DiResta, G. R., et al. (1997). Midazolam changes cerebral blood flow in discrete brain regions: an H2(15)O positron emission tomography study. *Anesthesiology, 87*, 1106–1117.

Veselis, R. A., Pryor, K. O., Reinsel, R. A., Mehta, M., Pan, H., & Johnson, R. (2008). Low-dose Propofol-induced amnesia is not due to a failure of encoding: Left inferior prefrontal cortex is still active. *Anesthesiology, 109*, 213–224.

White, N. S., & Alkire, M. T. (2003). Impaired thalamocortical connectivity in humans during general-anesthetic-induced unconsciousness. *NeuroImage, 19*, 402–411.

Wong, E. C., Buxton, R. B., & Frank, L. R. (1999). Quantitative perfusion imaging using arterial spin labeling. *Neuroimaging Clinics of North America, 9*, 333–342.

Xie, G., Deschamps, A., Backman, S. B., Fiset, P., Chartrand, D., Dagher, A., et al. (2011). Critical involvement of the thalamus and precuneus during restoration of consciousness with physostigmine in humans during propofol anaesthesia: A positron emission tomography study. *British Journal of Anaesthesia, 106*, 548–557.

CHAPTER SEVENTEEN

Brain Imaging Using Hyperpolarized ^{129}Xe Magnetic Resonance Imaging

Simrun Chahal*, Braedan R.J. Prete*, Alanna Wade*, Francis T. Hane*,†,1, Mitchell S. Albert*,†,‡

*Lakehead University, Thunder Bay, ON, Canada
†Thunder Bay Regional Health Research Institute, Thunder Bay, ON, Canada
‡Northern Ontario School of Medicine, Thunder Bay, ON, Canada
[1]Corresponding author: e-mail address: francishane@gmail.com

Contents

Abstract

Hyperpolarized (HP) ^{129}Xe magnetic resonance imaging (MRI) is a novel iteration of traditional MRI that relies on detecting the spins of ^{1}H. Since ^{129}Xe is a gaseous signal source, it can be used for lung imaging. Additionally, ^{129}Xe dissolves in the blood stream and can therefore be detectable in the brain parenchyma and vasculature. In this work, we provide detailed information on the protocols that we have developed to image ^{129}Xe within the brains of both rodents and human subjects.

1. INTRODUCTION

In addition to traditional medical imaging techniques such as computed tomography, positron emission tomography, and proton (^{1}H)

Methods in Enzymology, Volume 603
ISSN 0076-6879
https://doi.org/10.1016/bs.mie.2018.01.027

magnetic resonance imaging (MRI), hyperpolarized (HP) ^{129}Xe MRI has emerged as a potential MR imaging modality for noninvasive, functional, anatomical, and physiological studies of the mammalian lungs and brain (Albert et al., 1994; Bachert et al., 1996; Black et al., 1996; Fain et al., 2007; Hane, Imai, et al., 2017; Möller et al., 2002; Tooker et al., 2003).

Perhaps due to its direct relationship with the mammalian respiratory system, HP ^{129}Xe MRI has been used in research settings to study the feasibility of aiding in the diagnoses of a variety of pulmonary diseases, such as interstitial lung disease, asthma, cystic fibrosis, and chronic obstructive pulmonary disease (Couch et al., 2015). Recently, HP ^{129}Xe MRI has been used to study other biological environments of interest, including the mammalian brain. This success is primarily attributed to the lipophilicity of ^{129}Xe and its capability to dissolve in mammalian blood with a partition coefficient of 0.15 (Goto et al., 1998). Due to its lipophilicity, ^{129}Xe is capable of interacting with several biological media via hydrophobic interactions that are important for the formation of higher-level protein structures and lipid bilayer membranes (Cherubini & Bifone, 2003). Furthermore, ^{129}Xe can be used to sensitively detect its surrounding chemical environment because of its wide-ranging magnetic resonance chemical shifts. These chemical shifts have been reported to be between +212 ppm for xenon nuclei bound to red blood cells (Hane, Smylie, et al., 2016) and +79 ppm for xenon encapsulated by a cryptophane-based biosensing scaffold (Schröder, 2017). These unique properties make HP ^{129}Xe MRI a convenient imaging modality for studying various biological media, including lipid-rich neuronal membranes, such as those found in the mammalian brain, from both their anatomical and physiological perspectives (Albert et al., 1994; Barany et al., 1987; Hane, Imai, et al., 2017; Mazzanti et al., 2011; Oros & Shah, 2004; Zhou et al., 2011).

Due to the relatively low concentrations of gas under standard conditions of temperature and pressure, noble gas MRI requires nuclear hyperpolarization, most commonly achieved with spin-exchange optical pumping (SEOP), which results in an imbalance between the spin states of ^{129}Xe nuclei, where there is a greater population of spin-down than spin-up ^{129}Xe nuclei (Goodson, 2002; Goodson et al., 2017; Happer, 1972; Oros & Shah, 2004; Walker & Happer, 1997). By this methodology, the nuclear magnetic resonance signal-to-noise ratios (SNRs) obtained are up to 100,000 times greater than thermally polarized gas, which results in an increase in tissue sensitivity and thus contributes to greater MR image contrast (Albert et al., 1994). This signal enhancement makes HP ^{129}Xe MRI an ideal imaging modality for sensitive measurements of mammalian organ structure and function such as in brain (Fig. 1).

Xenon is also a known general anesthetic although its mechanism is not yet fully understood. Because of the detectability of HP ^{129}Xe in the brain, HP ^{129}Xe MRI may be a technique used in the near future to elucidate additional information about the mechanism of Xe anesthesia. In fact, the original research goal of Mitchell Albert when he coinvented HP ^{129}Xe MRI was this very problem. This research problem is not without its challenges. Since that time, however, the HP ^{129}Xe MRI technique has been adapted to solve a variety of other pressing research questions including high sensitivity fMRI and Alzheimer's disease (AD) studies (Li et al., 2017; Mazzanti et al., 2011).

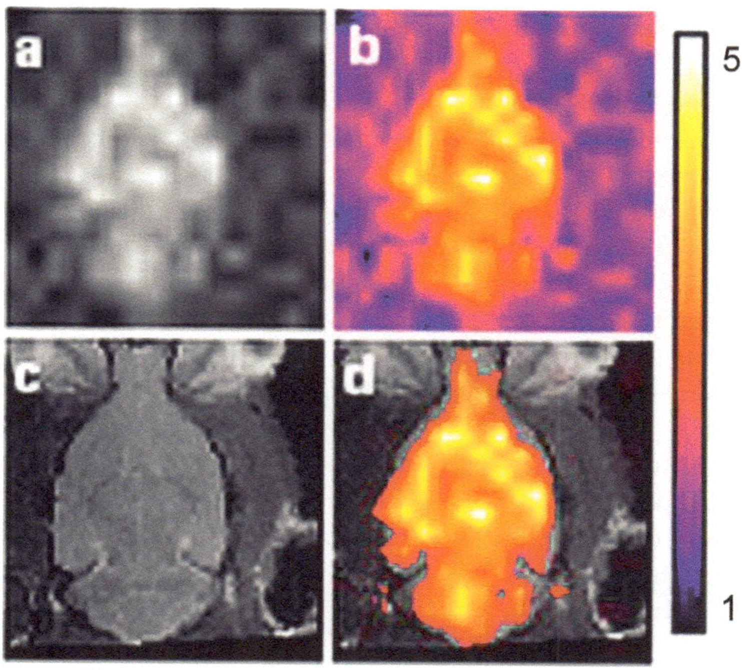

Fig. 1 (A) HP ^{129}Xe CSI acquired with a 2D CSI pulse sequence from rat head under normal breathing conditions (slice thickness 10 mm). (B) Identical image as (A) with false *color* applied. Warmer *colors* indicate increased HP ^{129}Xe signal intensity. (C) Proton MRI of a rat head showing a 1-mm coronal slice through the brain acquired with a RARE pulse sequence. (D) Proton image shown with overlay of HP ^{129}Xe MRI. *Figure reprinted with permission from Mazzanti, M. L., Walvick, R. P., Zhou, X., Sun, Y., Shah, N., Mansour, J., et al. (2011). Distribution of hyperpolarized xenon in the brain following sensory stimulation: Preliminary MRI findings.* PLoS One, 6, *e21607.*

In this work, we, in detail, document the methodology of conducting HP ^{129}Xe MRI of the mammalian brain. We apply this methodology to detail how to image both the rodent and the human brain.

2. EXPERIMENTAL PROTOCOL FOR HP ^{129}Xe BRAIN IMAGING IN RODENTS

Herein we describe the experimental protocol for imaging of the rodent brain using HP ^{129}Xe MRI. Our protocol involves anesthesia, a surgical intubation of the trachea followed by imaging. Obviously, these techniques must be approved by the institutional animal care committee and be carried out by experienced registered (also known as "licensed" in some jurisdictions) veterinarian technicians (RVT or LVT, respectively), research associates, or veterinarians. We have found that the benefit of utilizing the skills of an experienced RVT is well worth the added expense.

2.1 Rodent Anesthesia

2.1.1 Devices, Materials, and Pharmaceuticals

2.1.1.1 Devices

- Isoflurane induction chamber
- Mapleson E circuit with appropriate nose cone for rodent
- Ohmeda Isotec 3 vaporizer (Northern Vaporizers, Skipton, UK)
- Oxygen tank and hose
- Infusion pump
- Custom ventilator system (Nouls, Fanarjian, Hedlund, & Driehuys, 2011)
- Philips Achieva 3.0 T Clinical MR system
- Xemed polarizer (Xemed, Durham, NH)
- Custom quadrature RF rodent head coil (refer Section 2.4 for more details)
- 1 L Tedlar bag

2.1.1.2 Materials

- Tear gel or any other suitable lubricant
- Heating pad
- 6 m IV line
- 10-mL syringe
- 26 g catheter

- Gauze
- Warm water
- Tourniquet material
- 16 g IV catheter (to be used as an endotracheal tube (ETT))
- Blue surgical pad
- Silk sutures
- Surgical scissors and scalpel
- Tweezers
- 10 cm thread
- Curved hemostats
- Tape
- Tissue adhesive
- Retractor
- Alcohol wipes
- Permanent marker

2.1.1.3 Pharmaceuticals

- Isoflurane
- Lidocaine topical anesthetic cream
- 5 mL 0.9% saline
- 10 mg/mL propofol (45 mg/kg/h)
- 0.1 mL of 0.3 mg/mL buprenorphine
- 340 mg/mL Euthansol
- 1 L enriched ^{129}Xe (86%)
- 0.05 mL bupivacaine (to be diluted with 0.2 mL saline)
- Isoflurane vaporizer (Northern Vaporizers, Skipton, UK)
- Heating Pad
- 16 g IV catheter (to be used as an ETT)

2.1.2 Preanesthetic Preparation

1. Remove food and water from the rat cage 1 h prior to anesthesia to ensure no food is present in the mouth or in a location that could block the airway.
2. Perform preanesthetic evaluations on the rodent to ensure there are no health concerns, which may affect the study. Contact a veterinarian if there are any health concerns.
3. Weigh the rat so the appropriate dose of anesthetic can be injected.

2.1.3 Isoflurane Anesthesia

1. Ensure the anesthesia system is properly assembled and isoflurane chamber is full but not overfilled. Ensure the induction chamber, Mapleson E circuit, and scavenger chamber are connected properly.
2. Turn on the oxygen and open valve to 55 psi.
3. Place the animal in induction chamber and seal closed.
4. Turn on the oxygen to approximately 1–1.5 L/min and allow acclimatization for up to 1 min.
5. Turn on the isoflurane and dispense 4% isoflurane. Once the rat loses consciousness, reduce the isoflurane concentration to 1.5%–2%, titrating for effect.
6. Allow the rodent to rest until it does not respond to rocking of the induction chamber while monitoring breathing (breathing should go from rapid and shallow to slower, more regular breaths).
7. Flush the induction chamber with oxygen. Unplug oxygen hose from the outflow port on nose cone and plug into Mapleson E circuit. Unplug scavenger hose from induction chamber and attach to the Mapleson E circuit.
8. Remove the rodent from induction chamber and place the nose cone approximately 2 mm from the nose of the rat. This allows the rat to exhale into the surrounding air as opposed to fighting the positive pressure provided by the isoflurane vaporizer.
9. Keeping the rat on the heating pad, position and secure the rat while ensuring the nose is still in the cone. Titrate isoflurane to approximately 1%–2% to maintain anesthesia and adjust to ensure the rodent remains in the appropriate plane for anesthesia.
10. Monitor the breathing, heart rate, toe pink, eye blink, and mucous membrane color throughout time anesthetized (Table 1).
11. Inject 5 mL of normal saline and 0.1 mL of buprenorphine subcutaneously.

    Reference Chart

2.1.4 Propofol Anesthesia

Once the rat is sufficiently anesthetized using isoflurane, the rat can be anesthetized to the surgical plane using propofol anesthesia.

1. Connect a 6-m IV line and 10-mL syringe to infusion pump and fill with propofol so that there are no air bubbles. A long IV line is required because the IV infusion pump is located outside the MRI scan room.

Table 1 Rodent Anesthesia Vital Signs

	Too Light	Surgical Plane	Too Deep
Heart beat	Fast, regular	Slow, regular	Irregular, very slow, absent
Respirations	Fast, regular, shallow	Slower, regular, deep	Irregular, very slow, strained, absent
Toe pinch reflex	Present, animal will pull leg back	Absent	Absent
Eye blink reflex	Animal will blink	Absent	Absent
Mucous membrane color	Pink	Pink	Red, pale, gray, white, blue

2. Use the following equation to determine infusion rate:

$$\text{Infusion rate} = \frac{(\text{propofol dose}\,[45\text{mg/kg/h}])(\text{weight}\,[\text{kg}])}{60\left[\frac{\text{min}}{\text{h}}\right]\text{propofol concentration}\,[10\,\text{mg/mL}]}$$

3. Apply topical anesthetic to the tail of the rat. After 2 min, clean the tail of the rodent with an alcohol wipe.
4. Clamp tourniquet around the base of the tail and place the tail in warm water for several seconds.
5. Remove the tail from the water and insert the IV catheter into a lateral caudal vein.
6. Remove the tourniquet and secure the hub to the tail with tape and tissue adhesive.
7. Connect the catheter to the IV line of propofol. Turn on the syringe pump to the calculated infusion rate.
8. Allow the rat to enter the surgical plane. This may take up to 1 h. Rats are quite prone to spontaneously recovering if not under isoflurane anesthesia, while the propofol biodistributes. We have found that leaving the rat on 0.5% isoflurane is a good way to "balance" the anesthesia. A "stretching motion" of the extremities is often indicative of anesthesia to the surgical plane.

2.2 Surgical Intubation

Once the rat is anesthetized to the surgical plane, the rat must be surgically intubated with an ETT (14–16 g IV catheter, size dependent on the size of the rat). Given the small size of the rat, this procedure is quite delicate and experienced personnel are required to ensure the success of this procedure.

1. Once the rat is in the surgical plane, lay the rat down on a heating pad covered with blue surgical pads, taping the extremities to allow easy access to the neck area. Maintain oxygen/isoflurane delivery throughout.
2. Shave the neck of the rat.
3. Using a permanent "Sharpie" marker, mark the midline, and the top and the bottom of the sternum as well as the mandible.
4. Use approximately 0.05 mL lidocaine (bupivacaine) diluted with 0.2 mL saline to locally block the surgical site.
5. Using a scalpel or small surgical scissors, make a midline incision in the skin between the top of the sternum and just below the mandible. Continue to cut through the subcutaneous adipose tissue and fascia until the sternohyoid muscle is visualized.
6. Visualize the raphe (the line where the two sides of muscles meet). Using curved hemostats, blunt dissect the fascia joining the sternohyoid muscle at the raphe. The midline raphe can be difficult to visualize on some animals, and it is very possible that the sternohyoid muscle will be dissected not at midline making visualizing the trachea more difficult. The trachea should lie approximately 5 mm posterior to the surface of the sternohyoid muscle. It is identified by the presence of cartilaginous rings. To aid with visualization, a retractor may be used to hold the sides of the surgical site.
7. Once visualized, push tweezers under the trachea. Grab the 10 cm piece of thread with the tweezers. Pull the thread under the trachea so that 5 cm of thread is exposed on either side of the trachea. Place the tweezers under the trachea again.
8. Using the tweezers, lift the trachea slightly and use small surgical scissors to make a transverse incision in the trachea approximately ½ the diameter of the trachea (approximately 1 mm).
9. Without delay, insert the endotracheal catheter into the trachea to a depth so that the top of the catheter is approximately in line with the nose. Tie a double in the thread securing the ETT into the trachea. Place the nose cone close to the ETT so that the rat is able to breath via

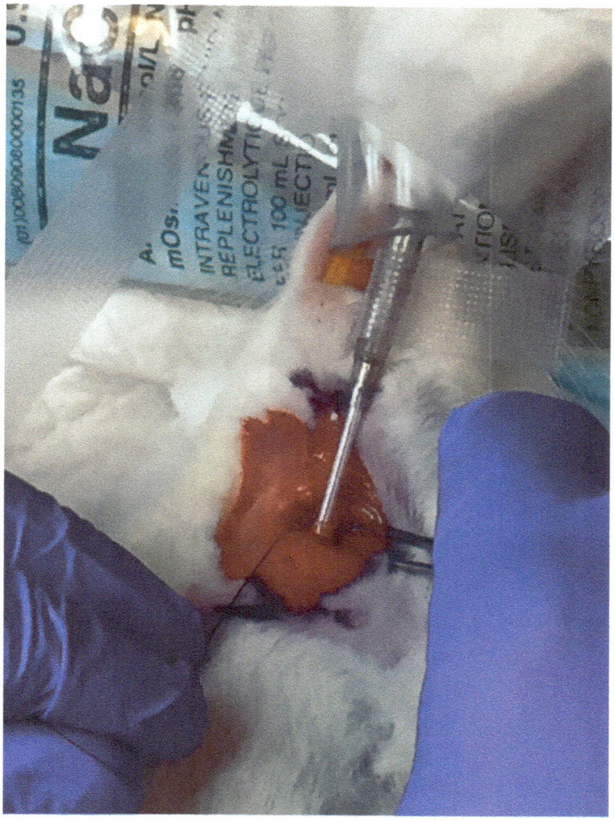

Fig. 2 Surgical intubation of endotracheal tube after tracheotomy procedure prior to suturing.

the ETT (Fig. 2). The rat may rapidly desaturate if intubated and breathing room air, or worse, extubated but with a severed trachea.

10. Suture the skin closed using silk or nylon sutures.

At this point you must prepare to move the rat from the animal surgical suite into the MRI room. This task requires the help of several assistants. One person must carry the rat, the other the IV line (now disconnected), and someone else the IV pump to the outside of the MRI room. The IV line must be run to the IV pump. Usually an additional person is needed to help out in case someone needs something rapidly. It is imperative that the rat be moved rapidly and placed onto the animal ventilator as it will rapidly desaturate on room air.

2.3 Animal Ventilation

Ventilation is necessary to accurately control the amount of xenon delivered to the rat. We utilized a custom built rodent ventilator (Nouls et al., 2011).

1. Turn on the ventilator dispensing 100% O_2.
2. Connect the ETT to the ventilator output. This may be somewhat difficult as the rat head needs to be placed inside of the RF coil, while the ventilator output is pushed onto the ETT. There is often limited room inside of the RF coil with the rat's head, and it is usually best to assign this task to someone with small fingers.
3. Generally, the tidal volume should be set to approximately 4 mL in rats.
4. Values of various parameters for rats are referenced Table 2 (Nouls et al., 2011):

2.3.1 *Breath-Hold vs Continuous Breathing Ventilation*

Both breath-hold and continuous breathing xenon ventilation can be used for brain imaging, each with their own advantages. Breath-hold ventilation can be used for imaging of the lung but can only be used for fast imaging sequences. Continuous ventilation provides a better SNR as the lungs can be saturated with the contrast agent providing rapid diffusion into the vasculature (Choi et al., 2013).

2.4 HP ^{129}Xe Brain Imaging: Rodent Model

All scans are acquired using a Philips Achieva 3T clinical MR System operating at the ^{129}Xe resonance frequency of 35.33 MHz.

1. Shim the magnetic field on the ^{1}H signal using a phantom approximately the same size as the rat.
2. Place the rat in a custom RF coil as described below (Fig. 3):

Table 2 Rodent Ventilator Settings

	Orifice Diameter (mm)	Supply Pressure (psig)	Breathing Rate (per min)	Inhalation Duration (ms)	Breath-Hold Duration (ms)	Expiration Duration (ms)	Tidal Volume (mL)
O_2	0.13	6.0	60	250	250	500	1.0
N_2	0.24	5.0					3.0
^{3}He	0.22	4.0					3.0
^{129}Xe	0.26	3.5					3.0

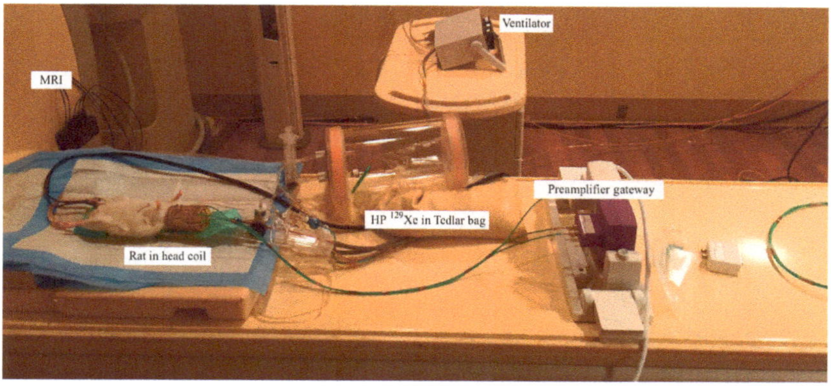

Fig. 3 Experimental setup of rat in custom quadrature head coil prior to image acquisition.

The animal RF transmit/receive coil is an eight-rung birdcage coil and adopted a four-end-ring structure in order to achieve dual frequency tuning (^{1}H and ^{129}Xe Larmor frequencies at 3 T). The dimension of the coil was designed to yield an optimized filling factor for the animal, and the coil works on quadrature-driven mode on both frequencies to obtain improved SNRs, especially for xenon scans.

3. Obtain a ^{1}H image of the rat brain using turbo spin echo (TSE) imaging. The following parameters are typically used: TR = 2 s, TE = 40 ms, flip angle = 12 degrees, slice thickness = 2 mm, with a field of view of 120 mm × 150 mm and a matrix size of 256 × 256 yielding in plane resolution of 0.586 mm.
4. Once the ^{1}H image is registered, obtain a ^{129}Xe scan of the rat brain using 2D xenon gradient echo imaging using 30% polarized ^{129}Xe at an 80% ^{129}Xe/20% oxygen ratio (polarized using a Xemed polarizer (Xemed, Durham, NH)).
   a. The following parameters are typically used for imaging the brain of the rat: TR = 197 ms, TE = 1.67 ms, flip angle of 40 degrees, slice thickness = 30 mm, with a field of view of 150 mm × 150 mm, and a matrix size of 64 × 64 yielding a plane resolution of 2.34 mm.
5. Analyze the images using custom MATLAB® image processing scripts. Sample scripts are found in the supplementary information in Hane, Li, et al. (2017).

   Overlay the xenon image on the proton image obtained using GIMP or similar imaging software.

3. EXPERIMENTAL PROTOCOL FOR HP ^{129}Xe BRAIN IMAGING IN HUMANS

HP ^{129}Xe is detectable in distal organs, including the brain, since it has a long relaxation time (T1) of 6–8 s in oxygenated blood and 3–4 s in deoxygenated blood (Albert, Balamore, Kacher, Venkatesh, & Jolesz, 2000; Albert, Kacher, Balamore, Venkatesh, & Jolesz, 1999; Norquay et al., 2015; Wolber, Cherubini, Dzik-Jurasz, Leach, & Bifone, 1999). Magnetic resonance spectroscopy (MRS) of ^{129}Xe in the human brain reveals five distinct peaks at 187, 192.5, 195, 198, and 215 ppm assigned to soft muscular tissue, white matter, gray matter, cerebrospinal fluid, and blood, respectively (Rao, Stewart, Norquay, Griffiths, & Wild, 2015). Conducting human research requires approval of the Institutional Research Ethics Board and likely also the services of a respiratory therapist and imaging technologist. High-resolution images with high SNR, such as displayed in Fig. 4, are possible using this technique.

3.1 HP ^{129}Xe Brain Imaging in Humans

3.1.1 Materials

- Philips Achieva 3T MRI scanner
- Clinical MR Solutions ^{1}H/^{129}Xe dual tuned head coil (MR Solutions, Guildford, UK)
- The clinical head coil is of a coaxial quadrature birdcage design to achieve dual frequency tuning (^{1}H and ^{129}Xe Larmor frequencies at 3 T). The coil has a diameter and length of 23 cm. Similar to the custom rodent head coil listed above, the CMRS coil works on a quadrature-driven mode to increase SNRs. Xemed ^{129}Xe gas polarizer (Xemed LLC, Durham)

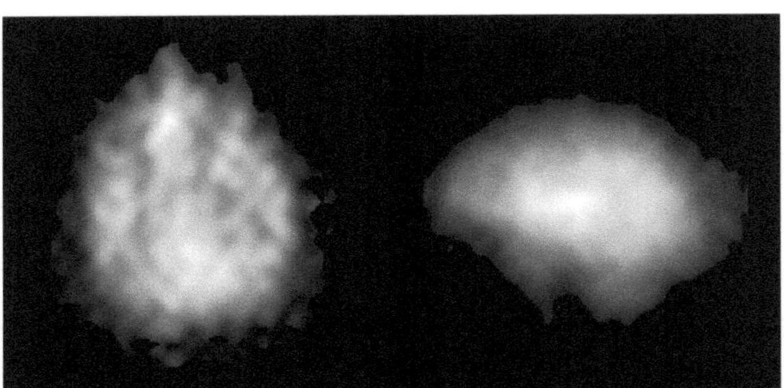

Fig. 4 HP ^{129}Xe MRI of the human brain at 3 T in the axial (*left*) and sagittal (*right*) plane.

- 1 L Tedlar bag
- ¼″ Tygon tubing (18″) with Tedlar adapter (Saint Gobain, Akron, OH)

1. Enriched (84%) ^{129}Xe gas is polarized using a Xemed polarizer (Xemed LLC, Durham) and dispensed into a 1-L Tedlar bag. The ^{129}Xe gas is also mixed with medical grade oxygen (O_2) or nitrogen (N_2).
2. The bag is transferred and stored within the bore of the MR scanner. Place the human participant inside the dual tuned head coil and ensure comfort. Provide them with ear protection and the unconnected breathing tube. This breathing tube can be held loosely in the participant's mouth or nearby. This breathing tube will be used by the participant to breath in the HP ^{129}Xe. This tube must be given to the participant at the appropriate time. Bearded participants or those with larger heads make it difficult to provide them the tube, hence the suggestion to have them hold the tube loosely in their mouth.
3. Begin by acquiring a ^{1}H fMRI data using a TSE imaging sequence. This image can either be T1 weighted or T2 weighted. We used the following imaging parameters to acquire our ^{1}H images: FOV = 250 mm × 250 mm, matrix = 256 × 256, TR/TE = 3 s/80 ms, NSA = 2, FA = 90 degrees.
4. The Tedlar bag is connected to the tubing. The participant is instructed to take three deep breaths of room air. The clip is removed from the Tedlar bag tubing, and the participant is instructed to inhale all of the gas from the bag and hold their breath for 20 s. A member of the research team should be counting to 20 out loud. Additionally, the participant should be encouraged to keep holding their breath if necessary.
5. A ^{129}Xe MRI is acquired using a dynamic 2D fast field echo pulse sequence. Image acquisition begins 10 s following inhalation, upon exhalation, and 10 s following exhalation. Typical imaging parameters are (Hane et al., 2017): FOV = 250 mm × 250 mm, matrix = 32 × 32, TR/TE = 250 ms/0.84 ms, NSA = 1, FA = 12 degrees, Bandwidth = 150 Hz/pixel.
6. At 20 s, the participant should be encouraged to take several deep breaths. It is common for the participant to experience mild adverse effects from the Xe, including dizziness, nausea, headaches, and blood oxygen desaturation. Participants who begin to desaturate should be encouraged to take several deep breaths. Those who desaturate to below 90% for more than 10 s should be placed on O_2 even though they will likely recover by the time the O_2 is administered.

4. CONCLUSION

Though generally used in the lungs, HP ^{129}Xe MRI is an exciting new imaging modality for studying the brain. As a result of its lipophilicity, ^{129}Xe is capable of hydrophobic interactions with lipid bilayer membranes in the mammalian brain (Barany et al., 1987; Cherubini & Bifone, 2003); furthermore, ^{129}Xe can dissolve in blood and thus be circulated to various mammalian organs, including the brain.

Using these methods of HP ^{129}Xe MRI, it may be possible to image a multitude of different brain pathologies, including stroke; also, this MRI technology is currently being investigated for studying neurodegenerative diseases, such as AD (Zhou et al., 2011). Although conventional ^{1}H MRI may have a greater signal than HP ^{129}Xe MRI, HP ^{129}Xe may still be able to obtain physiologically relevant information in brain regions which could provide an alternative to traditional blood-oxygen-level-dependent functional MRI. This, along with many other facets of HP ^{129}Xe MRI, has the potential to increase over time as factors such as SEOP improve with new technologies (Mazzanti et al., 2011).

ACKNOWLEDGMENTS

This work is partially supported by the Weston Brain Institute and the Natural Sciences and Engineering Research Council (NSERC). F.T.H. wishes to thank the generous donors of the BrightFocus Foundation and the Canadian Institutes for Health Research (CIHR) for their fellowship support. B.R.J.P. is supported by a NSERC Undergraduate Student Research Award (USRA).

REFERENCES

Albert, M. S., Balamore, D., Kacher, D. F., Venkatesh, A. K., & Jolesz, F. A. (2000). Hyperpolarized ^{129}Xe T1 in oxygenated and deoxygenated blood. *NMR in Biomedicine, 13*, 407–414.

Albert, M. S., Cates, G. D., Driehuys, B., Happer, W., Saam, B., Springer, C. S., et al. (1994). Biological magnetic resonance imaging using laser-polarized ^{129}Xe. *Nature, 370*, 199–201.

Albert, M. S., Kacher, D. F., Balamore, D., Venkatesh, A. K., & Jolesz, F. A. (1999). T1 of 129-Xe in blood and the role of oxygenation. *Journal of Magnetic Resonance, 140*, 264–273.

Bachert, P., Schad, L. R., Bock, M., Knopp, M. V., Ebert, M., Grobmann, T., et al. (1996). Nuclear magnetic resonance imaging of airways in humans with use of hyperpolarized 3He. *Magnetic Resonance in Medicine, 36*, 192–196. https://doi.org/10.1002/mrm.1910360204.

Barany, M., Spigos, D. G., Mok, E., Venkatasubramanian, P. N., Wilbur, A. C., & Langer, B. G. (1987). High resolution proton magnetic resonance spectroscopy of human brain and liver. *Journal of Magnetic Resonance, 5*, 393–398.

Black, R. D., Middleton, H. L., Cates, G. D., Cofer, G. P., Driehuys, B., Happer, W., et al. (1996). In vivo He-3 MR images of guinea pig lungs. *Radiology, 199*, 867–870. https://doi.org/10.1148/radiology.199.3.8638019.

Cherubini, A., & Bifone, A. (2003). Hyperpolarised xenon in biology. *Progress in Nuclear Magnetic Resonance Spectroscopy, 42*, 1–30.

Choi, J. S., Kim, M. J., Chung, Y. E., Kim, K. A., Choi, J. Y., Lim, J. S., et al. (2013). Comparison of breathhold, navigator-triggered, and free-breathing diffusion-weighted MRI for focal hepatic lesions. *Journal of Magnetic Resonance Imaging, 38*, 109–118. https://doi.org/10.1002/jmri.23949.

Couch, M. J., Blasiak, B., Tomanek, B., Ouriadov, A. V., Fox, M. S., Dowhos, K. M., et al. (2015). Hyperpolarized and inert gas MRI: The future. *Molecular Imaging and Biology, 17*, 149–162.

Fain, S. B., Korosec, F. R., Holmes, J. H., O'Halloran, R., Sorkness, R. L., & Grist, T. M. (2007). Functional lung imaging using hyperpolarized gas MRI. *Journal of Magnetic Resonance Imaging, 25*, 910–923. https://doi.org/10.1002/jmri.20876.

Goodson, B. M. (2002). Nuclear magnetic resonance of laser-polarized noble gases in molecules, materials, and organisms. *Journal of Magnetic Resonance, 155*, 157–216.

Goodson, B. M., Ranta, K., Skinner, J. G., Coffey, A. M., Nikolaou, P., Gemeinhardt, M., et al. (2017). The physics of hyperpolarized gas MRI. In M. S. Albert & F. T. Hane (Eds.), *Hyperpolarized and inert gas MRI: From technology to application in research and medicine* (pp. 23–46). New York: Elsevier.

Goto, T., Suwa, K., Uezono, S., Ichinose, F., Uchiyama, M., & Morita, S. (1998). The blood-gas partition coefficient of xenon may be lower than generally accepted. *British Journal of Anaesthesia, 80*, 255–256.

Hane, F. T., Imai, H., Kimura, A., Fujiwara, H., Rao, M., Wild, J. M., et al. (2017). chapter 16. *Brain imaging using hyperpolarized xenon MRI* (1st ed.). Elsevier.

Hane, F., Li, T., Lawrence-Dewar, J. M., Hassan, A., Granberg, K., Pellizzari, R., et al. (2017). Using hyperpolarized ^{129}Xe in human participants to perform functional magnetic resonance imaging (fMRI). In *International Society for Magnetic Resonance in Medicine* (p. 4875).

Hane, F. T., Li, T., Smylie, P., Pellizzari, R. M., Plata, J. A., DeBoef, B., et al. (2017). In vivo detection of a hyperpolarized xenon magnetic resonance molecular imaging contrast agent. *Scientific Reports*, 7, 41027.

Hane, F., Smylie, P., Li, T., Ruberto, J., Dowhos, K., Ball, I., et al. (2016). HyperCEST detection of cucurbit[6]uril in whole blood using an ultrashort saturation pre-pulse train. *Contrast Media & Molecular Imaging, 11*, 285–290. https://doi.org/10.1002/cmmi.1690.

Happer, W. (1972). Optical pumping. *Reviews of Modern Physics, 44*, 169–240.

Li, T., Hane, F. T., Lawrence-dewar, J. M., Hassan, A., Granberg, K., Pellizzari, R. M., et al. (2017). Using hyperpolarized 129Xe MRI to detect impaired cerebral perfusion in human brain with Alzheimer's disease results & discussion conclusion. In *Proceedings on International Society for Magnetic Resonance in Medicine* (p. 487). https://doi.org/10.1007/s00330-015-3834-9.

Mazzanti, M. L., Walvick, R. P., Zhou, X., Sun, Y., Shah, N., Mansour, J., et al. (2011). Distribution of hyperpolarized xenon in the brain following sensory stimulation: Preliminary MRI findings. *PLoS One, 6*, e21607.

Möller, H. E., Chen, X. J., Saam, B., Hagspiel, K. D., Johnson, G. A., Altes, T. A., et al. (2002). MRI of the lungs using hyperpolarized noble gases. *Magnetic Resonance in Medicine, 47*, 1029–1051. https://doi.org/10.1002/mrm.10173.

Norquay, G., Leung, G., Stewart, N. J., Tozer, G. M., Wolber, J., & Wild, J. M. (2015). Relaxation and exchange dynamics of hyperpolarized ^{129}Xe in human blood. *Magnetic Resonance in Medicine, 74*, 303–311.

Nouls, J., Fanarjian, M., Hedlund, L., & Driehuys, B. (2011). A constant-volume ventilator and gas recapture system for hyperpolarized gas MRI of mouse and rat lungs. *Concepts in Magnetic Resonance Part B: Magnetic Resonance Engineering*, *39B*, 78–88. https://doi.org/10.1002/cmr.b.

Oros, A.-M., & Shah, N. J. (2004). Hyperpolarized xenon in NMR and MRI. *Physics in Medicine and Biology*, *49*, R105–R153. https://doi.org/10.1088/0031-9155/49/20/R01.

Rao, M., Stewart, N., Norquay, G., Griffiths, P., & Wild, J. (2015). Imaging the human brain with dissolved xenon MRI at 1.5 T. In *International Society for Magnetic Resonance in Medicine Annual Meeting* (p 1254).

Schröder, L. (2017). Xenon biosensor HyperCEST MRI. In M. S. Albert & F. T. Hane (Eds.), *Hyperpolarized inert gas MRI from technology to application in research and medicine* (pp. 263–275). Elsevier Inc.

Tooker, A. C., Hong, K. S., McKinstry, E. L., Costello, P., Jolesz, F. A., & Albert, M. S. (2003). Distal airways in humans: Dynamic hyperpolarized ^{3}He MR imaging—Feasibility. *Radiology*, *227*, 575–579. https://doi.org/10.1148/radiol.2272012146.

Walker, T., & Happer, W. (1997). Spin-exchange optical pumping of noble-gas nuclei. *Reviews of Modern Physics*, *69*, 629–642.

Wolber, J., Cherubini, A., Dzik-Jurasz, A. S. K., Leach, M. O., & Bifone, A. (1999). Spin-lattice relaxation of laser-polarized xenon in human blood. *Proceedings of the National Academy of Sciences of the United States of America*, *96*, 3664–3669.

Zhou, X., Sun, Y., Mazzanti, M., Henninger, N., Mansour, J., Fisher, M., et al. (2011). MRI of stroke using hyperpolarized ^{129}Xe. *NMR in Biomedicine*, *24*, 170–175.

Kolibri 360 GmbH
Ferdinand-Jühlke-Straße 7
99095 Erfurt, Deutschland
produktsicherheit@kolibri360.de